Digital Systems Design with VHDL and Synthesis:

An Integrated Approach

K.C. Chang

IEEE
COMPUTER
SOCIETY

Los Alamitos, California

Washington • Brussels • Tokyo

Library of Congress Cataloging-in-Publication Data

Chang, K. C. (Kou-Chuan), 1957 –
    Digital systems design with VHDL and synthesis : an integrated approach / K.C. Chang.
        p.   cm.
    ISBN 0-7695-0023-4
    1. Electronic digital computers – Circuits – Design and construction – Data processing.   2. VHDL (Hardware description language).   3. System design – Data processing.   I. Title.
TK7888.4.C435   1999
621.39 ' 2 — dc21

99-24750
CIP

IEEE Computer Society Press Order Number BP00023
Library of Congress Number 99-24750
ISBN 0-7695-0023-4

Additional copies may be ordered from:

IEEE Computer Society Press
Customer Service Center
10662 Los Vaqueros Circle
P.O. Box 3014
Los Alamitos, CA 90720-1314
Tel: +1-714-821-8380
Fax: +1-714-821-4641
cs.books@computer.org

IEEE Service Center
445 Hoes Lane
P.O. Box 1331
Piscataway, NJ 08855-1331
Tel: +1-732-981-0060
Fax: +1-732-981-9667
mis.custserv@computer.org

IEEE Computer Society
Watanabe Building
1-4-2 Minami-Aoyama
Minato-ku, Tokyo 107-0062
JAPAN
Tel: +81-3-3408-3118
Fax: +81-3-3408-3553
tokyo.ofc@computer.org

Publisher: Matt Loeb
Manager of Production, CS Press: Deborah Plummer
Advertising/Promotions: Tom Fink
Production Editor: Denise Hurst
Printed in the United States of America

Dedicated to

my parents,

Chia-Shia and Jin-Swei,

my wife,

Tsai-Wei,

and my children,

Alan, Steven, and Jocelyn

Preface

The advance of very large-scale integration (VLSI) process technologies has made "system on a chip" a feasible goal for even the most complex system. The market has demanded shorter design cycles than ever before, while design complexity continues to increase. Traditional schematic capture design approaches are no longer sufficient to be competitive in the industry. New design processes and methodologies are required to design much more complex systems within a shorter design cycle. Furthermore, the gap between the design concept and implementation must be minimized. The goal of this book is to introduce an integrated approach to digital design principles, process, and implementation. This is accomplished by presenting the digital design concepts and principles, VHDL coding, VHDL simulation, synthesis commands, and strategies together.

It is the author's belief that it is better to learn digital design concepts and principles together with the high-level design language, verification, and synthesis. Overlooking key digital design concepts will result in inefficient designs that don't meet performance requirements. Without the high level design language, the design concepts cannot be captured fast enough. Lacking the verification techniques results in a design with errors. Not knowing the synthesis strategy and methodology does not allow the design to be synthesized with the desired results. Therefore, an integrated approach to design concepts, principles, VHDL coding, verification, and synthesis is crucial. To accomplish this in a book is challenging since there are so many potential combinations of design concepts with techniques for VHDL coding, verification, and synthesis. The author has approached this challenge by focusing on the ultimate end products of the design cycle: the implementation of a digital design. VHDL code conventions, synthesis methodologies, and verification techniques are presented as tools to support the final design implementation. By taking this approach, readers will be taught to apply and adapt techniques for VHDL coding, verification, and synthesis to various situations.

To minimize the gap between design concepts and implementation, design concepts are introduced first. VHDL code is presented and explained line by line to capture the logic behind the design concepts. The VHDL code is then verified using VHDL test benches and simulation tools. Simulation waveforms are shown and explained to verify the correctness of the design. The same VHDL code is synthesized. Synthesis commands and strategies are presented and discussed. Synthesized schematics and results are analyzed for area and timing. Variations on the design techniques and common mistakes are also addressed.

This book is the result of the author's practical experience as both a designer and an instructor. Many of the design techniques and design considerations illustrated throughout the chapters are examples of viable designs. The author's teaching experience has led to a step-by-step presentation that addresses common mistakes and hard-to-understand concepts in a way that makes learning easier.

Practical design concepts and examples are presented with VHDL code, simulation waveforms, and synthesized schematics so that readers can better understand their correspondence and relationships. Unique features of the book include the following:

1. More than 207 complete examples of 14,099 lines of VHDL code are developed. Every line of VHDL code is analyzed, simulated, and synthesized with simulation waveforms and schematics shown and explained. VHDL codes, simulation waveforms, and schematics are shown together, allowing readers to grasp the concepts in the context of the entire process.
2. Every line of VHDL code has an associated line number for easy reference and discussion. The Courier font is used for the VHDL code portion because each character of this font occupies the same width, which allows them to line up vertically. This is close to how an American Standard Code for Information Interchange (ASCII) file appears when readers type in their VHDL code with any text editor.
3. The VHDL code examples are carefully designed to illustrate various VHDL constructs, features, practical design considerations, and techniques for the particular design. The examples are complete so that readers can assimilate overall ideas more easily.
4. Challenging exercises are provided at the end of each chapter so that readers can put into practice the ideas and information offered in the chapter.
5. A complete Finite Impulse Response filter (700K+ transistors) ASIC design project from concept, VHDL coding, verification, synthesis, layout, to final timing verification is provided to demonstrate the entire semi-custom standard cell design process.
6. A complete microcontroller (AM2910) design project from concept, VHDL coding, verification, synthesis, to release to a gate array vendor for layout, to final postlayout timing verification with test vector generation is used to demonstrate the entire gate array application-specific integrated circuit (ASIC) design process.
7. A complete error correction and detection circuit (TI SN54ALS616) design is taken from concept, VHDL coding, verification, synthesis, to a field programmable gate array (FPGA) implementation.
8. Synthesis commands and strategies are presented and discussed for various design situations.
9. Basic digital design circuits of counters, shifters, adders, multipliers, dividers, and floating-point arithmetic are discussed with complete VHDL examples, verification, and synthesis.
10. Test benches VHDL coding and techniques are used and demonstrated along with the design examples.
11. Preferred practices for the effective management of designs are discussed.
12. VHDL design partitioning techniques are addressed and discussed with examples.
13. An entire chapter (Chapter 8) is dedicated to clock- and reset-related circuits. Clock skew, short path, setup and hold timing, and synchronization between clock domains are discussed with examples.

14. Commonly used custom blocks such as first in, first out (FIFO), dual port random access memory (RAM), and dynamic RAM models are discussed with examples.
15. All figures in the book are available through the Internet to assist instructors.
16. VHDL code examples are available through the Internet.

0.1 ORGANIZATION OF THE BOOK

This book is divided into 15 chapters and one appendix. Appendix A is a VHDL package example that has been referenced in many examples in the chapters. The details of each chapter are as follows:

Chapter 1 describes design process and flow.

Chapter 2 describes how to write VHDL to model basic digital circuit primitives or gates. Flip-flop, latch, and three-state buffer inferences are illustrated with examples. VHDL synthesis rules are presented to provide readers with an understanding of what circuits will be synthesized from the VHDL.

Chapter 3 presents the VHDL simulation and synthesis design environments. Synopsys simulation and synthesis design environments and design tools are introduced. Mentor QuickVHDL simulation environment is also discussed. A typical design process for a block is discussed to improve the debugging process.

Chapter 4 presents VHDL modeling, synthesis, and verification of several basic combination circuits such as selector, encoder, code converter, equality checker, and comparators. Each circuit is presented with different VHDL models and their differences and trade-offs are discussed.

Chapter 5 concentrates on several binary arithmetic circuits. Half adder, full adder, ripple adder, carry look ahead adder, countone, leading zero, and barrel shifter are presented with VHDL modeling, synthesis, and test bench.

Chapter 6 discusses sequential circuits such as counters, shift registers, parallel to serial converter, and serial to parallel converter with VHDL modeling, synthesis, and test bench.

Chapter 7 presents a framework to organize registers in the design. Registers are categorized and discussed. Partition, synthesis, and verification strategies of registers are discussed. VHDL modeling, synthesis, and verification techniques are illustrated.

Chapter 8 is dedicated to clock- and reset-related circuits. The synchronization between different clock domains is discussed. The clock tree generation, clock delay, and clock skew are presented and discussed with timing diagrams and VHDL code. The issues of gated clock and clock divider are also introduced.

Chapter 9 presents examples of dual-port RAM, synchronous and asynchronous FIFO, and dynamic RAM VHDL models. These blocks are commonly used as custom drop-in macros. They are used to interact with the rest of the design so that the complete design can be verified.

Chapter 10 illustrates the complete semicustom ASIC design process of a finite impulse response ASIC design through the steps of design description, VHDL coding, functional verification, synthesis, layout, and back-annotated timing verification.

Chapter 11 discusses the concept of a microprogram controller. The design of a AMD AM2910 is presented through the gate array design process from VHDL coding to postlayout back-annotated timing verification. The test vector generation is also illustrated.

Chapter 12 discusses the principles of error and correcting Hamming codes. An actual TI EDAC integrated circuit is used as an example to design from VHDL code to all steps of FPGA design process.

Chapter 13 presents the concepts of binary fixed-point multiplication algorithms such as Booth-Wallace multiplier. The VHDL coding, synthesis, and verification are presented.

Chapter 14 discusses the concepts of binary fixed-point division algorithms. VHDL coding, synthesis, and verification are presented.

Chapter 15 discusses the floating-point number representation. Floating-point addition and multiplication algorithms are discussed and implemented in VHDL. They are verified and synthesized.

Appendix A lists a package that is used and referenced by many examples.

0.2 AUTHOR'S NOTE

This book assumes that readers have the basic knowledge of VHDL syntax, modeling, synthesis concepts, and Boolean algebra. To acquire some background in VHDL, readers are referred to my previous book (*Digital Design and Modeling with VHDL and Synthesis*, IEEE Computer Society Press, 1997) which concentrates on the complete VHDL language, syntax and coding techniques. The recommended VHDL background can be obtained from Chapters 2, 3, 4, 5, 6, 7, and 8 of that book. This book concentrates on the digital design from a basic half adder to a complete ASIC design, using VHDL coding, simulation, and synthesis techniques. The interfaces to FPGA and ASIC layout tools are also addressed.

Synopsys and Mentor Modeltech (QuickVHDL, which is similar to Modelsim) VHDL simulators, Exemplar FPGA synthesis, and Synopsys ASIC synthesis tools are used in this book. These tools are widely used and available throughout university and industry settings.

The book is intended for both practicing engineers and as a college text. Chapters are divided according to specific designs, from simple to complex. Techniques for VHDL coding, test bench, verification, synthesis, design management, and actual design examples are good references for practicing engineers. The book also provides a robust introduction to digital design. It presents the basic design concepts of a half adder, full adder, ripple adder, carry look ahead adder, counter, shift register, multiplier, divider, floating-point arithmetic unit, error correction and detection unit, microcontroller, FIFO, dual port RAM, dynamic RAM, to a 700 thousands transistor finite impulse response filter ASIC design. These design concepts and disciplines are reinforced by actual VHDL code, verification strategies, simulation commands and waveforms, and synthesis commands and strategies. VHDL code is available for reference and practice. Previous experiences with VHDL, as suggested earlier, would be helpful. However, an understanding of the code conventions can also be acquired from this book. Most students will soon become practicing engineers. This book not only gives you a block diagram, but also, more importantly, shows you how each design concept can be captured, implemented, and realized in real life. An old Chinese saying is: "do not let your eyes and head reach higher than your hands," which is to say that it is better to be able to do it with your hands, rather than just think and see something that is out of your grasp.

I wrote this book to help the reader cope with both the design revolution that is taking place, as well as the increased demand for design complexity and a shortened design cycle.

0.3 ACKNOWLEDGMENTS

This book would have been impossible without the following people:

Ron Fernandes, who edited the entire first manuscript.

My wife, Tsai-Wei, and my children, Alan, Steven, and Jocelyn, who were patient and sacrificed many of our weekends and dinners together. I owe my children so many smiles and bedtime stories because I often got home late and tired.

My brothers, sisters, and many friends who encouraged and supported me.

Ramagopal Madamala and Tsipi Landen at Chip Express provided support in running through the place and route tools for the design in Chapter 11.

The Institute of Electrical and Electronics Engineers (IEEE) Computer Society Press editorial team, Mathew Loeb, Cheryl Baltes, and Denise Hurst contributed many valuable comments. Additionally, Joseph Pomerance did a skillful and thorough editing of the manuscript.

The author deeply appreciates all of these people.

Contents

Chapter 1 INTRODUCTION   1

1.1   Integrated Design Process and Methodology 1
1.2   Book Overview 2

Chapter 2 VHDL AND DIGITAL CIRCUIT PRIMITIVES   4

2.1   Flip Flop 4
2.2   Latch 15
2.3   Three-State Buffer 18
2.4   Combinational Gates 22
2.5   VHDL Synthesis Rules 26
2.6   Pads 30
2.7   Exercises 30

Chapter 3 VHDL SIMULATION AND SYNTHESIS ENVIRONMENT AND
               DESIGN PROCESS   32

3.1   Synopsys VHDL Simulation Environment Overview 32
3.2   Mentor QuickVHDL Simulation Environment 36
3.3   Synthesis Environment 39
3.4   Synthesis Technology Library 45
3.5   VHDL Design Process for a Block 47
3.6   Exercises 52

Chapter 4 BASIC COMBINATIONAL CIRCUITS 53

4.1   Selector 53

4.2   Encoder 68

4.3   Code Converter 71

4.4   Equality Checker 73

4.5   Comparator with Single Output 79

4.6   Comparator with Multiple Outputs 82

4.7   Exercises 89

Chapter 5 BASIC BINARY ARITHMETIC CIRCUITS 91

5.1   Half Adder and Full Adder 91

5.2   Carry Ripple Adder 97

5.3   Carry Look Ahead Adder 101

5.4   Countone Circuit 119

5:5   Leading Zero Circuit 123

5.6   Barrel Shifter 132

5.7   Exercises 137

Chapter 6 BASIC SEQUENTIAL CIRCUITS 143

6.1   Signal Manipulator 143

6.2   Counter 150

6.3   Shift Register 166

6.4   Parallel to Serial Converter 177

6.5   Serial to Parallel Converter 180

6.6   Exercises 186

Chapter 7 REGISTERS

7.1   General Framework for Designing Registers 187

7.2   Interrupt Registers 189

7.3   DMA and Control Registers 193

7.4   Configuration Registers 196

7.5   Reading Registers 201

7.6   Register Block Partitioning and Synthesis 202

7.7   Testing Registers 213

7.8   Microprocessor Registers 219

7.9   Exercises 221

Chapter 8 CLOCK AND RESET CIRCUITS

8.1   Clock Buffer and Clock Tree 222

8.2   Clock Tree Generation 225

8.3   Reset Circuitry 228

8.4   Clock Skew and Fixes 230

8.5   Synchronization between Clock Domains 238

8.6   Clock Divider 242

8.7   Gated Clock 246

8.8   Exercises 250

Chapter 9 DUAL-PORT RAM, FIFO, AND DRAM MODELING

9.1   Dual-Port RAM 251

9.2   Synchronous FIFO 260

9.3   Asynchronous FIFO 266

9.4   Dynamic Random Access Memory (DRAM) 274

9.5   Exercises 286

Chapter 10 A DESIGN CASE STUDY: FINITE IMPULSE RESPONSE FILTER ASIC DESIGN   **288**

10.1   Design Description 288

10.2   Design Partition 293

10.3   Design Verification 307

10.4   Design Synthesis 322

10.5   Worst-Case Timing Analysis 325

10.6   Best-Case Timing Analysis 329

10.7   Netlist Generation 331

10.8   Postlayout Verification 334

10.9   Design Management 337

10.10   Exercises 339

Chapter 11 A DESIGN CASE STUDY: A MICROPROGRAM CONTROLLER DESIGN 341

11.1   Microprogram Controller 341

11.2   Design Description and Partition 344

11.3   Design Verification 362

11.4   Design Synthesis 377

11.5   Postsynthesis Timing Verification 382

11.6   Preparing Release Functional Vectors 383

11.7   Postlayout Verification 387

11.8   Design Management 387

11.9   Exercises 389

Chapter 12 ERROR DETECTION AND CORRECTION   **390**

12.1   Error Detection and Correction Code 390

12.2   Single Error Detecting Codes 390

12.3   Single Error Correcting Codes 391

12.4   Single Error Correcting and Double Error Detecting Codes 393

12.5   Error Detecting and Correcting Code Design Example 394

12.6   Design Verification 400

12.7   Design Synthesis 403

12.8   Netlist Generation and FPGA Place and Route 407

12.9   Exercises 407

Chapter 13 FIXED-POINT MULTIPLICATION   **408**

13.1   Multiplication Concept 408

13.2   Unsigned Binary Multiplier 409

13.3   2's Complement Multiplication 419

13.4   Wallace Tree Adders 423

13.5   Booth-Wallace Tree Multiplier 425

13.6   Booth-Wallace Tree Multiplier Verification 429

13.7   Booth-Wallace Tree Multiplier Synthesis 431

13.8   Multiplication with Shift and Add 437

13.9   Exercises 442

13.10   References 444

Chapter 14 FIXED-POINT DIVISION   **445**

14.1   Basic Division Concept 445

14.2   32-Bit Divider 451

14.3   Design Partition 452

14.4   Design Optimization 453

14.5   Design Verification 458

14.6   Design Synthesis 462

14.7   Exercises 465

14.8   Reference 466

Chapter 15 FLOATING-POINT ARITHMETIC 467

15.1   Floating-Point Number Representation 467

15.2   Floating-Point Addition 468

15.3   Floating-Point Multiplication 479

15.4   Exercises 484

Appendix A PACKAGE PACK 485

Index 496

Chapter 1

Introduction

Welcome to the book!

The advance of very large-scale integration (VLSI) process technologies has made "system on a chip" feasible for very complex systems. This is partly responsible for the market demand for much shorter design cycles even though design complexity has continued to increase.

There are many design and VHDL code examples, simulation waveforms, and synthesized schematics shown in this book to illustrate their correspondence. VHDL code can be downloaded from the Internet for exercises. All examples have been verified with VHDL simulation and synthesis tools to ensure high fidelity of the VHDL code. All VHDL codes are listed line by line for reference and discussion in the text. Each VHDL example is complete, no fragments so that the complete picture is clear to you. Design techniques of VHDL coding, verification, and synthesis that implement the design concepts and principles are illustrated with actual examples throughout the book.

In the next section, an integrated design process and methodology is introduced to serve as our base line design flow and approach.

1.1 INTEGRATED DESIGN PROCESS AND METHODOLOGY

Figure 1-1 shows a flow chart that describes an integrated VHDL design process and methodology. The design is first described in VHDL. VHDL coding techniques will be applied to model the design efficiently and effectively. A preliminary synthesis for each VHDL code is done to check for obvious design errors such as unintended latches and insufficient sensitivity list before the design is simulated. This also ensures that all VHDL code can be synthesized. This process is discussed and illustrated in Chapter 2 as the VHDL design process for a block. A test bench is developed to facilitate the design verification. Test bench examples and verification techniques are discussed throughout the book for various designs. A more detailed synthesis can be started (in parallel with the test bench development) to calibrate design and timing constraints. The design architecture may require a change, based on the timing result. The complete design can be verified with the test bench. There will be iterations among the VHDL coding, synthesis, test bench, and simulation. After the design is free of design errors, the complete and detailed synthesis is done. A netlist can be generated for the layout tool to do the layout. The layout may be performed with Field Programmable Gate Array (FPGA), gate array, or standard cell layout tools. The layout tool can generate another VHDL netlist with the timing delay file in Standard Delay Format (SDF), which can be used for simulation with the test bench. Using the simulation results, test vectors can be generated for the foundry to test the fabricated ASIC design. A FPGA design does not need the test vector. The layout can be performed in-house or by a vendor. For the ASIC design, the layout tool can generate a GDSII file to describe the masks for fabrication.

The synthesis tool can also generate a VHDL netlist for postsynthesis and prelayout simulation. This is not common. The postlayout simulation is preferred since the timing generated from the layout tools is much closer to the actual hardware. Note also that the test bench should be used for both functional simulation and postlayout timing simulation without the need to change.

In this book, many examples are used to illustrate all the steps of the process described in Figure 1-1.

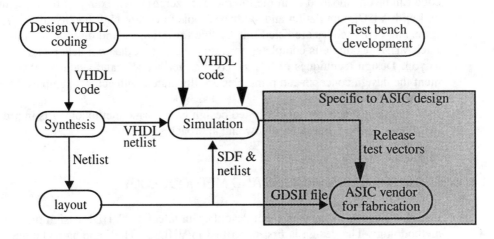

Figure 1-1 VHDL design process and methodology.

1.2 BOOK OVERVIEW

Chapter 2 describes how to write VHDL to model basic digital circuit primitives or gates. Flip-flops, latches, and three-state buffers inference is illustrated with examples. VHDL synthesis rules are presented to provide guidelines what circuits will be synthesized by synthesis tools.

Chapter 3 presents the VHDL simulation and synthesis design environments. Synopsys simulation and synthesis design environments and design tools are introduced. Mentor QuickVHDL simulation environment is also discussed. A typical design process for a block is discussed to improve the debugging process.

Chapter 4 presents VHDL modeling, synthesis, and verification of several basic combination circuits such as selector, encoder, code converter, equality checker, and comparators. Each circuit is presented with different VHDL models so that their differences and trade-offs are discussed.

Chapter 5 concentrates on several binary arithmetic circuits. Half adder, full adder, ripple adder, carry look ahead adder, count one, leading zero, and barrel shifter are presented with VHDL modeling, synthesis, and test bench.

Chapter 6 discusses sequential circuits such as counters, shift registers, parallel to serial converter, and serial to parallel converter with VHDL modeling, synthesis, and test bench.

Chapter 7 presents a framework to organize registers in the design. Registers are categorized and discussed. Partition, synthesis, and verification strategies of registers are discussed. VHDL modeling, synthesis, and verification are illustrated.

Chapter 8 is dedicated to clock- and reset-related circuits. The synchronization between different clock domains is discussed. Clock tree generation, clock delay, and clock skew are presented and discussed with timing diagrams and VHDL code. The issues of gated clock and clock divider are also introduced.

Chapter 9 presents examples of dual-port RAM, synchronous and asynchronous FIFO, and dynamic RAM VHDL models. These blocks are commonly used as custom drop-in macros. They are used to interact with the rest of the design so that the complete design can be verified.

Chapter 10 illustrates the complete semicustom ASIC design process of a Finite Impulse Response ASIC design through the steps of design description, VHDL coding, functional verification, synthesis, layout, back-annotated timing verification.

Chapter 11 discusses the concept of a microprogram controller. The design of a AMD AM2910 is presented through the gate array design process from VHDL coding to postlayout back-annotated timing verification. Test vector generation is also illustrated.

Chapter 12 discusses the principles of error and correcting Hamming codes. An actual TI EDAC integrated circuit is used as an example to design from VHDL code to all steps of FPGA design process.

Chapter 13 presents the concepts of binary fixed-point multiplication algorithms such as Booth-Wallace multiplier. The VHDL coding, synthesis, and verification are presented.

Chapter 14 discusses the concepts of binary fixed-point division algorithms. VHDL coding, synthesis, and verification are presented.

Chapter 15 discusses the floating-point number representation. Floating-point addition and multiplication algorithms are discussed and implemented with VHDL codes. They are verified and synthesized.

Chapter 2

VHDL and Digital Circuit Primitives

Digital circuit primitive gates are basic building components such as AND, NAND, NOR, and INVERTER used by synthesis tools to construct designs. In this chapter, we will use VHDL to describe these basic digital circuit primitive gates to show how basic digital circuit gates can be described in VHDL so that the synthesis tools can get the design we want. It is important to understand the correspondence between VHDL and basic digital circuit gates.

2.1 FLIP-FLOP

A D flip-flop is the most common sequential element used in the digital design. Figure 2-1 shows a D flip-flop schematic symbol. It has two inputs, D and CLK, which represent data and clock, respectively. It has a data output port Q. The following VHDL code can be used to model the D flip-flop. The IEEE library and std_logic_1164 package are referenced in lines 1 and 2. Input and output ports are described in the VHDL entity from lines 3 to 7.

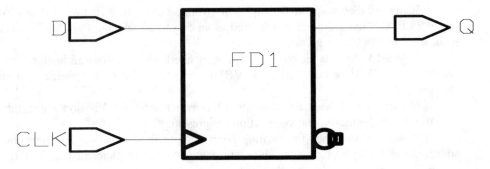

Figure 2-1 D flip-flop symbol.

```
1     library IEEE;
2     use IEEE.std_logic_1164.all;
3     entity DFF is
4       port(
5         CLK, D : in  std_logic;
6         Q      : out std_logic);
7     end DFF;
8     architecture RTL of DFF is
9     begin
10      seq0 : process
11      begin
12        wait until CLK'event and CLK = '1';
13        Q <= D;
14      end process;
15    end RTL;
```

The behavior of the D flip-flop is modeled inside the VHDL architecture from lines 8 to 15. A concurrent process statement (lines 10 to 14) is used. Recall that during the initialization phase of the simulation, every object (which is not initialized with a value) is initialized to its leftmost value of its basic type. Type std_logic is an enumerated type with values ('U', 'X', '0', '1', 'Z', 'W', 'L', 'H' and '-') to indicate certain conditions such as uninitialized, forcing unknown, forcing '0', forcing '1', high impedance, weak unknown, weak '0', weak '1', and don't care, respectively. Since D, CLK, and Q are declared as std_logic type, they are initialized to the uninitialized value 'U'. Every concurrent process statement is evaluated until it is suspended. In this case, after D, CLK, and Q are initialized to 'U', the process statement is evaluated from the beginning (line 11) until line 12. The process is then suspended as it waits for the rising edge of the clock signal CLK. Figure 2-2 shows the simulation waveform obtained by running the following QuickVHDL simulation commands: run 50, force D 0, force CLK 0, run 50, force CLK 1, run 50, force CLK 0, force D 1, run 50, force CLK 1, run 50, force CLK 0, run 50, force CLK 1, run 50, force CLK 0, force D 0, run 50, force CLK 1, run 50, force CLK 0, run 50, force CLK 1, run 50. Basically, we use the **run** command to advance the simulation time, and use the **force** command to change the input signal values.

The first command, run 50, advances the simulation time to 50 ns. Signals CLK, D, and Q have the same value 'U' since their values have not been assigned. At time 50 ns, D and CLK are both set to '0'. The simulation time is advanced to 100 ns where the clock signal CLK is assigned to '1'. At this time, the **process statement** proceeds through lines 12, 13, 14, and back to 11. It is again suspended at line 12. The output signal Q is updated from 'U' to '0'. Signal D is assigned to '1' at time 150 ns. At time 200 ns, clock signal CLK is assigned to '1', the **process statement** is executed again (line 12, 13, 14, back to 11), and suspended once more at line 12. Signal Q is updated to '1' and the simulation process is repeated as before.

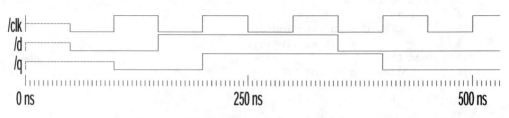

Figure 2-2 D flip-flop simulation waveform.

From the simulation waveform, it is easy to see that signal Q follows signal D with a delay. Whatever the value of signal D before the rising edge of the clock CLK, it is captured as the value of signal Q. This is the reason we often refer to a D flip-flop as a delay flip-flop.

Note that the DFF VHDL code is synthesized to the schematic shown in Figure 2-1. The D flip-flop has only one Q output; output Qn (inverted Q) is not connected. We can modify the DFF VHDL slightly to include the Qn output as shown in line 6. A temporary signal DFF is declared in line 9, which is used in line 14. Output signals Q and Qn are assigned in line 16. The DFFQN VHDL code is synthesized to the schematic as shown in Figure 2-3.

```
1     library IEEE;
2     use IEEE.std_logic_1164.all;
3     entity DFFQN is
4       port(
5         CLK, D : in  std_logic;
6         Q, Qn  : out std_logic);
7     end DFFQN;
8     architecture RTL of DFFQN is
9       signal DFF : std_logic;
10    begin
11      seq0 : process
12      begin
13        wait until CLK'event and CLK = '1';
14        DFF <= D;
15      end process;
16      Q <= DFF; Qn <= not DFF;
17    end RTL;
```

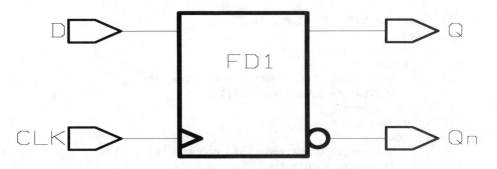

Figure 2-3 D flip-flop with Qn output.

The above D flip-flop is the simplest version. It does not have the synchronous or asynchronous reset input. Also, there are other ways to write VHDL to infer D flip-flops. The following VHDL code shows various ways to infer D flip-flops.

```
1    library IEEE;
2    use IEEE.std_logic_1164.all;
3    entity FFS is
4      port(
5         CLOCK : in  std_logic;
6         ARSTn : in  std_logic;
7         SRSTn : in  std_logic;
8         DIN   : in  std_logic_vector(5 downto 0);
9         DOUT  : out std_logic_vector(5 downto 0));
10   end FFS;
```

The ffs.vhd code infers six flip-flops with four **process statements**. Note that each **process statement** has the expression (CLOCK'event and CLOCK = '1') to check the rising edge of the clock signal CLOCK in lines 15, 20, 32, and 38. Every target signal (signals appearing on the left-hand side of the **signal assignment statements**) affected by the rising edge expression infers a number of flip-flops depending on the target signal width.

Processes seq0 (lines 13 to 17) and seq1 (lines 18 to 26) use a **wait statement** inside a **process statement** without a **sensitivity list**. The target signal to infer a flip-flop can have a Boolean expression on the right-hand side of the **signal assignment statements** as in lines 16, 24, and 26. The target signal value can also be affected by other **sequential statements** such as the **if statement** in lines 21 to 25. These Boolean expressions and **sequential statements** would result in combination circuits before the D input of the D flip-flop as shown in Figure 2-5 synthesized schematic. Processes seq2 (lines 28 to 35) and seq3 (lines 36 to 46) use a **process statement** with a **sensi-**

tivity list. There is no **wait statement** inside the **process statement.** As we recall, **a**
process statement **should have either a** *sensitivity list* **or at least a** *wait statement,*
but not both.

```
11    architecture RTL of FFS is
12    begin
13      seq0 : process
14      begin
15        wait until CLOCK'event and CLOCK = '1';
16        DOUT(0) <= DIN(0) xor DIN(1);
17      end process;
18      seq1 : process
19      begin
20        wait until CLOCK'event and CLOCK = '1';
21        if (SRSTn = '0') then
22          DOUT(1) <= '0';
23        else
24          DOUT(1) <= DIN(1) nand DIN(2);
25        end if;
26        DOUT(4) <= SRSTn and (DIN(1) nand DIN(2));
27      end process;
28      seq2 : process (ARSTn, CLOCK)
29      begin
30        if (ARSTn = '0') then
31          DOUT(2) <= '0';
32        elsif (CLOCK'event and CLOCK = '1') then
33          DOUT(2) <= DIN(2) nor DIN(3);
34        end if;
35      end process;
36      seq3 : process (CLOCK)
37      begin
38        if (CLOCK'event and CLOCK = '1') then
39          DOUT(5) <= SRSTn and DIN(3);
40          if (SRSTn = '0') then
41            DOUT(3) <= '0';
42          else
43            DOUT(3) <= DIN(3);
44          end if;
45        end if;
46      end process;
47    end RTL;
```

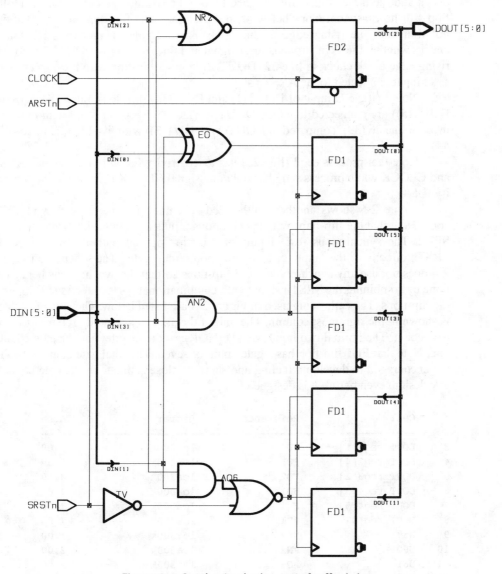

Figure 2-4 Synthesized schematic for ffs.vhd.

The input signals SRSTn and ARSTn are used to synchronously and asynchronously reset the flip-flops. SRSTn is used in lines 21, 26, 39, and 40 to show various ways to synchronously reset the flip-flops. The synchronous behavior is modeled by checking the rising edge before the SRSTn signal affects the target signal values. In other words, the event on signal SRSTn will not change the result until the clock rising edge is tested true. For example, lines 21 to 25 show the **if statement** using SRSTn to assign DOUT(1) after the clock rising edge check in line 20. Lines 26, 39, and 40 to 44 are also using SRSTn to synchronously reset the flip-flops. If the reset check and the target signal are assigned before the rising edge clock check, the flip-flop will have asynchronous behavior; the change of the flip-flop output value does not wait until the rising edge of the clock. For example, in process seq2, line 30 checks whether the asynchronous reset signal and line 31 assigns DOUT(2) before the rising edge clock check in line 32. DOUT(2) infers a flip-flop with the asynchronous reset input connected to the ARSTn signal.

Note that the D input of DOUT(1) and DOUT(4) flip-flops, (also DOUT(3) and DOUT(5)) are connected together. DOUT(4) (DOUT(5)) is modeled with one statement in line 26 (39), compared to DOUT(1) (DOUT(3)) with five lines 21 to 25 (40 to 44).

Note that process seq2 (line 28) has a **sensitivity list** with two signals ARSTn and CLOCK while process seq3 has only one signal CLOCK (line 36) in the **sensitivity list**.

Figure 2-5 shows another synthesized schematic for the same ffs.vhd VHDL code. Note that the flip-flops with synchronous inputs are synthesized with cell name SFD2. The synchronous reset inputs of the flip-flops are either connected to the SRSTn directly or the output of a combination circuit gate. The schematic in Figure 2-4 does not use any synchronous reset flip-flops so that the synchronous behavior is done by combining the SRSTn signal with combinational gates to feed the D inputs of the flip-flops. The cells being used in Figure 2-4 are listed below. The area of each cell is shown in the rightmost column. The unit of the area is commonly referred to as the gate count. The normal two input NAND (NOR), or the inverter has the gate count of one. Note that FD1 flip-flop has a gate count of seven. The total gate count is 52. You are encouraged to develop a feeling and sense of close estimate of the gate count for any design, even before it is designed.

```
1    Cell              Reference       Library           Area
2    ------------------------------------------------------------
3    DOUT_reg2[1]      FD1             lca300k           7.00
4    DOUT_reg2[4]      FD1             lca300k           7.00
5    DOUT_reg3[2]      FD2             lca300k           8.00
6    DOUT_reg4[3]      FD1             lca300k           7.00
7    DOUT_reg4[5]      FD1             lca300k           7.00
8    DOUT_reg[0]       FD1             lca300k           7.00
9    U59               NR2             lca300k           1.00
10   U60               AN2             lca300k           2.00
11   U61               AO6             lca300k           2.00
12   U62               IV              lca300k           1.00
13   U63               EO              lca300k           3.00
14   ------------------------------------------------------------
15   Total 11 cells                                      52.00
```

The following summarizes the gate count for the schematic in Figure 2-5.

```
1    Cell              Reference       Library         Area
2    -------------------------------------------------------------
3    DOUT_reg2[1]      SFD2            lca300k         8.00
4    DOUT_reg2[4]      FD1             lca300k         7.00
5    DOUT_reg3[2]      FD2P            lca300k         9.00
6    DOUT_reg4[3]      SFD2            lca300k         8.00
7    DOUT_reg4[5]      SFD2            lca300k         8.00
8    DOUT_reg[0]       FD1             lca300k         7.00
9    U50               AN2             lca300k         2.00
10   U51               ND2             lca300k         1.00
11   U52               NR2             lca300k         1.00
12   U53               EO              lca300k         3.00
13   -------------------------------------------------------------
14   Total 10 cells                                   54.00
```

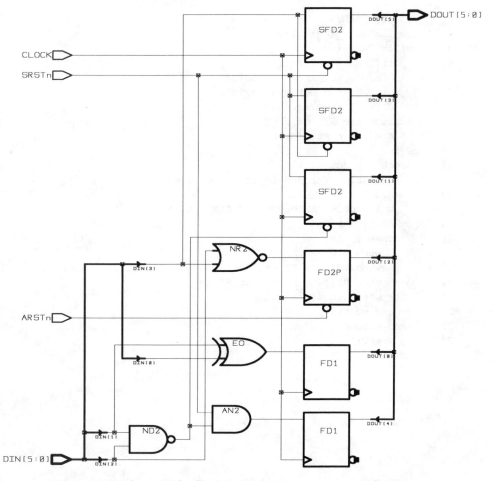

Figure 2-5 Schematic for ffs.vhd with synchronous reset D flip-flop.

Schematics Figure 2-4 and Figure 2-5 are synthesized from the same VHDL code. They function exactly the same except for their speed differences. Their total gate counts are also different. As we have stated above, it is a common practice to use synthesis tools to try to meet the speed requirements with the lowest possible gate count by using synthesis constraints. The synthesis constraints will be illustrated throughout later chapters.

A JK flip-flop can be described with the following VHDL code. When J&K is "10", the output is set to '1'. When J&K is "01", the output is cleared to '0'. When J&K is "11", the output toggles (reverses) the previous value. When J&K is "00", the output retains its previous value. Figure 2-6 shows the synthesized schematic. The JK flip-flop has a gate count of nine.

The function of a JK flip-flop can also be implemented with a D flip-flop. Figure 2-7 shows an example. It has a total gate count of 12. You can see that the same functionality (same VHDL code) can be synthesized with different schematics. This is achieved by synthesis commands to use only the D flip-flops.

```
1    library IEEE;
2    use IEEE.std_logic_1164.all;
3    entity JKFF is
4      port(
5        CLK, J, K : in  std_logic;
6        Q         : out std_logic);
7    end JKFF;
8    architecture RTL of JKFF is
9      signal FF : std_logic;
10   begin
11     seq0 : process
12       variable JK : std_logic_vector(1 downto 0);
13     begin
14       wait until CLK'event and CLK = '1';
15       JK := J & K;
16       case JK is
17         when "10" => FF <= '1';
18         when "01" => FF <= '0';
19         when "11" => FF <= not FF;
20         when others => null;
21       end case;
22     end process;
23     Q <= FF;
24   end RTL;
```

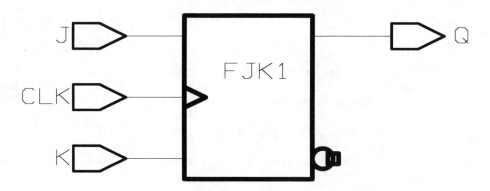

Figure 2-6 Synthesized schematic for jkff.vhd.

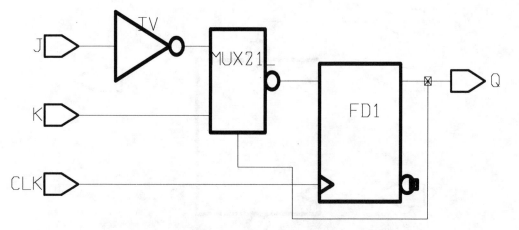

Figure 2-7 Synthesized schematic for jkff.vhd with D flip-flop.

The synchronous and asynchronous reset for a JK flip-flop is similar to the D flip-flop. It is also common to have synchronous and asynchronous preset D and JK flip-flops. What do we do when a flip-flop has both asynchronous preset and reset (clear)? The following VHDL code can be used. Figure 2-8 shows the synthesized schematic. What would happen if both SETn and CLRn are asserted to '0'? From a VHDL point of view, lines 12 to 15 imply that CLRn is checked first; it has higher priority over the SETn. Note that if statements always infer a priority of execution.

```
1     library IEEE;
```

```
2      use IEEE.std_logic_1164.all;
3      entity DFFSC is
4        port(
5          CLK, D, SETn, CLRn : in  std_logic;
6          Q                  : out std_logic);
7      end DFFSC;
8      architecture RTL of DFFSC is
9      begin
10       seq0 : process (SETn, CLRn, CLK)
11       begin
12         if (CLRn = '0') then
13           Q <= '0';
14         elsif (SETn = '0') then
15           Q <= '1';
16         elsif (CLK'event and CLK = '1') then
17           Q <= D;
18         end if;
19       end process;
20     end RTL;
```

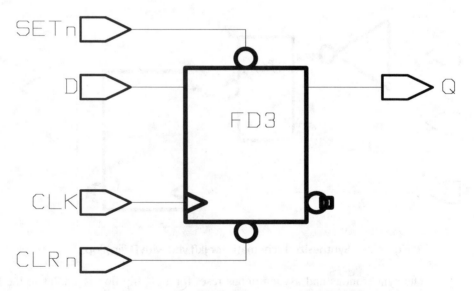

Figure 2-8 Synthesized schematic for dffsc.vhd.

The dffcs.vhd VHDL code is similar to dffsc.vhd, except the check of CLRn and SETn is reversed as shown in lines 12 to 15. Figure 2-9 shows the synthesized schematic. Note that extra gates of an inverter and a NAND gate are used to ensure the priority of CLRn and SETn described in the VHDL code. This is not the same as the priority of the cell FD3.

```
1    library IEEE;
2    use IEEE.std_logic_1164.all;
3    entity DFFCS is
4      port(
5        CLK, D, SETn, CLRn : in  std_logic;
6        Q                  : out std_logic);
7    end DFFCS;
8    architecture RTL of DFFCS is
9    begin
10     seq0 : process (SETn, CLRn, CLK)
11     begin
12       if (SETn = '0') then
13         Q <= '1';
14       elsif (CLRn = '0') then
15         Q <= '0';
16       elsif (CLK'event and CLK = '1') then
17         Q <= D;
18       end if;
19     end process;
20   end RTL;
```

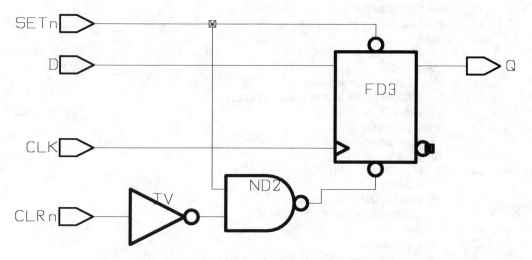

Figure 2-9 Synthesized schematic for dffcs.vhd.

2.2 LATCH

A basic latch has two inputs, LE (latch enable) and D (data), and an output Q. When the latch enable input LE is asserted, Q gets the same value as D. When LE is deasserted, Q retains its old value. The following VHDL code shows various ways to describe a latch. Lines 4 to 11 specify the VHDL entity. The behavior of a latch can be described in a VHDL **process statement** as shown in lines 14 to 19. When LE is asserted '1', the output (DOUT(0)) of the latch gets the input (DIN(0) xor DIN(1)). When LE is not asserted, DOUT(0) will not change its value because line 17 will not be executed. The **if statement** does not have the **else clause**. A latch is inferred whenever a target signal is not assigned a value (its old value is retained) in an execution path of a conditional statement such as the **if and case statement**. In process seq1 (lines 20 to 27), DOUT(1) is not assigned a value when both RSTn and LE are not asserted. In process seq2 (lines 28 to 36), DOUT(2) is not assigned a value in the **case statement** when LE is not asserted. In process seq3 (line 37 to 46), DOUT(3) is not assigned a value when LE is not asserted.

```
1     library IEEE;
2     use IEEE.std_logic_1164.all;
3     use work.pack.all;
4     entity LATCHES is
5       port(
6         RSTn, LE   : in  std_logic;
7         DVEC       : in  std_logic_vector(1 downto 0);
8         DIN        : in  std_logic_vector(4 downto 0);
9         DVOUT      : out std_logic_vector(1 downto 0);
10        DOUT       : out std_logic_vector(4 downto 0));
11    end LATCHES;
12    architecture RTL of LATCHES is
13    begin
14      seq0 : process (DIN, LE)
15      begin
16        if (LE = '1') then
17          DOUT(0) <= DIN(0) xor DIN(1);
18        end if;
19      end process;
20      seq1 : process (RSTn, LE, DIN)
21      begin
22        if (RSTn = '0') then
23          DOUT(1) <= '0';
24        elsif (LE = '1') then
25          DOUT(1) <= DIN(1) nand DIN(2);
26        end if;
27      end process;
28      seq2 : process (LE, DIN)
29      begin
30        case LE is
31          when '1' =>
32            DOUT(2) <= DIN(2) nor DIN(3);
33          when others =>
34            NULL;
35        end case;
36      end process;
37      seq3 : process (RSTn, LE, DIN)
```

```
38        begin
39          if (LE = '1') then
40            if (RSTn = '0') then
41              DOUT(3) <= '0';
42            else
43              DOUT(3) <= DIN(3);
44            end if;
45          end if;
46        end process;
47        LATCH_C(RSTn, LE, DIN(4), DOUT(4));
48        LATCH_C(RSTn, LE, DVEC, DVOUT);
49      end RTL;
```

The latches.vhd VHDL code can be synthesized as shown in the Figure 2-10 schematic. Note that processes seq0 and seq2 (lines 14 and 28) do not have the signal RSTn in the **process sensitivity list** and the target signal values DOUT(0) and DOUT(2) are not affected by the input signal RSTn. In process seq1, RSTn is checked in line 22, which is before the LE check in line 24. RSTn asynchronously resets DOUT(1) to '0' independently of the value of LE. On the other hand, in process seq3, the RSTn is checked in line 40, which is inside the **if statement** starting in line 39. DOUT(3) is reset to '0' when both LE and RSTn inputs are asserted. The latch is synchronously reset. Note that DOUT(1) is synthesized with the cell (LD3) while DOUT(3) is synthesized with the cell SLD3.

The **sequential statements** inside the **process statements** (as shown in process seq0, seq1, seq2, and seq3) to describe latches can be written as a VHDL procedure. Lines 47 and 48 use **concurrent procedure call statements** to call procedure LATCH_C that is in the package pack.vhd (Appendix A). The package named **pack** is referenced in line 3. The following summarizes the cells used in the schematic. Note that a latch has a gate count of five.

```
1     Cell              Reference      Library      Area
2     -------------------------------------------------------
3     DOUT_reg2[1]      LD3            lca300k      5.00
4     DOUT_reg3[2]      LD1            lca300k      5.00
5     DOUT_reg4[3]      SLD3           lca300k      5.00
6     DOUT_reg5[4]      LD3            lca300k      5.00
7     DOUT_reg[0]       LD1            lca300k      5.00
8     DVOUT_reg[0]      LD3            lca300k      5.00
9     DVOUT_reg[1]      LD3            lca300k      5.00
10    U85               NR2            lca300k      1.00
11    U86               ND2            lca300k      1.00
12    U87               EO             lca300k      3.00
13    -------------------------------------------------------
14    Total 10 cells                               40.00
```

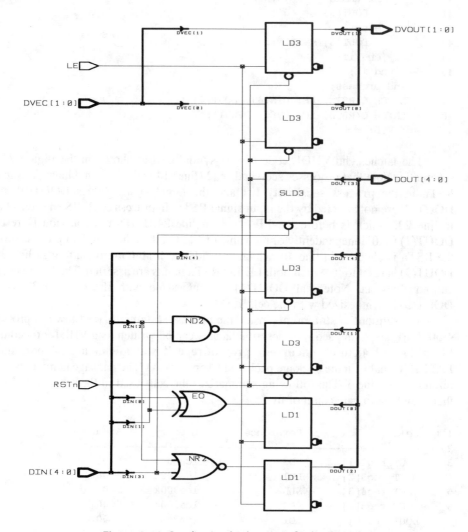

Figure 2-10 Synthesized schematic for latches.vhd.

2.3 THREE-STATE BUFFER

A three-state buffer has two inputs, EN (enable) and D (data), and one output Y. When EN is asserted, Y gets the same value as D. When EN is not asserted, Y gets the high-impedance state 'Z'. From a VHDL synthesis point of view, whenever a target signal is assigned a value 'Z', the target signal will be synthesized to a three-state buffer. The following VHDL code shows various ways that three-state buffers can be inferred. Lines 4 to 9 declare the VHDL entity. In process seq0 (lines 12 to 19), DOUT(0) is assigned 'Z' in line 17 whenever EN is not asserted in the **if statement**. In process seq1 (lines 20 to 28), DOUT(1) is assigned 'Z' in line 26 when En is not '1' in the **case statement**. Line 29 uses a **concurrent conditional signal assignment statement** to assign DOUT(2) to 'Z' when EN is not '1'. Lines 30 to 32 use a **concurrent selected signal assignment statement** to assign DOUT(3) to 'Z' when EN is not '1'. Lines 33 and 34 use a **concurrent procedure call** to call procedure TRIBUF declared in pack.vhd (Appendix A). The package pack is referenced in line 3. Figure 2-11 shows the synthesized schematic.

```
1     library IEEE;
2     use IEEE.std_logic_1164.all;
3     use work.pack.all;
4     entity STATE3 is
5       port(
6         EN     : in  std_logic;
7         DIN    : in  std_logic_vector(6 downto 0);
8         DOUT   : out std_logic_vector(6 downto 0));
9     end STATE3;
10    architecture RTL of STATE3 is
11    begin
12      seq0 : process (DIN, EN)
13      begin
14        if (EN = '1') then
15          DOUT(0) <= DIN(0) xor DIN(1);
16        else
17          DOUT(0) <= 'Z';
18        end if;
19      end process;
20      seq1 : process (EN, DIN)
21      begin
22        case EN is
23          when '1' =>
24            DOUT(1) <= DIN(2) nor DIN(3);
25          when others =>
26            DOUT(1) <= 'Z';
27        end case;
28      end process;
29      DOUT(2) <= DIN(1) nand DIN(2) when EN = '1' else 'Z';
30      with EN select
31        DOUT(3) <= DIN(3) when '1',
32                   'Z'    when others;
33      TRIBUF(DIN(4), EN, DOUT(4));
34      TRIBUF(DIN(6 downto 5), EN, DOUT(6 downto 5));
35    end RTL;
```

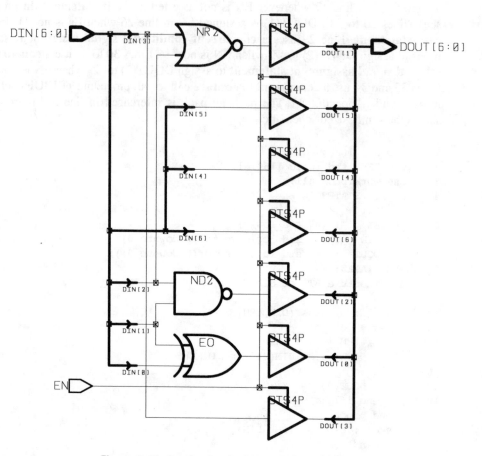

Figure 2-11 Synthesized schematic for state3.vhd.

The following list summarizes the cell area. Note that a three-state buffer has a gate count of four.

Cell	Reference	Library	Area
U66	ND2	lca300k	1.00
U67	NR2	lca300k	1.00
U68	EO	lca300k	3.00
U69	BTS4P	lca300k	4.00
U70	BTS4P	lca300k	4.00
U71	BTS4P	lca300k	4.00
U72	BTS4P	lca300k	4.00

10	U73	BTS4P	lca300k	4.00
11	U74	BTS4P	lca300k	4.00
12	U75	BTS4P	lca300k	4.00
13	--			
14	Total 10 cells			33.00

A three-stated signal is usually driven by multiple sources with three-state buffers. The following VHDL code shows a simple example. Signal DOUT is connected to both three-state buffer outputs as shown in Figure 2-12. This kind of connection is commonly used in an internal three-state bus and a bidirectional three-state bus which will be further discussed. The simulation of this example is left as Exercise 2.8.

```
1    library IEEE;
2    use IEEE.std_logic_1164.all;
3    use work.pack.all;
4    entity ST3EX is
5      port(
6        EN1, EN2, D1, D2 : in  std_logic;
7        DOUT             : out std_logic);
8    end ST3EX;
9    architecture RTL of ST3EX is
10   begin
11     TRIBUF(D1, EN1, DOUT);
12     TRIBUF(D2, EN2, DOUT);
13   end RTL;
```

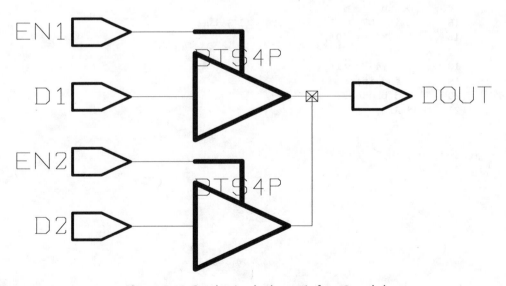

Figure 2-12 Synthesized schematic for st3ex.vhd.

2.4 COMBINATIONAL GATES

Basic combinational gates perform various logical operations such as NAND, NOR, AND, OR, and INVERTERS. These basic gates can be generated with various VHDL constructs. The following VHDL code shows four, two-input NOR gates which can be generated with different VHDL statements. Line 10 uses the VHDL Boolean operator 'nor' in a simple **signal assignment statement**. Lines 11 and 12 use the **conditional signal assignment statement** to generate Y2 and Y3. Lines 13 to 20 use a **process statement** with an **if statement** to generate Y4. Figure 2-13 shows the synthesized schematic. It is clear that two-input NOR gates are used for these four different types of VHDL constructs. The key is to write efficient good VHDL to model the circuit function correctly. Most of the rest are left to the synthesis tools to generate function-ally correct circuits that satisfy design requirements.

```
1     library IEEE;
2     use IEEE.std_logic_1164.all;
3     entity NORGATE is
4       port(
5         D1, D2, D3, D4, D5, D6, D7, D8 : in  std_logic;
6         Y1, Y2, Y3, Y4                 : out std_logic);
7     end NORGATE;
8     architecture RTL of NORGATE is
9     begin
10      Y1 <= D1 nor D2;
11      Y2 <= '0' when (D3 or  D4) = '1' else '1';
12      Y3 <= '1' when (D5 nor D6) = '1' else '0';
13      p0 : process (D7, D8)
14      begin
15        if (D7 = '1') or (D8 = '1') then
16          Y4 <= '0';
17        else
18          Y4 <= '1';
19        end if;
20      end process;
21    end RTL;
```

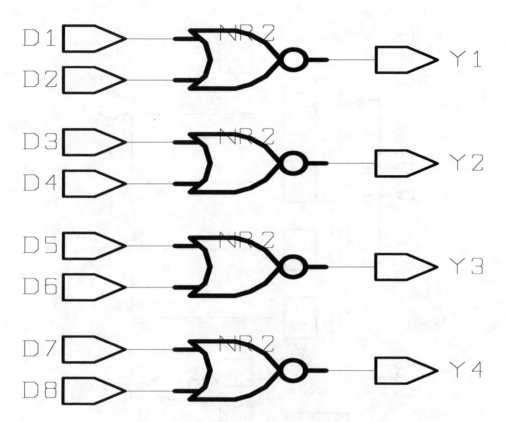

Figure 2-13 Synthesized schematic for norgate.vhd.

Other basic combinational circuit primitive gates include the multiplexer (or simply called MUX) and AND-NOR gates. However, they do not have a single VHDL Boolean operator to correspond. The following VHDL code shows examples to describe the multiplexer and AND-NOR functions. Line 13 uses VHDL Boolean operators 'and' and 'nor' to describe the AND-NOR gate function. Line 14 describes a multiplexer function. A **multiplexer** is also called a **selector** because the output is selected from the inputs based on the selection signal. In line 14, D4 is the selection signal. Y2 is assigned to D5 if D4 is '1', otherwise, Y2 is assigned to D6. Line 23 is similar except that the target signal is 2-bit wide and the concatenation operator '&' is used. Line 24 is a **concurrent procedure call statement**. Lines 15 to 22 use a **process statement** with an **if statement** inside. Note that the **sensitivity list** (line 15) should include all signals that are read within the **process statement**. Figure 2-14 shows the synthesized schematic with AND-NOR and 2-to-1 MUX gates.

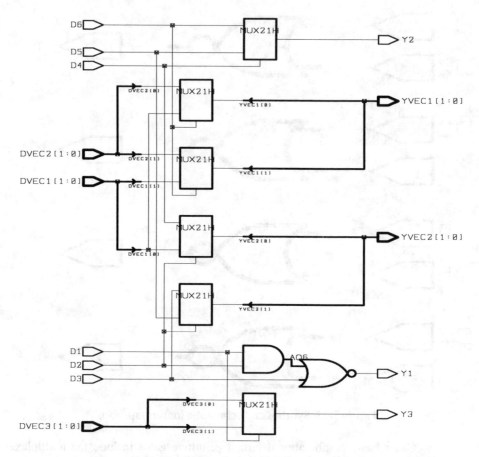

Figure 2-14 Synthesized schematic for complex.vhd.

```
1    library IEEE;
2    use IEEE.std_logic_1164.all;
3    use work.pack.all;
4    entity COMPLEX is
5      port(
6        D1, D2, D3, D4, D5, D6 : in  std_logic;
7        DVEC1, DVEC2, DVEC3    : in  std_logic_vector(1 downto 0);
8        YVEC1, YVEC2           : out std_logic_vector(1 downto 0);
9        Y1, Y2, Y3             : out std_logic);
10   end COMPLEX;
11   architecture RTL of COMPLEX is
12   begin
```

```
13        Y1 <= (D1 and D2) nor D3;
14        Y2 <= D5 when D4 = '1' else D6;
15        p0 : process (DVEC1, DVEC2, D6)
16        begin
17          if (D6 = '1') then
18            YVEC1 <= DVEC1;
19          else
20            YVEC1 <= DVEC2;
21          end if;
22        end process;
23        YVEC2 <= D3 & D4 when D2 = '0' else D5 & DVEC1(0);
24        KMUX21(D1, DVEC3(0), DVEC3(1), Y3);
25      end RTL;
```

The same complex.vhd can also be synthesized with the schematic shown in Figure 2-15 without using MUX gates. Figure 2-14 schematic has the following area statistic.

```
1     Number of ports:              19
2     Number of nets:               19
3     Number of cells:               7
4     Number of references:          2
5
6     Combinational area:        26.000000
7     Noncombinational area:      0.000000
8   . Net Interconnect area:     41.065342
9
10    Total cell area:           26.000000
11    Total area:                67.065338
```

Figure 2-15 schematic has the following area statistic. Based on these two schematics and area statistics, they have a different number of gates, interconnect nets, and total area measurements. Their timing delays are not shown here, but they are also different. However, they should have the same logical function. They conform to the golden rule of synthesis: meet the timing requirement (speed) with the smallest possible area. When the design is big and complex, there is no time to worry about what types of combination gates synthesis tools use for the design. The particular type of combinational gates used by the synthesis tools should be the lower priority that designers need to worry. Designers can spend more time on the architecture, functionality, speed, and area requirements by applying the synthesis tools to achieve their goals.

```
1     Number of ports:              19
2     Number of nets:               26
3     Number of cells:              14
4     Number of references:          5
5
6     Combinational area:        24.000000
```

```
7     Noncombinational area:        0.000000
8     Net Interconnect area:       60.131393
9
10    Total cell area:             24.000000
11    Total area:                  84.131393
```

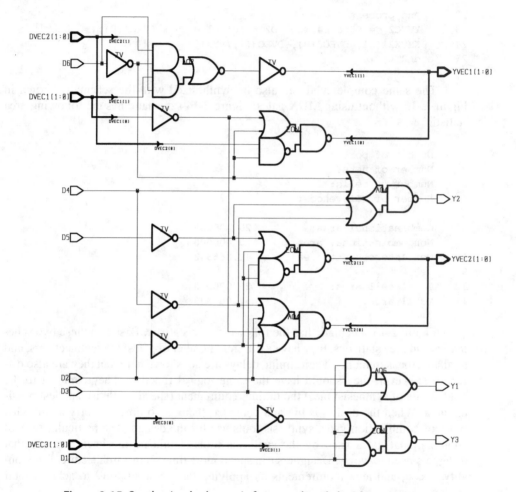

Figure 2-15 Synthesized schematic for complex.vhd without MUX gates.

2.5 VHDL SYNTHESIS RULES

We have discussed various ways to write VHDL code so that flip-flops, latches, three-state buffers, and combinational circuits can be generated by synthesis tools. It is easy to see the type of primitive gates generated by the synthesis tools when the VHDL code is independent. When they are mixed it is more difficult. Here are some general guidelines:

- When there is a signal edge expression such as (CLOCK'event and CLOCK = '1'), the target signals affected by the edge expression will be synthesized as flip-flops. The edge expression can also be a negative edge detection (CLOCK'event and CLOCK = '0'). It is a good practice to have only one edge detection expression within a **process statement**. Some VHDL codes have (CLK'event and CLK = '1' and CLK'last_value = '0') to ensure that the clock signal is actually a rising edge from '0' to '1'. This is not necessary. Also, some synthesis tools do not take this to infer flip-flops.

- If a target signal can be assigned a value 'Z', the target signal will infer three-state buffer(s).

- . If there is no edge detection expression and the target signal is not always assigned with a value, the target signal will infer a latch (or latches). It is a common mistake to unintentionally infer a latch.

- If none of the above situations exist, combinational circuits will be synthesized.

These guidelines are in priority order so that if the first condition is met it will drive the synthesis. If a target signal is synthesized to a flip-flop, it will not be synthesized as a latch even if the target signal is not fully assigned in every execution path. This is because a flip-flop is a sequential circuit that can remember its previous state. What happens when we start combining different types of inference? For example, what happens if a target signal is assigned value 'Z' and a latch or a flip-flop is inferred? The following VHDL code shows such a case. Line 14 has an edge detection expression so that the target signal FFOUT would be synthesized to a flip-flop. However, it can be assigned with value 'Z' in line 13. A three-state buffer will be inferred. Lines 18 to 25 show that the target signal LATCHOUT is not fully assigned in every condition (LE = '0'), therefore, a latch is inferred. It is also assigned value 'Z' in line 21 to infer a three-state buffer. Figure 2-16 shows the synthesized schematic. Note that FFOUT signal is implemented with a flip-flop followed by a three-state buffer. The three-state buffer enable signal also comes from a flip-flop. The same situation applies to the LATCHOUT signal. Two latches and one three-state buffer are synthesized.

```
1     library IEEE;
2     use IEEE.std_logic_1164.all;
3     entity TRIREG is
4       port (
5         CLOCK, LE, TRIEN, DIN : in  std_logic;
6         FFOUT, LATCHOUT       : out std_Logic);
7     end TRIREG;
8     architecture RTL of TRIREG is
9     begin
10      ff0 : process (TRIEN, CLOCK)
11      begin
12        if (TRIEN = '0') then
```

```
13              FFOUT <= 'Z';
14          elsif (CLOCK'event and CLOCK = '1') then
15              FFOUT <= DIN;
16          end if;
17        end process;
18        latch0 : process (LE, TRIEN, DIN)
19        begin
20          if (TRIEN = '0') then
21              LATCHOUT <= 'Z';
22          elsif (LE = '1') then
23              LATCHOUT <= DIN;
24          end if;
25        end process;
26      end RTL;
```

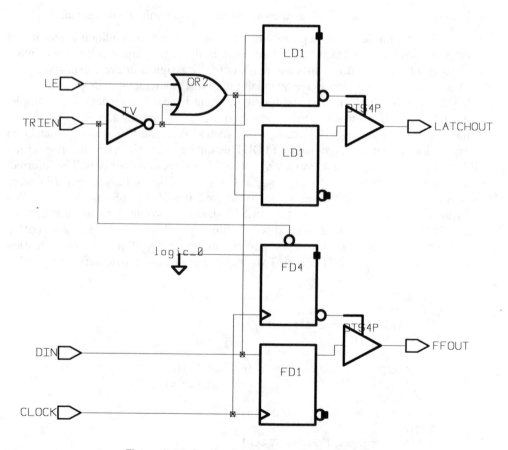

Figure 2-16 Synthesized schematic for trireg.vhd.

It is a good practice to isolate flip-flop, latch, and three-state buffer inferences to ensure design correctness. The following VHDL code shows the inferences of the flip-flop and the latch are separate from the three-state buffer. Figure 2-17 shows the synthesized schematic.

```
27    architecture RTLA of TRIREG is
28      signal DFF, LATCH : std_logic;
29    begin
30      ff0 : process
31      begin
32        wait until (CLOCK'event and CLOCK = '1') ;
33        DFF <= DIN;
34      end process;
35      FFOUT <= DFF when TRIEN = '1' else 'Z';
36      latch0 : process (LE, DIN)
37      begin
38        if (LE = '1') then
39          LATCH <= DIN;
40        end if;
41      end process;
42      LATCHOUT <= LATCH when TRIEN = '1' else 'Z';
43    end RTLA;
```

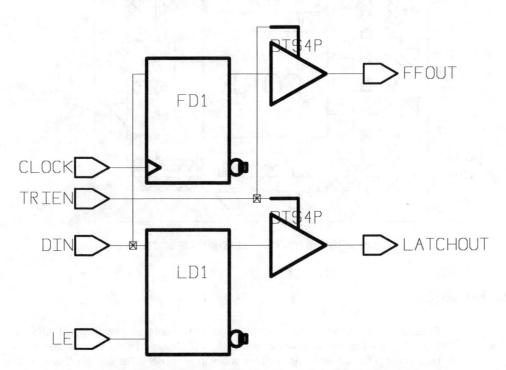

Figure 2-17 Synthesized schematic for trireg.vhd with RTLA architecture.

2.6 PADS

Pads are used in the ASIC and FPGA to physically interface between the design implemented with digital circuit primitives and the outside world. Figure 2-18 shows the general die configuration. Pads have a pad bonding to connect to the package, transistors for drivers, and electric static protection circuits. A VDD pad is used to connect to the power supply, VSS pads are used to connect to ground, and signal pads are used to interface between the circuit inside the CORE and outside signals. Note also that the power and ground inside the CORE are connected to the power and ground pads. There are a power ring and a ground ring along the pads on the boundary of the die. Generally, synthesis tools will be used as much as possible to implement the circuits inside the CORE. Pads are usually instantiated with **component instanti-ation statements**. Readers are encouraged to review Chang, *Digital Design and Modeling with VHDL and Synthesis,* IEEE Computer Press, 1997 Chapter 10 for modeling pads with VHDL. We will discuss pad issues further in later chapters with examples.

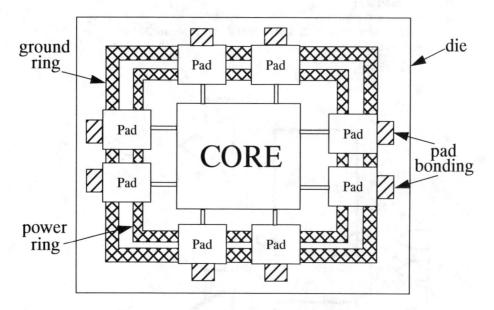

Figure 2-18 Die configuration with core and pads.

2.7 EXERCISES

2.1 Refer to ffs.vhd VHDL code seq2 process. What would happen if signal ARSTn or CLOCK is not included in the process sensitivity list? Signal DIN is used in process seq2, but it is not in the sensitivity list. Is it necessary to have DIN in the seq2 process sensitivity list? What would be the effect to include DIN in the seq2 process sensitivity list?

2.2 Refer to ffs.vhd VHDL code seq3 process. What would be the effect of including DIN and SRSTn in the seq3 process sensitivity list?

2.3 Refer to ffs.vhd VHDL code. Sketch a simulation waveform to show the timing relationship among SRSTn, ARSTn, DIN, and DOUT, especially when the value of each signal is changed. Run a VHDL simulator to verify your answer.

2.4 Synthesize ffs.vhd VHDL code with a library other than LSI lca300k. How many different flip-flops are used in your synthesized schematic?

2.5 Refer to dffqn.vhd. What circuit would be synthesized if we insert line 16 after line 14?

2.6 Refer to Figure 2-7. Draw a different schematic to use a D flip-flop and other combinational circuit gates to implement a JK flip-flop.

2.7 Refer to latches.vhd VHDL code and pack.vhd in Appendix A. VHDL procedure LATCH_C is called. How many LATCH_C procedure declarations in pack.vhd? Are both procedure calls called the same procedure? How do the VHDL tools know which procedure to use? What is the common name of this feature?

2.8 Refer to latches.vhd VHDL code. Sketch a simulation waveform to show the timing relationship among RSTn, LE, and DOUT, especially when the value of each signal is changed. Run a VHDL simulator to verify your answer.

2.9 Refer to pack.vhd VHDL code. What are the restrictions on the DIN and DOUT in LATCH_C procedure? Can we call the procedure with actual parameters DA(4 downto 3), and DB(2 downto 1) to map to DIN and DOUT to infer two latches? Explain the reasons. How do you modify the procedures to handle this situation? Use your synthesis tool to verify your answers.

2.10 Refer to latches.vhd VHDL code. What are the advantages and disadvantages of using **procedure calls** compared to using **process statements** to infer latches?

2.11 Refer to latches.vhd VHDL code. Justify the sensitivity list for every **process statement**. What would happen if any one or all signals were missing from the original sensitivity list?

2.12 Refer to state3.vhd VHDL code. Sketch a simulation waveform to show the timing relationship among DIN, EN, and DOUT, especially when the value of each signal is changed. Run a VHDL simulator to verify your answer.

2.13 Refer to pack.vhd VHDL code. What are the differences between procedures INVTRIBUF, INVTRIBUFn, TRIBUF, TRIBUFn?

2.14 Refer to pack.vhd VHDL code. What are the restrictions on the DIN and DOUT in the TRIBUF procedure? Can we call the procedure with actual parameters DA(4 downto 3), and DB(2 downto 1) to map to DIN and DOUT to infer two, three-state buffers? Explain the reasons. How do you modify the procedures to handle this situation? Use your synthesis tool to verify your answers.

2.15 Refer to st3ex.vhd VHDL code. Develop a truth table for inputs D1, D2, EN1, EN2 to have values '0', '1' and show the corresponding DOUT value. Use a VHDL simulator to verify your answer. What would happen if both EN1 and EN2 are asserted '1' and D1 and D2 have different values?

2.16 Refer to complex.vhd VHDL code. Line 13 is a signal assignment statement "Y1 <= (D1 and D2) nor D3;". Are the parentheses required? Explain the reason for this.

2.17 Write a functional behavioral model for each type of pad (input pad, output pad, bi-directional pad, three-state output pad, pull-up bidirectional pad, pull-down three-state output pad).

Chapter 3

VHDL Simulation and Synthesis Environment and Design Process

A basic VHDL design environment consists of a set of VHDL simulation tools and a set of synthesis tools. In this chapter, VHDL simulation tools, VHDL synthesis tools, synthesis technology component libraries, VHDL design libraries, and design directory structures are introduced with examples. Understanding their setups and relationships is critical to effectively and efficiently managing the design. Synthesis commands are used to direct synthesis tools to achieve design objectives and to detect design errors and report statistics. These concepts are presented in the context of a VHDL design process.

3.1 SYNOPSYS VHDL SIMULATION ENVIRONMENT OVERVIEW

Synopsys VHDL simulation tools include the following key items. Figure 3-1 illustrates their relationships.

- A VHDL analyzer parses VHDL code, checks and reports syntax errors, and stores the compiled object database in VHDL simulation libraries.

- A VHDL simulator gets the VHDL object database and executes a primary design unit (specified by the user) from a simulation library. It saves the simulation results such as text files, simulation waveforms, and simulation transcript recording simulation commands and messages (from VHDL assert statements) for users to verify.

- A VHDL waveform viewer displays simulation waveforms to show particular interested signals and variables in the design.

- A VHDL source code debugger combines a VHDL simulator, a VHDL waveform viewer, and a source code display. It gives users greater control of simulation for debugging purposes. For example, a debugger allows users to set a breakpoint, to step through VHDL codes, and to evaluate values of VHDL signals and variables.

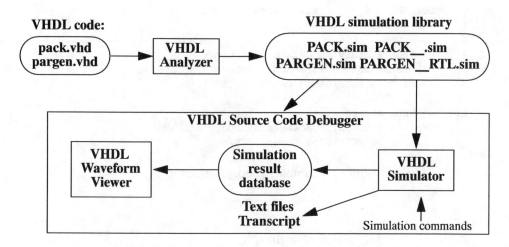

Figure 3-1 Synopsys VHDL simulation environment.

Figure 3-3 shows a sample directory structure starting at directory entitled **design**. It has subdirectories **vhdl, rtlsim, rtlsim.work, syn, syn.work, db,** and others (such as test vectors, etc.). Under directory **vhdl**, there are subdirectories **ch2** and **ch3** for keeping VHDL source codes. Directory **ch3** has two VHDL files, pack.vhd and pargen.vhd. Directory **rtlsim** is the directory where RTL VHDL simulation is run. For example, in directory **rtlsim**, Unix commands "**vhdlan ../vhdl/ch3/pack.vhd**," "**vhdlan ../vhdl/ch3/pargen.vhd**" run the SYNOPSYS VHDL analyzer. This generates object database files PACK.sim, PACK__.sim, PARGEN.sim, and PARGEN__RTL.sim in directory *rtlsim.work*, which is used as a VHDL simulation library. File .synopsys_vss.setup (as shown below) in directory **rtlsim** specifies the VHDL design library mapping in lines 3 and 4, that maps the working library WORK to ../rtlsim.work directory. Lines 5, 8, and 9 also map other VHDL simulation libraries. Line 1 specifies the time unit as nanosecond for running simulation. Lines 12 to 13 specify the VHDL source code directory for the source code debugger to find the VHDL code. Lines 8 and 9 use the Unix environment variable $SYNOPSYS and $ARCH. $SYNOPSYS variable is used to specify the physical directory where Synopsys software is stored. $ARCH variable is used to specify the machine type such as SUN, HP700, etc.

```
1      TIMEBASE    = NS
2      WAVEFORM    = WAVES + WIF
3      WORK        > DEFAULT
4      DEFAULT     : ../rtlsim.work
5      PACK_LIB    : ../pack.lib
6
7      -- VHDL library to UNIX dir mappings --
8      SYNOPSYS: $SYNOPSYS/$ARCH/packages/synopsys/lib
9      IEEE: $SYNOPSYS/$ARCH/packages/IEEE/lib
10
11     -- VHDL source files search path --
12     USE = . $SYNOPSYS/packages/synopsys/src \
13              $SYNOPSYS/packages/IEEE/src
```

Figure 3-2 Sample .synopsys_vss.setup file.

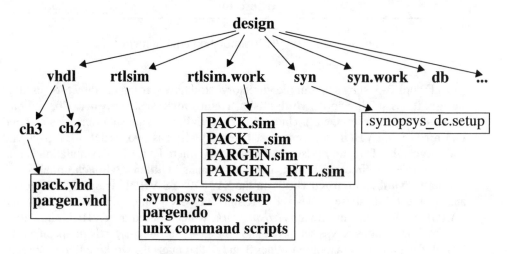

Figure 3-3 Sample directory structure — Synopsys simulation.

File pargen.vhd is as follows:
```
1      library IEEE;
2      use IEEE.std_logic_1164.all;
3      use work.pack.all;
4      entity PARGEN is
5        port(
6          CBEn0, CBEn1, CBEn2, CBEn3 : in  std_logic;
7          AD  : in  std_logic_vector(31 downto 0);
8          PAR : out std_logic);
9      end PARGEN;
10     architecture RTL of PARGEN is
11     begin
```

```
12      PAR <= REDUCE_XOR(AD) xor CBEn0 xor CBEn1 xor
13                               CBEn2 xor CBEn3;
14    end RTL;
```

In directory **rtlsim**, Unix command **vhdldbx** brings up the Synopsys VHDL VSS source code debugger. A primary design unit can be selected for simulation from a design library. For example, PARGEN with architecture RTL (PARGEN__RTL) can be selected for simulation. The following shows a sample of the simulation command file **pargen.do**. The command file can be executed by the vhdldbx command **include pargen.do**. Synopsys VHDL simulator **vhdlsim** can also be run as a batch job by taking the same command file. Each individual command can also be typed in interactively. Line 1 runs 30 time units (nanosecond as specified in .synopsys_vss.setup file). Line 2 changes the design hierarchy to the top with the primary design unit as the hierarchy name. Lines 3 to 4 specify signals to be displayed in the waveform. Note that a wildcard * can be used. The command **assign** puts a desired value into a signal. Note that the assigned value uses the VHDL syntax. For example, lines 6 to 9 use single quotes for single-bit signals. Lines 13 and 15 use double quotes for bus signals. Line 5 uses (others => '0') to assign '0' to every bit in **ad** bus.

```
1     run 30
2     cd /pargen
3     trace CBEn*
4     trace AD par
5     assign (others => '0') ad
6     assign '0' cben0
7     assign '0' cben1
8     assign '0' cben2
9     assign '0' cben3
10    run 30
11    assign '1' /pargen/cben1
12    run 30
13    assign "111" /pargen/ad(5 downto 3)
14    run 30
15    assign "11111" /pargen/ad(31 downto 27)
16    run 30
```

Figure 3-4 File pargen.do.

The aforementioned **vhdlan** command puts the compiled simulation database in the default VHDL **working library**. The default working library is mapped to ../rtlsim.work. Note that another simulation library PACK_LIB is mapped to ../pack.lib in line 5 of the .synopsys_vss.setup file. To put the compiled simulation database into the PACK_LIB, command "**vhdlan -w PACK_LIB filename**" can be used. For example, command "**vhdlan -w PACK_LIB ../vhdl/ch3/pack.vhd**" can be used to put the compiled pack.vhd into the PACK_LIB library which is mapped to the ../pack.lib directory. The PACK_LIB can be a **resource library** for another VHDL code. The following VHDL code shows a way to reference the PACK_LIB library. Line 2 declares the library PACK_LIB. Line 4 specifies that package pack will be referenced so that the function REDUCE_XOR (line 13) can be used.

```
1      library IEEE;
2      library PACK_LIB;
3      use IEEE.std_logic_1164.all;
4      use PACK_LIB.pack.all;
5      entity LIBREF is
6        port(
7          CBEn0, CBEn1, CBEn2, CBEn3 : in  std_logic;
8          AD : in  std_logic_vector(31 downto 0);
9          PAR : out std_logic);
10     end LIBREF;
11     architecture RTL of LIBREF is
12     begin
13       PAR <= REDUCE_XOR(AD) xor CBEn0 xor CBEn1 xor
14                              CBEn2 xor CBEn3;
15     end RTL;
```

Compared with the *pargen.vhd* example, the pack.vhd is compiled into the default working library so that *pargen.vhd* line 3 references the working library *work*. It is not necessary to declare library *work* as a library. By default, it is visible without declaration.

3.2 MENTOR QUICKVHDL SIMULATION ENVIRONMENT

The simulation environment of Mentor QuickVHDL is similar to the Synopsys. Both require methods to map the VHDL design libraries. To create a QuickVHDL design library, Unix command "**qvlib ../rtlsim.work**" can be used. The Unix command "**qvmap work ../rtlsim.work**" maps the working library WORK to a physical directory ../rtlsim.work. Unix commands "**qvcom ../vhdl/ch3/pack.vhd**," "**qvcom ../vhdl/ch3/pargen.vhd**" can be used to analyze VHDL files. Figure 3-5 shows a sample directory structure. The Modelsim VHDL simulator uses the commands vlib, vmap, vcom, and vsim which correspond to qvlib, qvmap, qvcom, and qvsim, respectively.

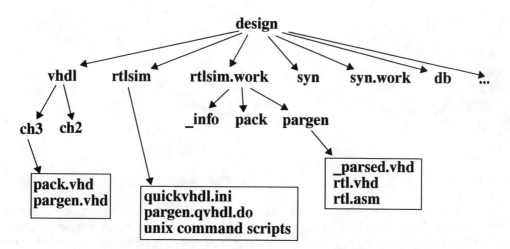

Figure 3-5 Sample directory structure — Mentor QuickVHDL simulation.

Figure 3-5 is very similar to Figure 3-3. File quickvhdl.ini is created and updated when Unix commands **qvlib** and **qvmap** are executed. Directory **../rtl-sim.work** is created by command **qvlib**. The QuickVHDL library mapping is done with the command **qvmap** while the Synopsys library mapping is specified in file **.synopsys_vss.setup**. The directory structure under **../rtlsim.work** has subdirectories with each design unit in a subdirectory. The Unix command **qvsim** brings up Quick-VHDL simulator. Simulation command file **pargen.qvhdl.do** (as shown below) can be executed with QuickVHDL command **do pargen.qvhdl.do**. Figure 3-7 shows the simulation waveform.

```
1     wave CBEn*
2     wave AD par
3     run 30
4     force ad 00000000000000000000000000000000
5     force cben0 0
6     force cben1 0
7     force cben2 0
8     force cben3 0
9     run 30
10    force cben1 1
11    run 30
12    force ad(5 downto 3) 111
13    run 30
14    force ad(31 downto 27) 11111
15    run 30
```

Figure 3-6 File pargen.qvhdl.do.

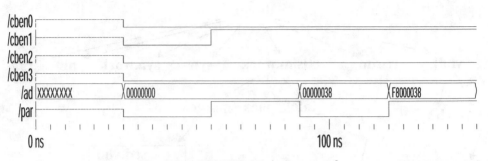

Figure 3-7 PARGEN simulation waveform.

Note that the simulation commands can be run one by one interactively or run as a batch (using QuickVHDL *do* command or Synopsys VSS *include* command) inside the VHDL simulators. Commands do and include are used to run batches of commands captured in a single file. These commands are used to check simple functions. In practice, a test bench is usually set up to run the simulation. The test bench will be discussed with more examples throughout the book.

Mentor also provides a VHDL simulator that can be run on a personal computer (PC). The commands and interfaces are similar to QuickVHDL.

The process to compile VHDL code to a VHDL simulation library in the Mentor QuickVHDL environment is similar to the Synopsys. As an example, we can create SYNOPSYS library and IEEE library using Synopsys VHDL codes. This enables the same VHDL code to be used both in Synopsys synthesis tools and Mentor Quick-VHDL simulation tools. The following shows the commands. Note that library SYN-OPSYS is generated before library IEEE is in line 14.

```
1    qvlib SYNOPSYS
2    qvmap SYNOPSYS SYNOPSYS
3    qvmap work SYNOPSYS
4    qvcom $SYNOPSYS/packages/synopsys/src/attributes.vhd
5    qvcom $SYNOPSYS/packages/synopsys/src/distributions.vhd
6    qvcom $SYNOPSYS/packages/synopsys/src/sdf_header.vhd
7    qvcom $SYNOPSYS/packages/synopsys/src/bvarithmetic.vhd
8    qvcom $SYNOPSYS/packages/synopsys/src/cyclone.vhd
9    qvcom $SYNOPSYS/packages/synopsys/src/mem_api.vhd
10   qvcom $SYNOPSYS/packages/synopsys/src/vhdlq.vhd
11   qvcom $SYNOPSYS/packages/synopsys/src/vhdlzq.vhd
12   qvcom $SYNOPSYS/packages/synopsys/src/qe.vhd
13     echo "SYNOPSYS library compiled"
14   qvlib IEEE
15   qvmap work IEEE
16   qvmap SYNOPSYS SYNOPSYS
17   qvmap IEEE IEEE
18   qvcom $SYNOPSYS/packages/IEEE/src/std_logic_1164.vhd
```

```
19    qvcom $SYNOPSYS/packages/IEEE/src/std_logic_arith.vhd
20    qvcom $SYNOPSYS/packages/IEEE/src/std_logic_misc.vhd
21    qvcom $SYNOPSYS/packages/IEEE/src/std_logic_signed.vhd
22    qvcom $SYNOPSYS/packages/IEEE/src/std_logic_unsigned.vhd
23    qvcom $SYNOPSYS/packages/IEEE/src/math_complex.vhd
24    qvcom $SYNOPSYS/packages/IEEE/src/std_logic_textio.vhd
25    qvcom $SYNOPSYS/packages/IEEE/src/math_real.vhd
26    qvcom $SYNOPSYS/packages/IEEE/src/std_logic_entities.vhd
27    qvcom $SYNOPSYS/packages/IEEE/src/std_logic_components.vhd
28    qvcom $SYNOPSYS/packages/IEEE/src/timing_p.vhd
29    qvcom $SYNOPSYS/packages/IEEE/src/timing_b.vhd
30    qvcom $SYNOPSYS/packages/IEEE/src/prmtvs_p.vhd
31    qvcom $SYNOPSYS/packages/IEEE/src/prmtvs_b.vhd
32    qvcom $SYNOPSYS/packages/IEEE/src/gs_types.vhd
33       echo "IEEE library compiled"
```

3.3 SYNTHESIS ENVIRONMENT

Figure 3-8 shows Synopsys synthesis tools and their relationship.

- *Design analyzer.* The Design Analyzer is a user interface tool. It has a command window for users to supply synthesis commands. Based on the command, other tools such as VHDL Compiler, Design Compiler, and Test Compiler can be invoked.

- *VHDL compiler.* The VHDL Compiler parses VHDL code, checks for syntax errors, and checks for VHDL constructs in the VHDL code that are synthesizable. It also generates a summary report with the number of inferred latches, flip-flops, and three-state buffers. It gives warning messages, such as when a signal is missing in the sensitivity list, and prepares an intermediate form for the Design Compiler to perform the optimization.

- *Design compiler.* The Design Compiler optimizes the design based on the synthesis constraints such as speed and area requirements. After the design is synthesized, a netlist can be generated for the place and route tools.

- *Test compiler.* The Test Compiler inserts scan paths, boundary scan cells, and a Test Access Port Controller (TAPC). It can also generate test patterns.

- *Design time.* The Design Time is a timing analyzer. It is used to find the critical paths and to perform timing checks.

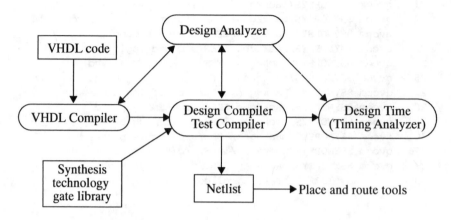

Figure 3-8 Synopsys synthesis tools.

Figure 3-9 shows a sample directory structure for synthesis. Directory *syn* is used to run the synthesis. The Unix command **design_analyzer** can be used to bring up the Synopsys Design Analyzer, which provides a user interface for running synthesis. The **design_analyzer** executes commands specified in file **.synopsys_dc.setup**, and brings up the a Design Analyzer window.

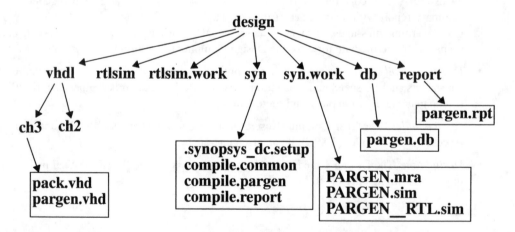

Figure 3-9 Sample directory structure — Synopsys synthesis.

The following is an example of a **.synopsys_dc.setup** file. Line 10 maps the working library WORK to **../syn.work**. Line 9 maps the IEEE library. Line 1 allows the Unix variable to be used in line 9 for the IEEE library. Line 2 declares a variable. Lines 6 and 7 declare the library path and their search sequence. Line 4 defines the target library which is the target technology library. Line 5 declares the symbol library for the schematics. Line 8 defines the synthetic library which has basic design building blocks such as adders and incrementers. Line 3 specifies a link library so that various synthesized designs can be found and linked. Line 11 defines the Synopsys synthesis variable to override its original default value. Lines 12 to 15 define user variables.

```
1     SYNOPSYS            = get_unix_variable ("SYNOPSYS")
2     comp_lib            = "lca300k"
3     link_library        = {"*" comp_lib + ".db" }
4     target_library      = comp_lib + ".db"
5     symbol_library      = comp_lib + ".sdb"
6     search_path         = {. /app/synopsys_1997.01/libraries/syn \
7                            /acct/synlib/lsi}
8     synthetic_library = {standard.sldb dw01.sldb dw02.sldb}
9     define_design_lib IEEE -path SYNOPSYS + "/hp700/packages/IEEE/
lib"
10    define_design_lib WORK -path ../syn.work
11    hdlin_report_inferred_modules = "verbose"
12    company = "STD LOGIC"
13    designer = "K. C. Chang"
14    VHDLSRC = "../vhdl/"
15    REPORT  = "../report/"
16    DBDIR   = "../db/"
```

Figure 3-10 Sample .synopsys_dc.setup file.

To run the synthesis, a synthesis script file is usually used. The following file compile.pargen is shown as an example. Line 1 executes a Unix command date during the synthesis. Line 2 declares the name of the design. Line 3 reads in the VHDL file with VHDLSRC variable defined in file **.synopsys_dc.setup** shown above. Line 4 includes another synthesis script file **compile.common**. Line 5 defines a clock with skew defined in line 6. Lines 7 to 8 specify the input and output delays for input and output ports. Line 9 synthesizes the design. Line 10 includes another synthesis script.

```
1     sh date
2     design = pargen
3     read -f vhdl VHDLSRC + "ch3/" + design + ".vhd"
4     include compile.common
5     create_clock -name clk -period CLKPERIOD -waveform {0.0
HALFCLK}
6     set_clock_skew -uncertainty CLKSKEW clk
7     set_input_delay  INPDELAY -add_delay -clock clk all_inputs()
8     set_output_delay CLKPER1  -add_delay -clock clk all_outputs()
9     compile -map_effort medium
10    include compile.report
```

Figure 3-11 File compile.pargen.

File **compile.common** is as follows: Line 1 defines the default input delay variable. Line 2 specifies the amount of clock skew. Line 3 defines a TIMEUNIT variable to be used in lines 4 to 15. Line 16 defines the wire load model to be used. Line 17 specifies the operating condition. Line 18 defines that all input ports have the same drive as an ND2 gate in the library. Line 19 defines the default loading of all output ports. Lines 20 to 22 declare different loading values. The purpose of this file is to define variables and defaults which are common to synthesis. For example, changing the TIMEUNIT (line 3) from 3 to 5 would change the clock period from 30 ns to 50 ns. Sometimes, the synthesis library has a time unit as a picosecond. The only change in this file would be the first three lines by a factor of 1,000. If there is a need to port the design to use another technology library, only files **.synopsys_dc.setup** and **compile.common** are required to change.

```
1     INPDELAY    = 2.5
2     CLKSKEW     = 1
3     TIMEUNIT    = 3
4     CLKPERIOD   = TIMEUNIT * 10
5     HALFCLK     = CLKPERIOD / 2
6     QUTRCLK     = CLKPERIOD / 4
7     CLKPER1     = TIMEUNIT
8     CLKPER2     = TIMEUNIT * 2
9     CLKPER3     = TIMEUNIT * 3
10    CLKPER4     = TIMEUNIT * 4
11    CLKPER5     = TIMEUNIT * 5
12    CLKPER6     = TIMEUNIT * 6
13    CLKPER7     = TIMEUNIT * 7
14    CLKPER8     = TIMEUNIT * 8
15    CLKPER9     = TIMEUNIT * 9
16    set_wire_load B9X9
17    set_operating_conditions "WCCOM"
18    set_driving_cell -cell ND2 all_inputs()
19    set_load  8 * load_of( comp_lib + "/ND2/A" ) all_outputs()
20    load8  =  8 * load_of( comp_lib + "/ND2/A" )
21    load16 = 16 * load_of( comp_lib + "/ND2/A" )
22    load32 = 32 * load_of( comp_lib + "/ND2/A" )
```

Figure 3-12 File compile.common.

Synthesis script file **compile.report** is shown below. It generates synthesis results to be placed in the REPORT directory. Save the synthesized data base in the DBDIR directory. Both variables REPORT and DBDIR are declared in file **.synopsys_dc.setup**.

```
1    check_design        > REPORT + design + ".rpt"
2    report_constraint  >> REPORT + design + ".rpt"
3    report_area        >> REPORT + design + ".rpt"
4    report_timing      >> REPORT + design + ".rpt"
5    create_schematic -all
6    write -f db -hier -o DBDIR  + design + ".db"
7    remove_design -design
8    remove_license VHDL-Compiler
```

Figure 3-13 File compile.report.

In Design Analyzer, the command "**include compile.pargen**" can be run. Figure 3-14 shows the synthesized schematic. Directory **../db** has a file **pargen.db**. Directory **../report** has a file **pargen.rpt**. It is strongly recommended that **../rtl-sim.work** and **../syn.work** be separate directories. When both Synopsys VHDL simulation and synthesis tools use the same working library, if the time stamp of a file is changed by one tool this would cause the problem of "out of date" issue for the other tool. Separating the two directories reduces the impact between the VHDL simulation process and the synthesis process.

Note that all design directories are relatively specified. This reduces the changes required when the design directory is moved (copied) into another location. Establishing a common directory structure (for the same project, design group, company) enables designers to find things much more easily. It also allows designers to support various projects without feeling things are out of place. However, setting up an agreed directory structure is sometimes a very personal matter. Guidelines should be established and disciplines should be maintained for the project supported by multiple designers.

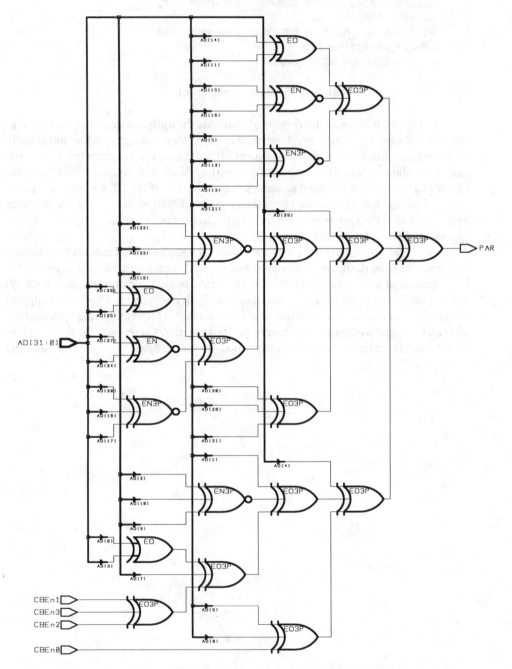

Figure 3-14 Synthesized schematic for PARGEN.

3.4 SYNTHESIS TECHNOLOGY LIBRARY

A synthesis technology library consists of definitions of basic components, operating conditions, and wire load models. Basic components are selected by the synthesis tool to implement the VHDL design based on functions, timing, area, and synthesis constraints. Each component has its logical function, timing, input pin loading, and output pin drive information. The library is usually provided by the ASIC or FPGA manufactures in a binary format understood by Synopsys synthesis tools. The synthesis technology library usually comes as **library_name.db**. The associated symbol library is usually named **library_name.sdb**. For example, the **.synopsys_dc.setup** file discussed above specifies **lca300k.db** and **lca300k.sdb**, which are stored in directory, /acct/synlib/lsi. Inside the Design Analyzer, command "**read /acct/synlib/lsi/ lca300k.db**" can be used to read in the synthesis technology library. Command **report_lib lca300k** can be used to report the library. The following shows part of the report. Lines 1 to 5 specify units used in the library. Lines 7 to 17 define several operating conditions. For example, operating condition WCMIL (worst case military) is 125 degrees with 4.5 volts. The derating factor is 1.4, which is 40 percent slower than the nominal operating condition at 25 degrees with 5.0 volts. Lines 19 to 36 declare a wire load model. Lines 27 to 36 are a lookup table. Each entry specifies the estimated wire length for a fanout number (the number of input pins connected to an output pin of a component). The wire length is then converted to a resistance and capacitance value to calculate the timing delay. Line 26 is the slope value which is used to calculate the wire length when the fanout is larger than the largest fanout number in the table.

```
1     Time Unit              : 1ns
2     Capacitive Load Unit   : 1.000000pf
3     Pulling Resistance Unit : 1kilo-ohm
4     Voltage Unit           : 1V
5     Current Unit           : 1mA
6
7     Operating Conditions:
8     Name   Process   Temp    Volt    Interconnect Model
9     ----------------------------------------------------------
10    NOM     1.00    25.00    5.00    best_case_tree
11    WCCOM   1.40    70.00    4.75    worst_case_tree
12    WCIND   1.40    85.00    4.75    worst_case_tree
13    WCMIL   1.40   125.00    4.50    worst_case_tree
14    BCCOM   0.70     0.00    5.25    best_case_tree
15    BCIND   0.70   -40.00    5.25    best_case_tree
16    BCMIL   0.70   -55.00    5.50    best_case_tree
17    TST     1.40    25.00    5.00    best_case_tree
18
19    Name          :   BBXX *
20    Location       :   LLCC
21    Resistance     :   0.020929
22    Capacitance    :   0.14221
23    Area           :   8.28598
```

```
24  Slope           :   0.398
25  Fanout   Length   Points Average Cap Std Deviation
26  -------------------------------------------------------
27      1    0.40
28      2    0.80
29      3    1.19
30      4    1.59
31      5    1.99
32      6    2.39
33      7    2.79
34      8    3.18
35      9    3.58
36     10    3.98
```

The following table shows a summary of five actual synthesis libraries. The table gives a quick summary about the library. The bigger the number is in each column, the greater the likelihood that the library is more calibrated. If the library has only one wire load model as shown in Library E, call the manufacture and ask when the wire load model should be used. If the answer is either "do not know" or "10 mils block," purge the synthesis library and save the disk space, and avoid the grief from a design that did not work.

	Number of operating conditions	Number of wire load models	Maximum number of fanout in the wire load model
Library A	8	62	10
Library B	14	40	99
Library C	6	4	24
Library D	3	7	40
Library E	3	1	5

Figure 3-15 Synthesis libraries summary.

It is important to select a good synthesis library (and manufacturer) to match your design objectives. For example, speed requirements, cost, production schedule, and design tools are all factors that need to be considered. Assume that a synthesis library has been selected. Find out the following:

- Use the **report_lib** command to list the component names in the library. Are input/output (I/O) pads included in the library? Some manufacturers provide I/O pads in a separate library. Check whether the three-state output pads and bidirectional pads enable input is low or high active (also use the data book). In general, I/O pads are instantiated with component instantiation. By providing the enable signal at the right level with synthesis, the signal can then be used directly in the I/O pad component instantiations.

- Are there clock buffer components? How is the clock buffer (or clock buffer tree) implemented? Is it inserted in the place / route process?

- Are there delay components? The design may need to use a delay component.

- How many different types of flip-flops? Are there any scan flip-flops? Are reset and present synchronous or asynchronous?

- Ask the vendor whether he has a recommended common synthesis script to set up before synthesizing any block. For example, a vendor suggests the following commands to set the maximum capacitance and maximum transition in line 1 and 4. Line 2 sets the input to port to have the same drive as an inv_1x output drive. Line 3 sets up all output ports to drive the input pin IN1 of an inv_3x component. Note that component names inv_1x and inv_3x are used which can be found by the **report_lib** command.

```
1       set_max_capacitance get_attribute(hx2000_wc/inv_1x/
OUT1,max_capacitance) all_inputs()
2       set_driving_cell -cell inv_1x all_inputs()
3       set_load get_attribute(hx2000_wc/inv_3x/IN1,capacitance)
all_outputs()
4       set_max_transition get_attribute(hx2000_wc/inv_1x/
IN1,max_transition) all_outputs()
```

- Prepare files **.synopsys_dc.setup** and **compile.common**. Bring up **design_analyzer**. Try a simple synthesis to ensure that **.synopsys_dc.setup** and **compile.common** are executed correctly. Now we are ready to go.

3.5 VHDL DESIGN PROCESS FOR A BLOCK

Now, we know how to write VHDL. We have set up the design directory structure, mapped VHDL libraries, and established synthesis libraries. We are anxious to start the design. Assume that we have blocks partitioned and are ready to design the first block. The question is when to do the VHDL simulation and when to do the synthesis. Do we wait until all blocks are designed with VHDL, and the whole design is verified before any synthesis is attempted? The answer is "no" for the following reasons:

- The synthesis process can catch obvious design errors that are harder and slower to find with VHDL simulation.

- If the design (schematic) is not close to what was intended, it is pointless to run the VHDL simulation.

- It may take too much time to get all related blocks designed before a meaningful VHDL simulation can be run.

- The synthesis result may demand design changes. For example, a long path may need to be broken down to two clock cycles. The associated control signals need to be changed.

    For example, the following VHDL code has passed the VHDL syntax error check by running VHDL analyzer (Synopsys vhdlan or Mentor QuickVHDL qvcom).

It may take some time to verify this block, not to mention to set up the test bench or to enter simulation commands.

```
1    library IEEE;
2    use IEEE.std_logic_1164.all;
3    entity WRONG is
4      port (
5        RSTn, CLK                 : in  std_logic;
6        A, B, C, D, E, F, G, H : in  std_logic;
7        FF1, FF2                  : out std_logic;
8        X, Y, Z                   : out std_logic);
9    end WRONG;
10   architecture RTL of WRONG is
11     signal EN1, EN2 : std_logic;
12   begin
13     p0 : process (A, B, C, D, E, E, G, H)
14     begin
15       if (A = '0') or (B = '1') then
16         X <= D nor (E nor F);
17       elsif (C = '1') then
18         Y <= G xor H;
19       end if;
20     end process;
21     p1 : process (RSTn, CLK)
22     begin
23       if (RSTn = '1') then
24         EN1 <= '0';
25         FF1 <= '1';
26         FF2 <= '1';
27       elsif (CLK'event and CLK = '1') then
28         EN1 <= E nand F;
29         EN2 <= A nor   H;
30         if (EN1 = '1') then
31           FF1 <= B and D;
32         end if;
33         if (EN2 = '1') then
34           FF2 <= A nor not B;
35         end if;
36       end if;
37     end process;
38   end RTL;
```

The design can be read in the Design Analyzer with the command "**read -f vhdl ../vhdl/ch3/wrong.vhd**." The following is a summary of the message. Lines 3 to 6 indicate that signal F is used in the process, but not in the sensitivity list. This is just a typo that shows two 'E's instead of one 'E' and one 'F' in line 13. It will take many simulations before this error is discovered.

```
1    Reading in the Synopsys vhdl primitives.
2        ../vhdl/ch3/wrong.vhd:
3    Warning: Variable 'F' is being read
4    in routine WRONG line 13 in file '../vhdl/ch3/wrong.vhd',
5    but is not in the process sensitivity list of the block
6        which begins there.   (HDL-179)
7
8    Inferred memory devices in process 'p0'
9    in routine WRONG line 13 in file
10           '../vhdl/ch3/wrong.vhd'.
11   ==============================================================
12   |Register Name| Type  |Width| Bus | AR | AS | SR | SS | ST |
13   ==============================================================
14   |X_reg        | Latch | 1   |  -  | N  | N  | -  | -  | -  |
15   |Y_reg        | Latch | 1   |  -  | N  | N  | -  | -  | -  |
16   ==============================================================
17   X_reg ----- reset/set: none
18   Y_reg ----- reset/set: none
19
20   Inferred memory devices in process 'p1'
21   in routine WRONG line 21 in file
22           '../vhdl/ch3/wrong.vhd'.
23   ==============================================================
24   |Register Name|  Type   |Width|Bus| AR | AS | SR | SS | ST |
25   ==============================================================
26   |EN1_reg      |Flip-flop| 1   | - | Y  | N  | N  | N  | N  |
27   |EN2_reg      |Flip-flop| 1   | - | N  | N  | N  | N  | N  |
28   |FF1_reg      |Flip-flop| 1   | - | N  | Y  | N  | N  | N  |
29   |FF2_reg      |Flip-flop| 1   | - | N  | Y  | N  | N  | N  |
30   ==============================================================
31
32   EN1_reg ------- Async-reset: RSTn
33   EN2_reg ------- set/reset/toggle: none
34   FF1_reg ------- Async-set: RSTn
35   FF2_reg ------- Async-set: RSTn
36
37   Warning: In design 'WRONG',
38       port 'Z' is not connected to any nets. (LINT-28)
```

Lines 8 to 18 indicate that two latches are inferred in process p0 that are not intended. This is caused by the incomplete assigning of a target object in an if or case statement. Lines 21 to 35 summarize the flip-flops. Note that EN2_reg is reported not asynchronously reset or preset in lines 27 and 33. This may be just a miss. Lines 37 and 38 is a message from the **check_design** command after the VHDL code is read in. Output port Z is not used. Up to this point, the design has not even been synthesized. It is just read in and checked. These are common mistakes that can be found much more easily with the synthesis tool. The following VHDL code shows that the above errors have been corrected. Line 9 is replaced with line 10 to remove output port Z. Line 15 is replaced with line 16 to include F in the sensitivity list. Line 18 is added to set default values so that signals X and Y are completely assigned in every execution path, and the latches are not inferred. Line 29 is added to reset signal EN2.

```
1        ------------------- file wrong2right.vhd
2     library IEEE;
3     use IEEE.std_logic_1164.all;
4     entity WRONG is
5        port (
6          RSTn, CLK              : in  std_logic;
7          A, B, C, D, E, F, G, H : in  std_logic;
8          FF1, FF2               : out std_logic;
9        --X, Y, Z                 : out std_logic); -- remove Z
10         X, Y                   : out std_logic);
11    end WRONG;
12    architecture RTL of WRONG is
13       signal EN1, EN2 : std_logic;
14    begin
15    --p0 : process (A, B, C, D, E, E, G, H) -- typo E -> F
16       p0 : process (A, B, C, D, E, F, G, H)
17       begin
18         X <= '0'; Y <= '0'; -- set value to remove latches
19         if (A = '0') or (B = '1') then
20           X <= D nor (E nor F);
21         elsif (C = '1') then
22           Y <= G xor H;
23         end if;
24       end process;
25       p1 : process (RSTn, CLK)
26       begin
27         if (RSTn = '1') then
28           EN1 <= '0';
29           EN2 <= '0';   -- added EN2
30           FF1 <= '1';
31           FF2 <= '1';
32         elsif (CLK'event and CLK = '1') then
33           EN1 <= E nand F;
34           EN2 <= A nor  H;
35           if (EN1 = '1') then
36             FF1 <= B and D;
37           end if;
38           if (EN2 = '1') then
39             FF2 <= A nor not B;
40           end if;
41         end if;
42       end process;
43    end RTL;
```

The following is a summary message after the corrected VHDL file is read in. Note that only flip-flops (no latches) are inferred with asynchronous reset and set. Figure 3-16 shows the synthesized schematic. Note that the asynchronous reset and set pins of the four flip-flops are connected to an inverter output of the RSTn input. This is due to line 27, which checks RSTn = '1'. This may be intended for RSTn = '0'. Lines 16 to 19 do not tell which way. It is a good practice to review the synthesized schematic to get a quick sense of obvious mistakes.

```
1      Reading in the Synopsys vhdl primitives.
2          ../vhdl/ch3/wrong2right.vhd:
3
4      Inferred memory devices in process 'p1'
5      in routine WRONG line 24 in file
6          '../vhdl/ch3/wrong2right.vhd'.
7      =======================================================
8      |Register Name|   Type    |Width|Bus| AR | AS | SR | SS | ST |
9      =======================================================
10     |EN1_reg      |Flip-flop|  1  | - | Y  | N  | N  | N  | N  |
11     |EN2_reg      |Flip-flop|  1  | - | Y  | N  | N  | N  | N  |
12     |FF1_reg      |Flip-flop|  1  | - | N  | Y  | N  | N  | N  |
13     |FF2_reg      |Flip-flop|  1  | - | N  | Y  | N  | N  | N  |
14     =======================================================
15
16     EN1_reg ------- Async-reset: RSTn
17     EN2_reg ------- Async-reset: RSTn
18     FF1_reg ------- Async-set: RSTn
19     FF2_reg ------- Async-set: RSTn
```

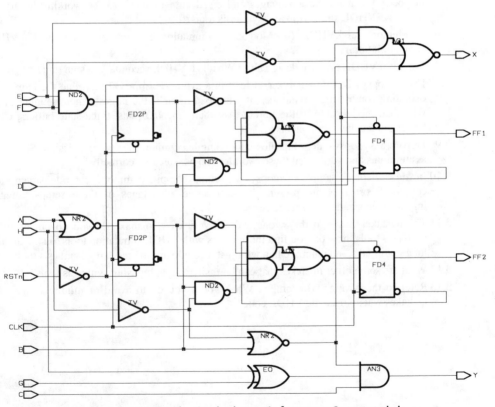

Figure 3-16 Synthesized schematic for wrong2correct.vhd.

3.6 EXERCISES

3.1 Find how the $SYNOPSYS and $ARCH environment variables are set by using Unix commands echo $SYNOPSYS and echo $ARCH. They are usually set in the .cshrc (C Shell) file or .kshrc (Korn Shell) file. If they are not set, ask your system administrator how to set them up. Visit the $SYNOPSYS directory to find out how subdirectories are set up. How many files are in $SYNOPSYS/packages/IEEE/src directory? View each file and write down the file name and its purpose. Use Unix command echo $PATH to see how the $PATH environment variable is set. Which directory contains the command: vhdlsim, vhdlan, vhdldbx, design_analyzer, dc_shell?

3.2 Define a design directory structure for a design for which you alone are responsible. State your assumptions, guidelines, usages, advantages, and purposes.

3.3 Define a design directory structure for a design by two or more designers. State your assumptions, guidelines, usages, advantages, and purposes.

3.4 The directory structure shown in this chapter is more or less flat. Is this an advantage compared to a deeper directory structure? For example, directory **rtlsim.work** can be embedded inside directory **rtlsim**.

3.5 In Synopsys VHDL simulation environment, how is the VHDL simulation library mapped? What are the advantages and disadvantages to share the working library for Synopsys VHDL simulation and synthesis tools?

3.6 In Mentor QuickVHDL (or Modelsim) simulation environment, how is the VHDL simulation library mapped?

3.7 What is a VHDL primary design unit? What is a VHDL secondary design unit?

3.8 Use Synopsys Design Analyzer to read in a synthesis library. Determine the number of operating conditions, the number of wire load models, the largest fanout number in the wire load model table in the library. Are there any I/O pads, three-state buffers, clock buffers, and delay components?

3.9 Prepare a .synopsys_dc.setup file and a compile.common file for your design. State your assumptions and verify that these two files can be executed correctly.

3.10 In this chapter, sample .synopsys_dc.setup, compile.common, compile.pargen, and compile.report files are presented. Discuss the advantages of these setups. Suggest improvements and state your reasons.

3.11 The mistakes made in the wrong.vhd are very easy to make. If you do not have the synthesis tools, how do you find those errors with VHDL simulation tools? Estimate how long it will take and compare that with just a couple of minutes using synthesis tools.

3.12 What are some other mistakes that can be easily caught by using synthesis tools?

3.13 Refer to the wrong.vhd example. What would you do to avoid or minimize the same mistakes when you write VHDL code?

Chapter 4

Basic Combinational Circuits

In this chapter, we will use VHDL to describe several basic combinational circuits. There may be more than one way to write VHDL to describe the same circuit. These different approaches to VHDL implementation are discussed and compared with their corresponding synthesis scripts and synthesized results. For completeness, we also show how combinational circuits can be verified with test benches.

4.1 SELECTOR

A selector circuit selects an input to pass data to the output. For example, an input with 16 bits can have any one of the inputs identified and selected by using a 4-bit identifier input. The following VHDL code shows a selector with inputs A, SEL and output Y in lines 6 to 8. Note that there are four architectures (RTL1, RTL2, RTL3, and RTL4) for the same entity in the same VHDL text file. As a convention of this book, the continuously increasing line numbers associated with the VHDL code indicate that they are in the same text file. This allows descriptions of the code to be inserted so that the descriptions are close to the VHDL code they refer to. It is important to note this even where the code is broken by description comments. For example, descriptions are inserted between lines 9 and 10. Note also that the library clauses in lines 1 to 3 can be extended for the scope of the **primary design unit** (entity SELECTOR in this case). The library clauses do not need to be repeated before each architecture, as long as they are in sequence in the same text file. For example, there is no need to repeat the library clauses before lines 10, 49, 73, and 93 architecture declarations.

```
1    library IEEE;
2    use IEEE.std_logic_1164.all;
3    use IEEE.std_logic_unsigned.all;
4    entity SELECTOR is
5      port (
6        A   : in  std_logic_vector(15 downto 0);
7        SEL : in  std_logic_vector( 3 downto 0);
8        Y   : out std_logic);
9    end SELECTOR;
```

In the following architecture RTL1, the selector function is modeled with a **process statement**. Input signals A and SEL are the **process sensitivity list**. Inside the process statement, an **if statement** is used as shown in lines 14 to 46.

```
10    architecture RTL1 of SELECTOR is
11    begin
12      p0 : process (A, SEL)
13      begin
14        if      (SEL = "0000") then
15          Y <= A(0);
16        elsif (SEL = "0001") then
17          Y <= A(1);
18        elsif (SEL = "0010") then
19          Y <= A(2);
20        elsif (SEL = "0011") then
21          Y <= A(3);
22        elsif (SEL = "0100") then
23          Y <= A(4);
24        elsif (SEL = "0101") then
25          Y <= A(5);
26        elsif (SEL = "0110") then
27          Y <= A(6);
28        elsif (SEL = "0111") then
29          Y <= A(7);
30        elsif (SEL = "1000") then
31          Y <= A(8);
32        elsif (SEL = "1001") then
33          Y <= A(9);
34        elsif (SEL = "1010") then
35          Y <= A(10);
36        elsif (SEL = "1011") then
37          Y <= A(11);
38        elsif (SEL = "1100") then
39          Y <= A(12);
40        elsif (SEL = "1101") then
41          Y <= A(13);
42        elsif (SEL = "1110") then
43          Y <= A(14);
44        else
45          Y <= A(15);
46        end if;
47      end process;
48    end RTL1;
```

The following architecture RTL2 is similar to architecture RTL1, except for the **case statement** used in lines 53 to 70.

```
49    architecture RTL2 of SELECTOR is
```

```
50    begin
51      p1 : process (A, SEL)
52      begin
53        case SEL is
54          when "0000" => Y <= A(0);
55          when "0001" => Y <= A(1);
56          when "0010" => Y <= A(2);
57          when "0011" => Y <= A(3);
58          when "0100" => Y <= A(4);
59          when "0101" => Y <= A(5);
60          when "0110" => Y <= A(6);
61          when "0111" => Y <= A(7);
62          when "1000" => Y <= A(8);
63          when "1001" => Y <= A(9);
64          when "1010" => Y <= A(10);
65          when "1011" => Y <= A(11);
66          when "1100" => Y <= A(12);
67          when "1101" => Y <= A(13);
68          when "1110" => Y <= A(14);
69          when others => Y <= A(15);
70        end case;
71      end process;
72    end RTL2;
```

The following architecture RTL3 does not use a **process statement** as in architectures RTL1 and RTL2. Lines 75 to 91 are a **concurrent selected signal assignment statement**.

```
73    architecture RTL3 of SELECTOR is
74    begin
75      with SEL select
76        Y <= A(0)  when "0000",
77              A(1)  when "0001",
78              A(2)  when "0010",
79              A(3)  when "0011",
80              A(4)  when "0100",
81              A(5)  when "0101",
82              A(6)  when "0110",
83              A(7)  when "0111",
84              A(8)  when "1000",
85              A(9)  when "1001",
86              A(10) when "1010",
87              A(11) when "1011",
88              A(12) when "1100",
89              A(13) when "1101",
90              A(14) when "1110",
91              A(15) when others;
92    end RTL3;
```

In the following architecture RTL4, a predefined function conv_integer in package IEEE.std_logic_unsigned is used to convert 4-bit std_logic_vector to an integer. The integer is then used as an index to the array A as shown in line 95.

```
93    architecture RTL4 of SELECTOR is
94    begin
95       Y <= A(conv_integer(SEL));
96    end RTL4;
```

Architectures RTL1, RTL2, RTL3, and RTL4 take 39, 24, 20, and 4 lines of VHDL code respectively. Before we synthesize each architecture, let's make sure that all four architectures implement the selector function exactly. Note that architectures RTL1, RTL2, and RTL3 have something in common. The output Y is taking a value of signal A with an indexed integer value based on the signal SEL 4-bit value. It is suggested that you cut and paste the code since the syntax is similar among lines in the three architectures. For example, line 78 can be copied from line 77 by updating A(1) to A(2), and "0001" to "0010". However, it is easy to forget to change the value or to have a wrong value caused by typos. Several errors may surface:

• The integer index does not correspond to the desired 4-bit binary value. This does not cause any syntax error. The selected value of A is not intended.

• There may be duplicates of integer indexes so that different 4-bit binary values may select the same value in A. This again does not cause any syntax error. It causes more than one 4-bit value to select the same indexed value from A.

• There may be duplicates of 4-bit binary values. Luckily, this causes syntax error, since the case statement choices are required to cover all unique combinations.

It is strongly recommended that errors be caught as early as possible in the design process. Since the last error cannot pass the syntax check, we will concentrate on developing a test to ensure that the first two types of errors do not pass through our fingers. Otherwise, the result will be hardware that works in almost all cases except a few—the designer's worst nightmare!

The following VHDL code is a test bench example. Lines 1 to 4 declare library and packages used. Lines 5 and 6 are the entity declaration without anything inside. Lines 8 to 13 declare the SELECTOR component. Lines 14 to 16 declare signals used. Lines 17 to 18 define time constants. Since there are only 16 combinations of signal SEL, we can enumerate all of them. The concept is to automatically generate input signals A and SEL. These input signals are connected to all four instantiated components, each representing one architecture. All four component outputs are connected with signals Y1, Y2, Y3, and Y4, respectively. In the test bench, an expected correct result, Y, is computed so that Y can be compared with Y1, Y2, Y3, and Y4 automatically.

```
1    library IEEE;
2    use IEEE.std_logic_1164.all;
3    use IEEE.std_logic_arith.all;
```

```
4     use IEEE.std_logic_unsigned.all;
5     entity TBSELECTOR is
6     end TBSELECTOR;
7     architecture BEH of TBSELECTOR is
8       component SELECTOR
9       port (
10        A   : in  std_logic_vector(15 downto 0);
11        SEL : in  std_logic_vector( 3 downto 0);
12        Y   : out std_logic);
13      end component;
14      signal A    : std_logic_vector(15 downto 0);
15      signal SEL  : std_logic_vector( 3 downto 0);
16      signal Y, Y1, Y2, Y3, Y4 : std_logic;
17      constant PERIOD : time := 50 ns;
18      constant STROBE : time := 45 ns;
```

The following process **p0** uses a loop statement in lines 24 to 31. Index ranges from 0 to 31. Line 25 converts the integer value to a 5-bit std_logic_vector. The least significant 4-bits are used for signal SEL as shown in line 25. The most significant bit is used as the expected value Y, as shown in line 26. Line 27 first assigns all values in A to be the complement of the expected value. In line 28, the correct index of the 4-bit binary number in signal A is changed to the expected value. The result is that only the correct index of signal A is the same as the expected value; the other values in A are reversed. The first two types of errors will be caught. Line 29 waits for the time PERIOD before it changes another set of inputs. After the loop statement is executed 32 times, the process waits forever in line 31.

```
19    begin
20      p0 : process
21        variable cnt : std_logic_vector( 4 downto 0);
22      begin
23        for j in 0 to 31 loop
24          cnt  := conv_std_logic_vector(j,5);
25          SEL  <= cnt(3 downto 0);
26          Y    <= cnt(4);
27          A    <= (A'range => not cnt(4));
28          A(conv_integer(cnt(3 downto 0))) <= cnt(4);
29          wait for PERIOD;
30        end loop;
31        wait;
32      end process;
```

The purpose of the following process **check** is to compare the results Y1, Y2, Y3, and Y4 with expected value Y. Line 36 waits for STROBE time after the inputs are changed to compare the results. Note that input signals are changed every PERIOD (50 ns). The results are compared 5 ns before the inputs are changed so that the STROBE is declared as 45 ns. This means that after inputs are changed, we wait for 45 ns for the combinational circuits to settle down before the output values are compared. After they are compared, the next set of input values can be used 5 ns later. In a synchronous design, the strobe is usually done just before the rising edge of the clock. Lines 37 to 44 compare all 32 sets of output values. A counter err_cnt is used to count the number of errors in lines 34 and 41. Lines 45 and 46 print out a message to indicate whether the test is passed or failed. Line 47 then waits forever.

```
33    check : process
34      variable err_cnt : integer := 0;
35    begin
36      wait for STROBE;
37      for j in 0 to 31 loop
38        assert FALSE report " comparing ..." severity NOTE;
39        if (Y /= Y1)or (Y /= Y2) or (Y /= Y3) or (Y /= Y4) then
40          assert FALSE report "not compared" severity WARNING;
41          err_cnt := err_cnt + 1;
42        end if;
43        wait for PERIOD;
44      end loop;
45      assert (err_cnt = 0) report "test failed" severity ERROR;
46      assert (err_cnt /= 0) report "test passed" severity NOTE;
47      wait;
48    end process;
```

Lines 49 to 56 instantiate four SELECTOR components. They share the same inputs A and SEL. The outputs are mapped to Y1, Y2, Y3, and Y4, respectively.

```
49    selector1 : SELECTOR
50      port map ( A   => A, SEL => SEL, Y   => Y1);
51    selector2 : SELECTOR
52      port map ( A   => A, SEL => SEL, Y   => Y2);
53    selector3 : SELECTOR
54      port map ( A   => A, SEL => SEL, Y   => Y3);
55    selector4 : SELECTOR
56      port map ( A   => A, SEL => SEL, Y   => Y4);
57    end BEH;
```

Lines 58 to 73 configure each SELECTOR component corresponding to each architecture.

```
58    configuration CFG_TBSELECTOR of TBSELECTOR is
59      for BEH
60        for selector1 : SELECTOR
61          use entity work.SELECTOR(RTL1);
62        end for;
63        for selector2 : SELECTOR
64          use entity work.SELECTOR(RTL2);
65        end for;
66        for selector3 : SELECTOR
67          use entity work.SELECTOR(RTL3);
```

```
68          end for;
69          for selector4 : SELECTOR
70            use entity work.SELECTOR(RTL4);
71          end for;
72        end for;
73      end CFG_TBSELECTOR;
```

A portion of the simulation waveform for the TBSELECTOR is shown in Figure 4-1. Note that value A has only one value that is the same as Y. The rest are reversed. Signal A is shown as a hexadecimal value while signal SEL is shown as a 4-bit binary number. Note also that Y, Y1, Y2, Y3, and Y4 are the same.

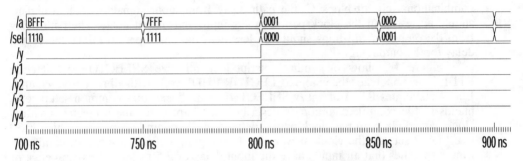

Figure 4-1 Simulation waveform for TBSELECTOR.

Test bench TBSELECTOR is written in a way that the test can be run and the results be compared automatically. There is no need to examine the waveform to ensure that values Y, Y1, Y2, Y3, and Y4 are exactly the same. TBSELECTOR is a small test case with only 32 sets of values. It is easy to check the waveform to see the correctness of the design. When the design gets bigger, through more signals and sets of values, it becomes time consuming and error prone to verify all tests by manually examining the waveform. Develop tests, if possible, that can automatically verify themselves. More simulations can be run as a batch which takes advantage of overnight and weekend computing time, while you enjoy family time and outdoor activities.

The following message is part of the simulation transcript. Note that each time line 38 of TBSELECTOR is executed, 2 lines are generated. Input values are changed every 50 ns starting at time 0. The results are compared 45 ns after input values are changed. Line 5 and 6 indicate that all 32 tests are compared.

```
1    # ** Note:  comparing ...
2    #    Time: 1545 ns  Iteration: 0  Instance:/
3    # ** Note:  comparing ...
4    #    Time: 1595 ns  Iteration: 0  Instance:/
5    # ** Note: test passed
6    #    Time: 1645 ns  Iteration: 0  Instance:/
```

Now, we have verified that all four architectures of SELECTOR are functionally equivalent. Let's synthesize the VHDL code into schematics. Figure 4-2 shows the general framework of a synchronous digital circuit and the relationship among input delay, output delay, and clock period. Note that the input delay value is the amount of time after the rising edge of the clock. The output delay value is the amount of time before the rising edge of the clock. Now assume that we have the ideal clock situation without any clock skew, and ignore the setup time and clock to Q-output for the flip-flop. In Figure 4-2, combinational circuit block B has input connections from the external input and from the output of flip-flops. The timing delay from external input signals, going through combinational circuit B, must arrive at the D input of the flip-flop before the rising edge of the clock. The path delay should be less than the clock period minus the input delay of each input signal. The path delay from the output of flip-flops should arrive at the D input of flip-flops within a full clock period. Combinational circuit block A has all input connections from external inputs. The path delay should be less than the clock period minus the input delay of each input signal. In combinational circuit block C, the path delay from input signals to output signals should be less than the clock period minus input and output delay. The path delay from the output of flip-flops should be less than the clock period minus the output delay for the output signal.

Figure 4-3 shows a synthesis script to synthesize SELECTOR architecture RTL1. Line 3 analyzes the whole SELECTOR VHDL code with all four architectures. Line 4 elaborates RTL1 of entity SELECTOR. Line 5 executes common setup commands in file compile.common as described in Chapter 3. Line 6 creates a virtual clock named clk with the clock period defined in file compile.common (which is 30 ns). Line 7 sets up the clock skew to be 1 ns (again, defined in compile.common). Line 8 specifies that all inputs have the input delay of 2.5 ns after the rising edge of the clock clk. Line 9 specifies that all outputs have the constraint to be CLKPER1 (10 percent of the clock period or 3 ns) amount of time before the rising edge of the clock. Line 10 synthesizes the circuit. Line 11 produces a synthesized report which summarizes area and timing. File compile.report and compile.common were presented in Chapter 3. Figure 4-4 shows the synthesized schematic for architecture RTL1.

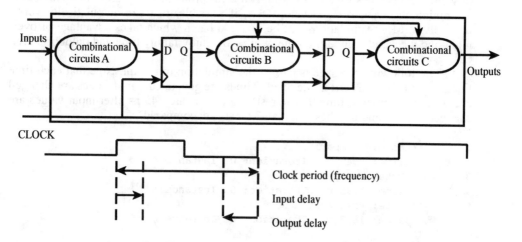

Figure 4-2 Input delay and output delay relationship to clock.

```
1      sh date
2      design = selector
3      analyze -f vhdl VHDLSRC + "ch4/" + design + ".vhd"
4      elaborate SELECTOR -architecture RTL1
5      include compile.common
6      create_clock -name clk -period CLKPERIOD -waveform {0.0
HALFCLK}
7      set_clock_skew -uncertainty CLKSKEW clk
8      set_input_delay  INPDELAY -add_delay -clock clk all_inputs()
9      set_output_delay CLKPER1  -add_delay -clock clk all_outputs()
10     compile -map_effort medium
11     include compile.report
```

Figure 4-3 Synthesis script file compile.selector.rtl1.

The following shows portions of the report file. The area report summarizes the numbers of ports, nets, cells, and references in lines 12 to 15. The combinational area is 52 units in line 17. There are no sequential circuits so that the sequential area is 0 unit as shown in line 18. Line 19 shows the area for the interconnect wires. Line 21 shows the total cell area. Line 22 shows the total of the cell area and the interconnect area.

```
1      ****************************************
2      Report : area
3      Design : SELECTOR
4      Version: 1997.01-44683
5      Date   : Sat Dec 13 11:25:25 1997
6      ****************************************
7
8      Library(s) Used:
9
10         lca300k (File: /acct/synlib/lsi/lca300k.db)
11
12     Number of ports:              21
13     Number of nets:               48
14     Number of cells:              28
15     Number of references:          5
16
17     Combinational area:      52.000000
18     Noncombinational area:    0.000000
19     Net Interconnect area:  230.847549
20
21     Total cell area:         52.000000
22     Total area:             282.847534
```

The following timing report summarizes the operating condition and the synthesis library used in line 33. Line 38 shows that the wire_load model B9X9 is used. Starting at line 49, the worst case timing path is shown. Line 49 starts with the rising edge of the clock clk at time 0. Lines 51 and 52 shows that the input signal SEL has 2.5 ns input delay. Lines 53 to 68 show the timing delay through the path to output Y. Note that the last character in the path delay timing indicates a rise ("r") or a fall ("f") time.

```
23    *****************************************
24    Report : timing
25            -path full
26            -delay max
27            -max_paths 1
28    Design : SELECTOR
29    Version: 1997.01-44683
30    Date   : Sat Dec 13 11:25:26 1997
31    *****************************************
32
33    Operating Conditions: WCCOM    Library: lca300k
34    Wire Loading Model Mode: enclosed
35
36    Design            Wire Loading Model        Library
37    ---------------------------------------------------
38    SELECTOR                B9X9                 lca300k
39
40
41    Startpoint: SEL[1] (input port clocked by clk)
42    Endpoint: Y (output port clocked by clk)
43    Path Group: clk
44    Path Type: max
45
46    Point                                  Incr      Path
47    ---------------------------------------------------
48    clock clk (rise edge)                  0.00      0.00
49    clock network delay (ideal)            0.00      0.00
50    input external delay                   2.50      2.50 f
51    SEL[1] (in)                            0.00      2.50 f
52    U81/Z (NR2)                            1.81      4.31 r
53    U86/Z (MUX31L)                         0.92      5.24 r
54    U87/Z (IV)                             0.57      5.81 f
55    U89/Z (MUX31L)                         0.73      6.54 r
56    U90/Z (IV)                             0.57      7.11 f
57    U93/Z (MUX31L)                         0.73      7.84 r
58    U94/Z (IV)                             0.57      8.41 f
59    U96/Z (MUX31L)                         0.73      9.14 r
60    U97/Z (IV)                             0.57      9.71 f
61    U100/Z (MUX31L)                        0.73     10.45 r
62    U101/Z (IV)                            0.57     11.02 f
63    U103/Z (MUX31L)                        0.73     11.75 r
64    U104/Z (IV)                            0.57     12.32 f
65    U107/Z (MUX31L)                        0.73     13.05 r
66    U108/Z (IV)                            0.82     13.87 f
67    Y (out)                                0.00     13.88 f
```

```
68        data arrival time                                          13.88
69
70        clock clk (rise edge)                       30.00          30.00
71        clock network delay (ideal)                  0.00          30.00
72        clock uncertainty                           -1.00          29.00
73        output external delay                       -3.00          26.00
74        data required time                                         26.00
75        ----------------------------------------------------------------
76        data required time                                         26.00
77        data arrival time                                         -13.88
78        ----------------------------------------------------------------
79        slack (MET)                                                12.12
```

Line 68 shows that the circuit path delay is 13.88 ns (including the input delay) from signal SEL to output signal Y. Line 70 depicts that the clock period is 30 ns. The clock skew is 1 ns in line 72. The output delay constraint is 10 percent (3 ns) before the rising edge of the clock. These constraints indicate that the output signal is required within 26 ns. The circuit takes only 13.88 ns as shown in line 68. Line 79 shows the difference between the data required time and the data arrival time as the slack. A positive slack means that the synthesized circuit meets the synthesis timing constraint.

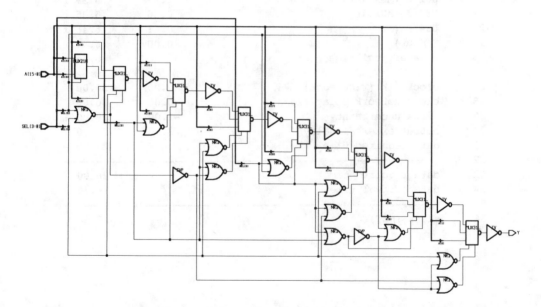

Figure 4-4 Synthesized schematic for SELECTOR RTL1.

Line 4 of the Figure 4-3 synthesis script file can be changed from RTL1 to RTL2 to elaborate architecture RTL2 of entity SELECTOR. Figure 4-5 shows the synthesized schematic. The following shows the summary report. Note that the cell area is reduced from 52 to 34 units. The total area is reduced from 282.85 to 142.83. The data required time is reduced from 13.88 ns down to 6.46 ns. Why?

```
1      ****************************************
2
3      Number of ports:             21
4      Number of nets:              26
5      Number of cells:              6
6      Number of references:         4
7
8      Combinational area:      34.000000
9      Noncombinational area:    0.000000
10     Net Interconnect area:  108.828125
11
12     Total cell area:         34.000000
13     Total area:             142.828125
14
15     Point                            Incr      Path
16     ---------------------------------------------------------
17     clock clk (rise edge)            0.00      0.00
18     clock network delay (ideal)      0.00      0.00
19     input external delay             2.50      2.50 f
20     SEL[1] (in)                      0.00      2.50 f
21     U82/Z (MUX51HP)                  2.69      5.19 f
22     U83/Z (MUX31L)                   0.66      5.85 r
23     U84/Z (IVP)                      0.61      6.46 f
24     Y (out)                          0.00      6.46 f
25     data arrival time                          6.46
26
27     clock clk (rise edge)           30.00     30.00
28     clock network delay (ideal)      0.00     30.00
29     clock uncertainty               -1.00     29.00
30     output external delay           -3.00     26.00
31     data required time                        26.00
32     ---------------------------------------------------------
33     data required time                        26.00
34     data arrival time                         -6.46
35     ---------------------------------------------------------
36     slack (MET)                               19.54
```

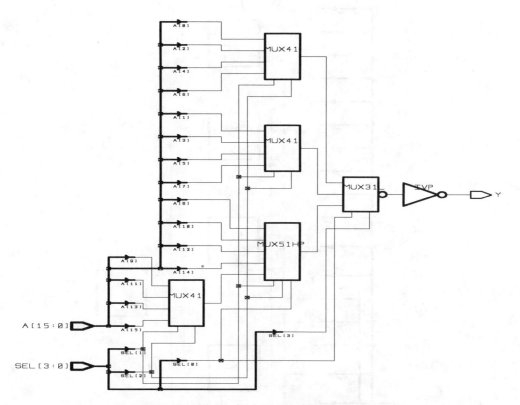

Figure 4-5 Synthesized schematic for SELECTOR RTL2.

Architecture RTL1 uses an **if statement**. It forms a series of comparison chains. Line 16 will not be executed unless line 14 is evaluated not true. Line 45 will not be executed until all previous 15 comparisons have failed. For line 45 to be evaluated, each preceeding conditions (in the elsif clause) must first be evaluated false. This means that lots of circuits need to be used to prevent line 45 from being executed in some conditions. Condition "0000" seems to have higher priority than "0001", etc. It forms a chain of priority. In architecture RTL2, all conditions can be implemented as multiplexers in parallel as shown in Figure 4-5. This results in a faster and smaller circuit. As a matter of fact, architectures RTL2, RTL3, and RTL4 are synthesized to the same schematic. Whenever possible, a long chain of **if elsif statement** should be examined carefully to see whether they are really necessary.

Examine the schematic in Figure 4-5. Is the schematic close to what you have in mind? A straightforward implementation would use multiplexers as shown in Figure 4-6. How?

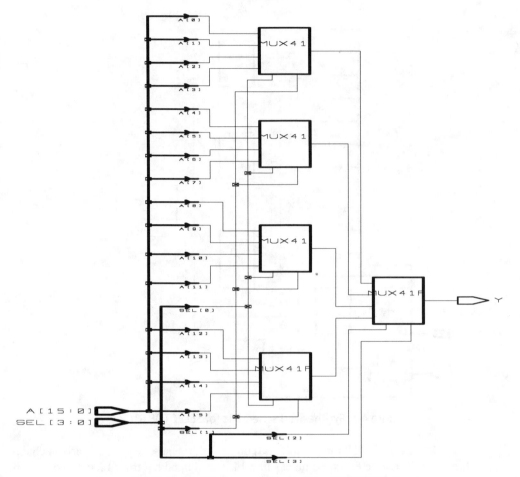

Figure 4-6 Using multiplexers to implement SELECTOR circuit.

In general, synthesis tools know the logical function of the desired circuit. There are just too many combinations to implement a circuit with the same logical function. Synthesis tools need some direction to help them to generate a circuit that is close to what we want. In this case, many combinations of 4-bits from A can be used for any of the first stage 4-to-1 multiplexers, and the results are logically equivalent. Let's help the synthesis tools to determine which 4-bits of signal A will go together with the following updated synthesis script file from file compile.selector.rtl2. The difference is in lines 9, 10, and 11. Note that A[4] (square bracket) is used in line 9, not A(4) (which is VHDL syntax for the array element). Each of them specifies 4-bits of input, and signal A has an input delay of CLKPER5, CLKPER7, and CLKPER8, respectively. Note that line 8 specifies that all input signals have the same default input delay. Lines 9, 10, and 11 overwrite A(15 downto 4) input delays. The input delay for

A(3 downto 0) remains as the default value. The synthesis script indeed generates the schematic shown in Figure 4-6.

```
1     sh date
2     design = selector
3     analyze -f vhdl VHDLSRC + "ch4/" + design + ".vhd"
4     elaborate SELECTOR -architecture RTL2
5     include compile.common
6     create_clock -name clk -period CLKPERIOD -waveform {0.0
HALFCLK}
7     set_clock_skew -uncertainty CLKSKEW clk
8     set_input_delay   INPDELAY -add_delay -clock clk all_inputs()
9     set_input_delay   CLKPER5  -clock clk {A[4] A[5] A[6] A[7]}
10    set_input_delay   CLKPER7  -clock clk {A[8] A[9] A[10] A[11]}
11    set_input_delay   CLKPER8  -clock clk {A[12] A[13] A[14] A[15]}
12    set_output_delay CLKPER1  -add_delay -clock clk all_outputs()
13    compile -map_effort medium
14    include compile.report
```

Figure 4-7 Synthesis script file compile.selector.rtl2i.

The following summarizes the timing and area statistics for the Figure 4-6 schematic. Note that A(15) takes about 2.03 ns to go through two MUX41P gates. SEL takes 3.16 ns (not shown in the report). This implementation in Figure 4-6 is faster and smaller than that in Figure 4-5.

```
1     Number of ports:            21
2     Number of nets:             25
3     Number of cells:             5
4     Number of references:        2
5
6     Combinational area:         32.000000
7     Noncombinational area:       0.000000
8     Net Interconnect area:     102.232483
9
10    Total cell area:            32.000000
11    Total area:                134.232483
12
13    Point                                    Incr      Path
14    -------------------------------------------------------------
15    clock clk (rise edge)                    0.00      0.00
16    clock network delay (ideal)              0.00      0.00
17    input external delay                    24.00     24.00 f
18    A[15] (in)                               0.00     24.00 f
19    U82/Z (MUX41P)                           1.12     25.12 f
20    U79/Z (MUX41P)                           0.91     26.03 f
21    Y (out)                                  0.00     26.04 f
22    data arrival time                                 26.04
23
```

```
24      clock clk (rise edge)                          30.00    30.00
25      clock network delay (ideal)                     0.00    30.00
26      clock uncertainty                              -1.00    29.00
27      output external delay                          -3.00    26.00
28      data required time                                      26.00
29      -------------------------------------------------------------
30      data required time                                      26.00
31      data arrival time                                      -26.04
32      -------------------------------------------------------------
33      slack (VIOLATED)                                        -0.04
```

The above shows a way to somewhat direct synthesis tools to generate a circuit close to what we prefer. Synthesis tools by Synopsys also provide synthesis directives inside the VHDL code as comments to achieve similar goals.

4.2 ENCODER

An encoder simply converts an input pattern to an output pattern. For example, a seven-segment encoder converts a 4-bit input to a 7-bit output to control the seven-segment display, which is used to form the number "8" as shown in a calculator display. The following example shows an encoder with a 4-bit input SEL and a 16-bit output Y. The integer value represented by the 4-bit input SEL forms an index to the output Y. All the bits in output Y will be '1', except for the one with the integer index value represented by SEL. The following shows four architectures to implement the encoder function. Architecture RTL1 uses an **if statement** inside a **process statement** with input signal SEL as the **process sensitivity list**. Note that line 13 sets the default value for every bit in Y to be '1'. The corresponding index represented by input signal SEL will be used to update the value to '0'.

```
1     library IEEE;
2     use IEEE.std_logic_1164.all;
3     use IEEE.std_logic_unsigned.all;
4     entity ENCODER is
5       port (
6         SEL : in  std_logic_vector(3 downto 0);
7         Y   : out std_logic_vector(15 downto 0));
8     end ENCODER;
9     architecture RTL1 of ENCODER is
10    begin
11      p0 : process (SEL)
12      begin
13        Y <= (Y'range => '1');
14        if    (SEL = "0000") then
15          Y(0) <= '0';
16        elsif (SEL = "0001") then
17          Y(1) <= '0';
18        elsif (SEL = "0010") then
```

```
19            Y(2) <= '0';
20          elsif (SEL = "0011") then
21            Y(3) <= '0';
22          elsif (SEL = "0100") then
23            Y(4) <= '0';
24          elsif (SEL = "0101") then
25            Y(5) <= '0';
26          elsif (SEL = "0110") then
27            Y(6) <= '0';
28          elsif (SEL = "0111") then
29            Y(7) <= '0';
30          elsif (SEL = "1000") then
31            Y(8) <= '0';
32          elsif (SEL = "1001") then
33            Y(9) <= '0';
34          elsif (SEL = "1010") then
35            Y(10) <= '0';
36          elsif (SEL = "1011") then
37            Y(11) <= '0';
38          elsif (SEL = "1100") then
39            Y(12) <= '0';
40          elsif (SEL = "1101") then
41            Y(13) <= '0';
42          elsif (SEL = "1110") then
43            Y(14) <= '0';
44          else
45            Y(15) <= '0';
46          end if;
47        end process;
48      end RTL1;
```

The following architecture RTL1 is similar to architecture RTL1, except for the **case statement** used inside the **process statement**.

```
49      architecture RTL2 of ENCODER is
50      begin
51        p1 : process (SEL)
52        begin
53          Y <= (Y'range => '1');
54          case SEL is
55            when "0000" => Y(0)  <= '0';
56            when "0001" => Y(1)  <= '0';
57            when "0010" => Y(2)  <= '0';
58            when "0011" => Y(3)  <= '0';
59            when "0100" => Y(4)  <= '0';
60            when "0101" => Y(5)  <= '0';
61            when "0110" => Y(6)  <= '0';
62            when "0111" => Y(7)  <= '0';
63            when "1000" => Y(8)  <= '0';
64            when "1001" => Y(9)  <= '0';
```

```
65              when "1010" => Y(10) <= '0';
66              when "1011" => Y(11) <= '0';
67              when "1100" => Y(12) <= '0';
68              when "1101" => Y(13) <= '0';
69              when "1110" => Y(14) <= '0';
70              when others => Y(15) <= '0';
71        end case;
72      end process;
73    end RTL2;
```

The following architecture RTL3 uses a **concurrent selected signal assignment statement**.

```
74    architecture RTL3 of ENCODER is
75    begin
76      with SEL select
77        Y <= "1111111111111110" when "0000",
78              "1111111111111101" when "0001",
79              "1111111111111011" when "0010",
80              "1111111111110111" when "0011",
81              "1111111111101111" when "0100",
82              "1111111111011111" when "0101",
83              "1111111110111111" when "0110",
84              "1111111101111111" when "0111",
85              "1111111011111111" when "1000",
86              "1111110111111111" when "1001",
87              "1111101111111111" when "1010",
88              "1111011111111111" when "1011",
89              "1110111111111111" when "1100",
90              "1101111111111111" when "1101",
91              "1011111111111111" when "1110",
92              "0111111111111111" when others;
93      end RTL3;
```

The following architecture RTL4 uses a **process statement** (in the same way as RTL1 and RTL2). In line 99, a function conv_integer from package IEEE std_logic_unsigned (line 3) is used.

```
94    architecture RTL4 of ENCODER is
95    begin
96      p3 : process (SEL)
97      begin
98        Y <= (Y'range => '1');
99        Y(conv_integer(SEL)) <= '0';
100     end process;
101   end RTL4;
```

The verification and synthesis process are similar to the SELECTOR circuit as presented in the last section. Therefore, they are left as an exercise (see Exercise 4.12).

4.3 CODE CONVERTER

A *code* is a binary sequence that represents certain values. For example, 4 binary bits can represent 16 possible values. The most common codes are the binary code, excess-3, and Gray code as shown in Figure 4-8. Binary code has the bit position weighted as 8-4-2-1, respectively. Excess-3 is similar to the binary code except that the corresponding integer value is three more than the binary code. The Gray code has the property to be exactly one bit difference between adjacent codes. The first eight codes are mirrored to the last eight codes with the leftmost bit reversed. This property exists recursively. For example, the first set of four codes (of the first eight codes) is mirrored to the second set of four codes with the second leftmost bit reversed.

```
1      Binary    Excess-3    Gray code
2      ----      ----        ----
3      0000      0011        0000
4      0001      0100        0001
5      0010      0101        0011
6      0011      0110        0010
7      0100      0111        0110
8      0101      1000        0111
9      0110      1001        0101
10     0111      1010        0100
11     1000      1011        1100
12     1001      1100        1101
13     1010      1101        1111
14     1011      1110        1110
15     1100      1111        1010
16     1101      0000        1011
17     1110      0001        1001
18     1111      0010        1000
```

Figure 4-8 4-bit Binary, Excess-3, and Gray code.

In digital design, a code conversion may be required. A *code conversion* is a type of encoder designed to convert an input pattern to an output pattern. The following BIN2EX3 VHDL code shows three different architectures to convert a binary code to an excess-3 code. By this time, you may be convinced that there is no need to use the **if statement** here.

```
1      ----------------------- file bin2ex3.vhd
2      library IEEE;
3      use IEEE.std_logic_1164.all;
4      use IEEE.std_logic_unsigned.all;
5      entity BIN2EX3 is
6        port (
7          BINARY  : in  std_logic_vector(3 downto 0);
8          EXCESS3 : out std_logic_vector(3 downto 0));
9      end BIN2EX3;
```

```
10    ----------------------------------------------------------
11    architecture RTL1 of BIN2EX3 is
12    begin
13      with BINARY select
14      EXCESS3 <= "0011" when "0000", -- 0
15                 "0100" when "0001", -- 1
16                 "0101" when "0010", -- 2
17                 "0110" when "0011", -- 3
18                 "0111" when "0100", -- 4
19                 "1000" when "0101", -- 5
20                 "1001" when "0110", -- 6
21                 "1010" when "0111", -- 7
22                 "1011" when "1000", -- 8
23                 "1100" when "1001", -- 9
24                 "1101" when "1010", -- 10
25                 "1110" when "1011", -- 11
26                 "1111" when "1100", -- 12
27                 "0000" when "1101", -- 13
28                 "0001" when "1110", -- 14
29                 "0010" when others; -- 15
30    end RTL1;
31    architecture RTL2 of BIN2EX3 is
32    begin
33      p0 : process (BINARY)
34      begin
35        case BINARY is
36          when "0000" => EXCESS3 <= "0011"; -- 0
37          when "0001" => EXCESS3 <= "0100"; -- 1
38          when "0010" => EXCESS3 <= "0101"; -- 2
39          when "0011" => EXCESS3 <= "0110"; -- 3
40          when "0100" => EXCESS3 <= "0111"; -- 4
41          when "0101" => EXCESS3 <= "1000"; -- 5
42          when "0110" => EXCESS3 <= "1001"; -- 6
43          when "0111" => EXCESS3 <= "1010"; -- 7
44          when "1000" => EXCESS3 <= "1011"; -- 8
45          when "1001" => EXCESS3 <= "1100"; -- 9
46          when "1010" => EXCESS3 <= "1101"; -- 10
47          when "1011" => EXCESS3 <= "1110"; -- 11
48          when "1100" => EXCESS3 <= "1111"; -- 12
49          when "1101" => EXCESS3 <= "0000"; -- 13
50          when "1110" => EXCESS3 <= "0001"; -- 14
51          when others => EXCESS3 <= "0010"; -- 15
52        end case;
53      end process;
54    end RTL2;
55    architecture RTL3 of BIN2EX3 is
56    begin
57      EXCESS3 <= BINARY + "0011";
58    end RTL3;
```

Architecture RTL1 uses a **case statement** inside a **process statement**. Architecture RTL2 uses a **concurrent selected signal assignment statement**. Architecture RTL3 uses operator "+" in line 57. Again, the verification and synthesis are left as Exercise 4.14.

4.4 EQUALITY CHECKER

It is common to check whether two signals have the same value. An equality checker serves this purpose. The following EQUAL VHDL code shows three different architectures to model the circuit. Entity EQUAL has two input signals, A and B. Both of them have the same width N defined in line 5 as a generic with a default value of 8. Line 7 specifies the input signals A and B based on the value of generic N. The output signal SAME is '1' when A is equal to B, '0' otherwise.

```
1      ---------------------- file equal.vhd
2      library IEEE;
3      use IEEE.std_logic_1164.all;
4      entity EQUAL is
5        generic (N : in integer := 8);
6        port (
7           A, B  : in  std_logic_vector(N-1 downto 0);
8           SAME  : out std_logic);
9      end EQUAL;
```

The following architecture RTL1 uses a **process statement** with signals A and B as the **process sensitivity list**. A variable SAME_SO_FAR is declared in line 13. Variable SAME_SO_FAR is set to '1'. Lines 16 to 21 is a **loop statement**. The loop index i goes left to right to compare the corresponding bit position of A and B. When the corresponding bit is not the same, Variable SAME_SO_FAR is set to '0' and the loop is exited. Line 22 assigns the output signal SAME with variable SAME_SO_FAR. Note that attribute range of A (A'range) is used in line 16, rather than 0 to N-1 or N-1 downto 0. When A and B are changed with a fixed value, there is no need to change the architecture.

```
10     architecture RTL1 of EQUAL is
11     begin
12       p0 : process (A, B)
13         variable SAME_SO_FAR : std_logic;
14       begin
15         SAME_SO_FAR := '1';
16         for i in A'range loop
17          if (A(i) /= B(i)) then
18             SAME_SO_FAR := '0';
19            exit;
20           end if;
21         end loop;
```

```
22          SAME <= SAME_SO_FAR;
23        end process;
24      end RTL1;
```

The above architecture RTL1 is a software algorithm to have a **loop statement** to compare bit by bit. When two corresponding bits are not the same, there is no need to compare further, the loop can be exited. From the hardware point of view, to check whether 2 bits are the same, an exclusive NOR gate (inverse of XOR gate, or XNOR gate) can be used. The following architecture RTL2 lines 28 to 30 is a **generate statement** to generate these XNOR gates to signal EACHBIT. Lines 31 to 39 will set output SAME to the ANDed result of all bits in EACHBIT. This is exactly the function of REDUCE_AND function in package pack.vhd as shown in Appendix A.

```
25      architecture RTL2 of EQUAL is
26        signal EACHBIT : std_logic_vector(A'length-1 downto 0);
27      begin
28        xnor_gen : for i in A'range generate
29          EACHBIT(i) <= not (A(i) xor B(i));
30        end generate;
31        p0 : process (EACHBIT)
32          variable BITSAME : std_logic;
33        begin
34          BITSAME := EACHBIT(0);
35          for i in 1 to A'length-1 loop
36            BITSAME := BITSAME and EACHBIT(i);
37          end loop;
38          SAME <= BITSAME;
39        end process;
40      end RTL2;
```

The following architecture RTL3 simply uses operator "=" to compare A and B as shown in line 43.

```
41      architecture RTL3 of EQUAL is
42      begin
43        SAME <= '1' when A = B else '0';
44      end RTL3;
```

To verify the EQUAL VHDL code, the following test bench is developed. Lines 7 to 12 declare the EQUAL component. There will be too many combinations of A and B values when A and B have more bits (when N is bigger). For example, if N is 10, there will be 1024 × 1024 combinations. When N is 32, the process of verifying all combinations will take a long time. The idea here is to develop the test bench that can work for any value of N. For a smaller value of N, all combinations can be enumerated and verified. The code should work for a bigger value of N. Line 13 declares a constant W to be 5. Line 14 declares signals A and B based on the value of W.

```
1      library IEEE;
```

```
2    use IEEE.std_logic_1164.all;
3    use IEEE.std_logic_arith.all;
4    entity TBEQUAL is
5    end TBEQUAL;
6    architecture BEH of TBEQUAL is
7      component EQUAL
8      generic (N : in integer := 8);
9      port (
10       A, B  : in  std_logic_vector(N-1 downto 0);
11       SAME  : out std_logic);
12     end component;
13     constant W   : integer := 5; -- signal width
14     signal  A, B : std_logic_vector(W-1 downto 0);
15     signal  Y, Y1, Y2, Y3 : std_logic;
16     constant PERIOD : time := 50 ns;
17     constant STROBE : time := PERIOD - 5 ns;
```

Process **p0** (lines 19 to 29) enumerates all possible values of A and B based on the value of W. Line 30 calculates the expected correct value. The **check** process (lines 31 to 45 is similar to the previous TBSELECTOR example. The loop is executed $2**(W*2)$ times.

```
18   begin
19     p0 : process
20     begin
21       for j in 0 to 2**W-1 loop
22         A  <= conv_std_logic_vector(j,W);
23         for k in 0 to 2**W-1 loop
24           B  <= conv_std_logic_vector(k,W);
25           wait for PERIOD;
26         end loop;
27       end loop;
28       wait;
29     end process;
30     Y <= '1' when A = B else '0';
31     check : process
32       variable err_cnt : integer := 0;
33     begin
34       wait for STROBE;
35       for j in 1 to 2**(W*2) loop
36         if (Y /= Y1) or (Y /= Y2) or (Y /= Y3) then
37           assert FALSE report "not compared" severity WARNING;
38           err_cnt := err_cnt + 1;
39         end if;
40         wait for PERIOD;
41       end loop;
42       assert (err_cnt = 0) report "test failed" severity ERROR;
43       assert (err_cnt /= 0) report "test passed" severity NOTE;
44       wait;
45     end process;
```

```
46    equal1 : EQUAL
47      generic map ( N => W)
48      port map ( A => A, B => B, SAME => Y1);
49    equal2 : EQUAL
50      generic map ( N => W)
51      port map ( A => A, B => B, SAME => Y2);
52    equal3 : EQUAL
53      generic map ( N => W)
54      port map ( A => A, B => B, SAME => Y3);
55    end BEH;
56    configuration CFG_TBEQUAL of TBEQUAL is
57      for BEH
58        for equal1 : EQUAL
59          use entity work.EQUAL(RTL1);
60        end for;
61        for equal2 : EQUAL
62          use entity work.EQUAL(RTL2);
63        end for;
64        for equal3 : EQUAL
65          use entity work.EQUAL(RTL3);
66        end for;
67      end for;
68    end CFG_TBEQUAL;
```

Lines 46 to 54 instantiate EQUAL components with generic map of N => W. Lines 56 to 68 configure EQUAL components to each architecture. All 32 × 32 combinations are verified.

To synthesize, the EQUAL VHDL code can be analyzed as before. During the elaboration, Synopsys synthesis command *elaborate EQUAL -arch RTL3 -param "N => 8"* can be used. Figure 4-9 shows the area and timing summary. Figure 4-10 shows the synthesized schematic. The synthesis of all architectures is left as Exercise 4.18.

```
1     ****************************************
2     Report : area
3     Design : EQUAL_N8
4     Number of ports:              17
5     Number of nets:               27
6     Number of cells:              11
7     Number of references:          3
8     Combinational area:        29.000000
9     Net Interconnect area:     89.041191
10    Total cell area:           29.000000
11    Total area:               118.041191
12    ****************************************
13    Operating Conditions: WCCOM   Library: lca300k
14    --------------------------------------------------------------
15    clock clk (rise edge)                    0.00      0.00
16    clock network delay (ideal)              0.00      0.00
```

17	input external delay	2.50	2.50 f
18	A[3] (in)	0.00	2.50 f
19	U55/Z (EN)	0.94	3.44 r
20	U60/Z (ND4)	0.61	4.05 f
21	U62/Z (NR2)	1.19	5.24 r
22	SAME (out)	0.00	5.25 r
23	data arrival time		5.25

Figure 4-9 Area and timing summary for EQUAL RTL3.

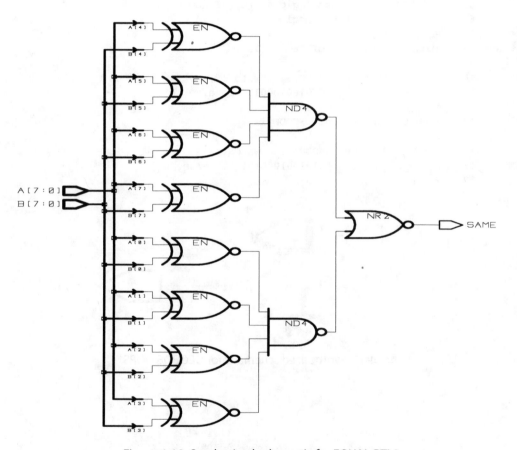

Figure 4-10 Synthesized schematic for EQUAL RTL3.

It is common to compare a signal with a known value. For example, an address bus is compared with a certain value to indicate whether a certain address is used. The following code shows an example. Sixteen bits of signal ADDR should be compared with "0101001000101001". However, the code is mistakenly written as "010100100010001" which is 1 bit short in architecture OOPS line 11. Comparing a 16-bit signal with a 15-bit value will never be equivalent. Figure 4-11 shows the synthesized schematic. This error may not be easy to see when this is embedded inside a large circuit. Caution is recommended. Figure 4-12 shows the correct schematic with architecture RTL synthesized.

```
1     library IEEE;
2     use IEEE.std_logic_1164.all;
3     entity DECODE is
4       port (
5         ADDR  : in  std_logic_vector(31 downto 0);
6         HIT   : out std_logic);
7     end DECODE;
8     architecture OOPS of DECODE is
9     begin
10      HIT <= '1' when ADDR(31 downto 16) =
11                     "010100100010001" else '0';
12    end OOPS;
13    architecture RTL of DECODE is
14    begin
15      HIT <= '1' when ADDR(31 downto 16) =
16                     "0101001000101001" else '0';
17    end RTL;
```

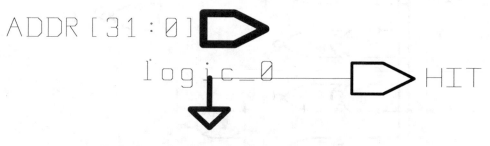

Figure 4-11 Synthesized schematic for DECODE OOPS.

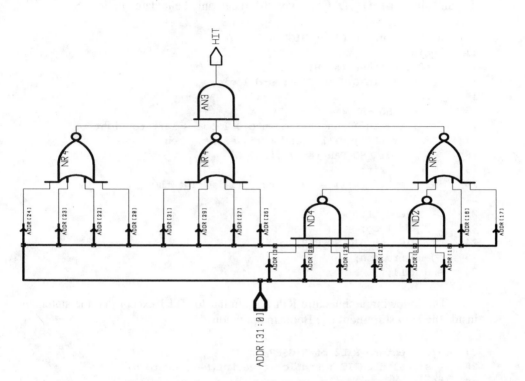

Figure 4-12 Synthesized schematic for DECODE RTL.

4.5 COMPARATOR WITH SINGLE OUTPUT

A comparator compares two input signals. The following VHDL shows a comparator with two input signals A and B. Output signal BIG is '1' when the unsigned value of A is greater than the unsigned value of B.

```
1       ---------------------- file bigger.vhd
2       library IEEE;
3       use IEEE.std_logic_1164.all;
4       entity BIGGER is
5         generic (N : in natural := 8);
6         port (
7           A, B  : in  std_logic_vector(N-1 downto 0);
8           BIG   : out std_logic);
9       end BIGGER;
```

The following architecture RTL1 is similar to the equality checker EQUAL VHDL code. It goes left to right. Whenever the corresponding bits of A and B are not the same, either A is bigger or B is bigger, the loop is exited. Line 15 sets up the default value for BIG_SO_FAR to be '0'. It can only be updated in line 18.

```
10    architecture RTL1 of BIGGER is
11    begin
12      p0 : process (A, B)
13        variable BIG_SO_FAR : std_logic;
14      begin
15        BIG_SO_FAR := '0';
16        for i in A'length-1 downto 0 loop -- left to right
17          if (A(i) = '1') and (B(i) = '0') then
18            BIG_SO_FAR := '1';
19            exit;
20          elsif (A(i) = '0') and (B(i) = '1') then
21            exit;
22          end if;
23        end loop;
24        BIG  <= BIG_SO_FAR;
25      end process;
26    end RTL1;
```

The following architecture RTL2 is similar to RTL1 except that the statements inside the **loop statement** use Boolean operations.

```
27    architecture RTL2 of BIGGER is
28      signal EACHBIT : std_logic_vector(N-1 downto 0);
29    begin
30      xnor_gen : for i in A'range generate
31        EACHBIT(i) <= not (A(i) xor B(i));
32      end generate;
33      p0 : process (EACHBIT, A, B)
34        variable EQUAL_SO_FAR, BIG_SO_FAR : std_logic;
35      begin
36        EQUAL_SO_FAR := '1';
37        BIG_SO_FAR   := '0';
38        for i in A'range loop
39          BIG_SO_FAR   :=
40            (EQUAL_SO_FAR and A(i) and (not B(i))) or
41            BIG_SO_FAR;
42          EQUAL_SO_FAR := EQUAL_SO_FAR and EACHBIT(i);
43        end loop;
44        BIG  <= BIG_SO_FAR;
45      end process;
46    end RTL2;
```

The following architecture RTL3 simply uses operator ">" in line 49.

```
47    architecture RTL3 of BIGGER is
48    begin
```

```
49      BIG  <= '1' when A > B else '0';
50      end RTL3;
```

The verification of the BIGGER VHDL code is left as Exercise 4.20. Figure 4-13 shows an example to synthesize BIGGER architecture RTL1. Note that line 4 specifies the parameter which maps the generic N to 8. Figure 4-14 shows the synthesized schematic.

```
1       sh date
2       design = bigger
3       analyze -f vhdl VHDLSRC + "ch4/" + design + ".vhd"
4       elaborate BIGGER -architecture RTL1 -param "N => 8"
5       include compile.common
6       create_clock -name clk -period CLKPERIOD -waveform {0.0
HALFCLK}
7       set_clock_skew -uncertainty CLKSKEW clk
8       set_input_delay  INPDELAY -add_delay -clock clk all_inputs()
9       set_output_delay CLKPER1  -add_delay -clock clk all_outputs()
10      compile -map_effort medium
11      include compile.report
```

Figure 4-13 Synthesis script file compile.bigger.rtl1.

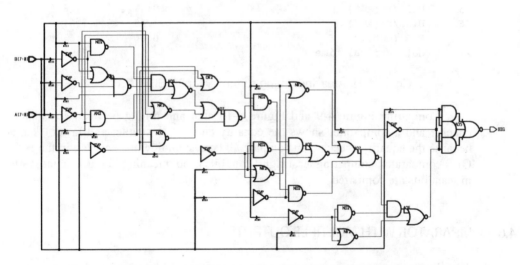

Figure 4-14 Synthesized schematic for BIGGER RTL1.

The following shows the area and timing for the Figure 4-14 schematic. Note that line 2 indicates the design name as BIGGER_N8. "_N8" is appended where 'N' is the generic name and '8' is the mapped value.

```
1    Report : area
2    Design : BIGGER_N8
3    Number of ports:              17
4    Number of nets:               43
5    Number of cells:              27
6    Number of references:          9
7    Combinational area:        37.000000
8    Net Interconnect area:    181.380203
9    Total cell area:           37.000000
10   Total area:               218.380203
11   ****************************************
12   Operating Conditions: WCCOM   Library: lca300k
13   ----------------------------------------------------------------
14     clock clk (rise edge)                    0.00      0.00
15     clock network delay (ideal)              0.00      0.00
16     input external delay                     2.50      2.50 f
17     B[1] (in)                                0.00      2.50 f
18     U128/Z (IVP)                             0.41      2.91 r
19     U135/Z (AO3)                             0.55      3.46 f
20     U139/Z (AO6)                             0.53      4.00 r
21     U141/Z (AO7)                             0.56      4.56 f
22     U142/Z (ND2)                             0.28      4.84 r
23     U146/Z (AO6)                             0.40      5.24 f
24     U148/Z (AO7)                             0.42      5.66 r
25     U152/Z (AO6)                             0.47      6.13 f
26     U153/Z (AO5)                             1.48      7.61 r
27     BIG (out)                                0.00      7.61 r
28     data arrival time                                  7.61
```

Figure 4-15 Area and timing summary for BIGGER RTL1.

Comparing Figure 4-9 and Figure 4-15 area and timing between an equality checker and a comparator shows the equality checker as smaller and faster. This is because the equality checker can compare all corresponding bits parallel. In the BIGGER comparator, comparing least significant bits is not meaningful until the most significant bits are compared.

4.6 COMPARATOR WITH MULTIPLE OUTPUTS

Suppose we want a comparator to compare two 32-bit inputs and three outputs. Each output indicates whether A is greater than, equal to, or less than B. The following shows the VHDL entity.

```
1    --------------------- file gle_32.vhd
2    library IEEE;
3    use IEEE.std_logic_1164.all;
```

```
4      use IEEE.std_logic_unsigned.all;
5      entity GLE is
6        port (
7          A, B : in  std_logic_vector(31 downto 0);
8          EQUAL, GREATER, LESS : out std_logic);
9      end GLE;
```

The following architecture RTL1 first sets up the default value in lines 7 to 9. It then goes through the loop to check bits from left to right.

```
1      ---------------------- file gle_rtl.vhd
2      architecture RTL1 of GLE is
3      begin
4        p0 : process (A, B)
5          variable EQ, GT, LT : std_logic;
6        begin
7          EQUAL   <= '1';
8          GREATER <= '0';
9          LESS    <= '0';
10         for j in A'range loop
11           if (A(j) and (not B(j))) = '1' then
12             GREATER <= '1';
13             EQUAL   <= '0';
14             exit;
15           elsif ((not A(j)) and B(j)) = '1' then
16             LESS    <= '1';
17             EQUAL   <= '0';
18             exit;
19           end if;
20         end loop;
21       end process;
22     end RTL1;
```

The following architecture RTL2 first sets up the default values in lines 27 to 29. Operators ">" and "<" are used in lines 31 and 33, respectively.

```
23     architecture RTL2 of GLE is
24     begin
25       p1 : process (A, B)
26       begin
27         EQUAL   <= '0';
28         GREATER <= '0';
29         LESS    <= '0';
30         if (A > B) then
31           GREATER <= '1';
32         elsif (A < B) then
33           LESS    <= '1';
34         else
35           EQUAL   <= '1';
```

```
36        end if;
37      end process;
38    end RTL2;
```

The following architecture RTL3 sets up the default values in lines 43 to 45.
Operators "=" and "<" are used in lines 47 and 49.

```
39    architecture RTL3 of GLE is
40    begin
41      p1 : process (A, B)
42      begin
43        EQUAL   <= '0';
44        GREATER <= '0';
45        LESS    <= '0';
46        if (A = B) then
47          EQUAL   <= '1';
48        elsif (A < B) then
49          LESS    <= '1';
50        else
51          GREATER <= '1';
52        end if;
53      end process;
54    end RTL3;
```

The following architecture RTL4 uses operators "=", ">", and "<" in lines 57 to
59.

```
55    architecture RTL4 of GLE is
56    begin
57      EQUAL   <= '1' when (A = B) else '0';
58      GREATER <= '1' when (A > B) else '0';
59      LESS    <= '1' when (A < B) else '0';
60    end RTL4;
```

The following architecture RTL5 uses operators "=" and ">" in lines 64 and 65.
The outputs are computed in lines 66 to 68 based on temporary values EQ_TEMP and
GT_TEMP.

```
61    architecture RTL5 of GLE is
62      signal EQ_TEMP, GT_TEMP : std_logic;
63    begin
64      EQ_TEMP <= '1' when (A = B) else '0';
65      GT_TEMP <= '1' when (A > B) else '0';
66      LESS    <= EQ_TEMP nor GT_TEMP;
67      GREATER <= GT_TEMP;
68      EQUAL   <= EQ_TEMP;
69    end RTL5;
```

The following architecture RTL6 uses operators "=" and "-" in lines 74 to 75. Note that line 75 infers a subtracter. Both A and B are extended 1 bit. The leftmost bit of SUB is used in lines 76 to 77 to generate outputs LESS and GREATER.

```
70    architecture RTL6 of GLE is
71      signal EQ_TEMP : std_logic;
72      signal SUB : std_logic_vector(A'length downto 0);
73    begin
74      EQ_TEMP <= '1' when (A = B) else '0';
75      SUB     <= ('0' & A) - ('0' & B);
76      LESS    <= SUB(A'length) and (not EQ_TEMP);
77      GREATER <= (not SUB(A'length)) and (not EQ_TEMP);
78      EQUAL   <= EQ_TEMP;
79    end RTL6;
```

To completely verify all combinations of A and B with all 32 bits will take a long time. To reduce the verification time, we have written the VHDL code in architectures independent of the length of A and B. There is no hard-coded value to be changed when their signal widths are changed. This enables us to write another VHDL entity gle_gen.vhd as shown below. Note that the entity has the same name GLE. It uses a generic with a default value of 32.

```
1     ----------------------- file gle_gen.vhd
2     library IEEE;
3     use IEEE.std_logic_1164.all;
4     use IEEE.std_logic_unsigned.all;
5     entity GLE is
6       generic ( N : integer := 32);
7       port (
8         A, B : in  std_logic_vector(N-1 downto 0);
9         EQUAL, GREATER, LESS : out std_logic);
10    end GLE;
```

The test bench is shown below. Lines 7 to 12 declare the GLE component. This allows the test bench to enumerate all possible combinations of input values when the generic value is mapped with a smaller value. Line 13 declares a constant W to be 5. W is then mapped to N in lines 53, 57, etc. The basic checking, component instantiation, and configurations are similar to previous examples.

```
1     library IEEE;
2     use IEEE.std_logic_1164.all;
3     use IEEE.std_logic_arith.all;
4     entity TBGLE is
5     end TBGLE;
6     architecture BEH of TBGLE is
7       component GLE
8       generic ( N : integer := 32);
9       port (
```

```
10        A, B : in  std_logic_vector(N-1 downto 0);
11        EQUAL, GREATER, LESS : out std_logic);
12      end component;
13      constant W   : integer := 5; -- signal width
14      signal  A, B : std_logic_vector(W-1 downto 0);
15      signal  E1, E2, E3, E4, E5, E6 : std_logic;
16      signal  G1, G2, G3, G4, G5, G6 : std_logic;
17      signal  L1, L2, L3, L4, L5, L6 : std_logic;
18      constant PERIOD : time := 50 ns;
19      constant STROBE : time := PERIOD - 5 ns;
20    begin
21      p0 : process
22      begin
23        for j in 0 to 2**W-1 loop
24          A  <= conv_std_logic_vector(j,W);
25          for k in 0 to 2**W-1 loop
26            B  <= conv_std_logic_vector(k,W);
27            wait for PERIOD;
28          end loop;
29        end loop;
30        wait;
31      end process;
32      check : process
33        variable err_cnt : integer := 0;
34      begin
35        wait for STROBE;
36        for j in 1 to 2**(W*2) loop
37          if (E1 /= E2) or (E2 /= E3) or (E3 /= E4) or
38             (E4 /= E5) or (E5 /= E6) or
39             (G1 /= G2) or (G2 /= G3) or (G3 /= G4) or
40             (G4 /= G5) or (G5 /= G6) or
41             (L1 /= L2) or (L2 /= L3) or (L3 /= L4) or
42             (L4 /= L5) or (L5 /= L6) then
43            assert FALSE report "not compared" severity WARNING;
44            err_cnt := err_cnt + 1;
45          end if;
46          wait for PERIOD;
47        end loop;
48        assert (err_cnt = 0) report "test failed" severity ERROR;
49        assert (err_cnt /= 0) report "test passed" severity NOTE;
50        wait;
51      end process;
52      GLE1 : GLE
53        generic map ( N => W)
54        port map ( A => A, B => B, EQUAL => E1,
55          GREATER => G1, LESS => L1);
56      gle2 : GLE
57        generic map ( N => W)
58        port map ( A => A, B => B, EQUAL => E2,
59          GREATER => G2, LESS => L2);
```

```
60      gle3 : GLE
61        generic map ( N => W)
62        port map ( A => A, B => B, EQUAL => E3,
63          GREATER => G3, LESS => L3);
64      gle4 : GLE
65        generic map ( N => W)
66        port map ( A => A, B => B, EQUAL => E4,
67          GREATER => G4, LESS => L4);
68      gle5 : GLE
69        generic map ( N => W)
70        port map ( A => A, B => B, EQUAL => E5,
71          GREATER => G5, LESS => L5);
72      gle6 : GLE
73        generic map ( N => W)
74        port map ( A => A, B => B, EQUAL => E6,
75          GREATER => G6, LESS => L6);
76    end BEH;
77    configuration CFG_TBGLE of TBGLE is
78      for BEH
79        for gle1 : GLE
80          use entity work.GLE(RTL1);
81        end for;
82        for gle2 : GLE
83          use entity work.GLE(RTL2);
84        end for;
85        for gle3 : GLE
86          use entity work.GLE(RTL3);
87        end for;
88        for gle4 : GLE
89          use entity work.GLE(RTL4);
90        end for;
91        for gle5 : GLE
92          use entity work.GLE(RTL5);
93        end for;
94        for gle6 : GLE
95          use entity work.GLE(RTL6);
96        end for;
97      end for;
98    end CFG_TBGLE;
```

Figure 4-16 can be used to synthesize the 32-bit GLE comparator. Note that line 3 analyzes gle_32.vhd rather than gle_gen.vhd. Line 4 analyzes all architectures in file gle_rtl.vhd. Line 5 elaborates GLE architecture RTL4. Figure 4-17 shows a portion of the synthesized schematic. Note that there are two comparator blocks at the bottom of the schematic. This is caused by lines 58 and 59 in GLE architecture RTL4. This is a Synopsys DesignWare block that has been designed and called out. There are also unconnected outputs of the DesignWare block since the block is designed with more outputs such as (GT, LT, EQ, GE, LE, NE) for (greater, less, equal, greater or equal, less or equal, and not equal, respectively). The synthesis command **check_design** can be used. Warning messages such as "**Pin 'TC' is connected to**

logic 0" and "**Warning: In design 'GLE_DW01_cmp6_32_2', port 'EQ' is not connected to any nets**" are generated. Synthesis command "**ungroup -all**" can be used to flatten the hierarchy. The **check_design** command flags that there are two ground symbols in the schematic with warning message "Warning: In design 'GLE', net 'n249' driven by pin 'U58/**logic_0**' has no loads." Another **compile** command will reoptimize the schematic and remove the ground symbols. The command **check_design** returns no warning messages. The verification and synthesis for other architectures are left as exercises.

```
1      sh date
2      design = gle
3      analyze -f vhdl VHDLSRC + "ch4/" + design + "_32.vhd"
4      analyze -f vhdl VHDLSRC + "ch4/" + design + "_rtl.vhd"
5      elaborate GLE -architecture RTL4
6      include compile.common
7      create_clock -name clk -period CLKPERIOD -waveform {0.0
HALFCLK}
8      set_clock_skew -uncertainty CLKSKEW clk
9      set_input_delay  INPDELAY -add_delay -clock clk all_inputs()
10     set_output_delay CLKPER6  -add_delay -clock clk all_outputs()
11     compile -map_effort medium
12     include compile.report
```

Figure 4-16 Synthesis script file compile.gle.rtl4.

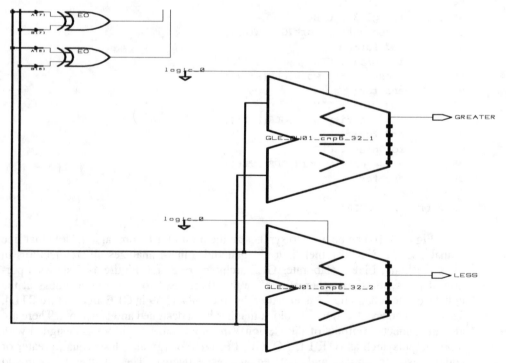

Figure 4-17 Synthesized schematic for GLE RTL4.

4.7 EXERCISES

4.1   In entity TBSELECTOR VHDL code line 18, will the TBSELECTOR simulation produce a "test passed" message when STROBE is set to 1 ns? What is the advantage of specifying the STROBE as 45 ns rather than 1 ns?

4.2   What would happen if line 31 alone in entity TBSELECTOR is deleted? What would happen if line 47 alone in entity TBSELECTOR is deleted? What would happen if both lines 31 and 47 in entity TBSELECTOR are deleted?

4.3   Architectures RTL1, RTL2, RTL3, and RTL4 VHDL code and entity SELECTOR VHDL code are put in the same text file. What needs to be done if they are split into five separate files? Are there any advantages to doing this? What are your conventions and reasons for the five filenames?

4.4   In entity TBSELECTOR VHDL code line 21 and 24, **cnt** is declared as a variable. Change it to a signal. Rerun the simulation. What are the differences?

4.5   Refer to entity SELECTOR architectures RTL1, RTL2, RTL3, and RTL4 VHDL code. Which architecture has the least chance of having a design error? Which architecture will synthesize to the largest circuit? What are the advantages of architecture RTL3 compared with architecture RTL2?

4.6   Refer to entity SELECTOR synthesis reports. The slack is positive and large. This indicates that the synthesis tool does not need to try very hard to synthesize the circuit. Change the output delay constraint (line 9 in Figure 4-3) CLKPER1 to CLKPER3, CLKPER5, CLKPER7, CLKPER8, and CLKPER9. Synthesize again for architecture RTL1 and RTL2. Compare the results. What is your conclusion?

4.7   Refer to entity SELECTOR synthesis. The input delay for all inputs are set to be the same. Set the input delay for the SEL as CLKPER3, and A as CLKPER5. Synthesize architectures RTL1 and RTL2. Compare with the results shown in Figure 4-4 and Figure 4-5. What are your observations?

4.8   Refer to Figure 4-6 schematic. Two MUX41P cells are used. One of them is used for inputs A(15 downto 12). Explain the reasons why other first stage multiplexers are MUX41, not MUX41P. What are the differences?

4.9   Refer to Figure 4-5 schematic. Use commands "report_timing -from SEL" and "report_timing -from A[0]" to compare the timing delay from all inputs of the circuit.

4.10 In Figure 4-7 synthesis script file, we have shown that synthesis commands will direct synthesis tools to generate different schematics. Use the PARGEN VHDL code in Chapter 3. Apply different input delays to various bits of input signal. Synthesize and compare the area and timing results.

4.11 If you have access to Synopsys synthesis tools, find out how to setup the synthesis directives so that SELECTOR architecture RTL2 can be synthesized with all multiplexers. Compare the synthesis results with the results of Figure 4-6.

4.12 Refer to ENCODER VHDL code. Develop a test bench to verify the correctness of all four architectures. Develop synthesis scripts to synthesize each architecture. Compare the synthesis results. Which architecture is synthesized with the best result? Explain why.

4.13 Refer to ENCODER VHDL code. What would happen if line 53 is removed? Simulate and synthesize architecture RTL2 after line 53 is removed.

4.14 Refer to BIN2EX3 VHDL code. Develop a test bench to verify the correctness of all three architectures. Estimate the number of gates that will be used in each architecture. Develop synthesis scripts to synthesize each architecture. Compare the synthesis results. Which architecture is synthesized with the best result? Explain why. Are the synthesized results close to what you have estimated?

4.15 Write VHDL code to convert a 4-bit binary code to a Gray code. Develop a test bench to verify the correctness of your implementation. Justify the reasons that your test bench covers all cases. Estimate the number of gates to be used. Synthesize the VHDL code. Are the synthesized results close to what you have estimated?

4.16 Write VHDL code to convert a 4-bit Excess-3 code to a Gray code. Develop a test bench to verify the correctness of your implementation. Justify the reasons that your test bench covers all cases. Estimate the number of gates to be used. Synthesize the VHDL code. Are the synthesized results close to what you have estimated?

4.17 Refer to EQUAL architecture RTL2. As mentioned in the text, function REDUCE_AND from the package pack.vhd (Appendix A) can be used. Write another architecture RTL4 to use the REDUCE_AND function. Update the test bench TBEQUAL to verify all four architectures with another value of W.

4.18 Synthesize EQUAL VHDL code for all architectures with N => 32. Compare the synthesis results.

4.19 A zero detector is similar to the equality checker except that the constant has every bit as '0'. Write VHDL code for a zero detector. Write a test bench to verify its correctness. Synthesize the VHDL code.

4.20 Refer to BIGGER VHDL code. Develop a test bench to verify all three architectures.

4.21 Write synthesis script files to synthesize all three architectures of BIGGER VHDL code to map N with 32. Compare your synthesis results for all three architectures. Compare the results with Exercise 4.18.

4.22 Refer to BIGGER VHDL code. It compares the unsigned values of A and B. Write VHDL code to compare A and B which are treated as 2's complement value. Develop a test bench to verify the VHDL code. Synthesize the VHDL code and compare with Exercise 4.21.

4.23 Refer to TBGLE VHDL code. Run the test bench to see whether the test bench enumerates all possible combinations of A and B 5-bit values.

4.24 Refer to GLE VHDL code (gle_32.vhd and gle_rtl.vhd). Before you synthesize the VHDL code, take a guess at the area and timing for all architectures. Which architecture will produce the smallest area? Which architecture will produce the fastest circuit? Write synthesis script files to synthesize all six architectures. Compare the area and timing among all six architectures. Do they match your presynthesis estimates?

Chapter 5

Basic Binary Arithmetic Circuits

In this chapter we will use VHDL to describe several basic binary arithmetic combinational circuits such as half adder, full adder, ripple carry adder, carry look ahead adder, incrementer, and 2's complementer. Other commonly used circuits such as counting the number of ones, counting the number of leading zero positions, and the barrel shifter are also discussed. There are generally many ways to describe the same circuit in VHDL. We will look at some of these different constructs and how they impact the synthesized circuit. We also show test bench techniques to verify these circuits.

5.1 HALF ADDER AND FULL ADDER

A half adder adds 2 binary bits A, B and generates 2 output bits S (sum) and C (carry) as shown in the following VHDL code. The sum bit output is '1' when input bits A and B are not the same. In line 10, this is accomplished using an xor operator. The carry bit output is '1' when both inputs A and B are '1'. Line 11 uses operator "and" for this purpose. There are many different ways to represent this functionality in VHDL. The half adder (HA) logic is straightforward, so we use this simplified coding approach to generate the schematic in Figure 5-1. because of the simplicity of the schematic, we are assured of the required output, and the simulation step can be omitted.

```
1     library IEEE;
2     use IEEE.std_logic_1164.all;
3     entity HA is
4       port (
5         A, B  : in  std_logic;
6         S, C  : out std_logic);
7     end HA;
8     architecture RTL of HA is
9     begin
10      S <= A xor B;
11      C <= A and B;
12    end RTL;
```

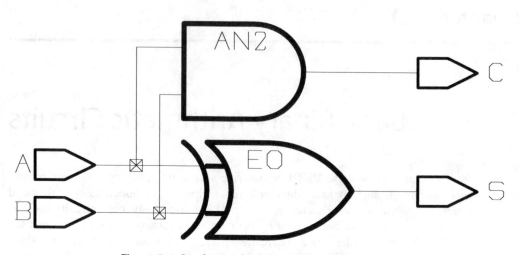

Figure 5-1 Synthesized schematic for HA VHDL code.

The full adder (FA) adds a level of complexity by supporting a carry input bit. In short, it adds 3 bits and generates 2 output bits. The following VHDL code describes a full adder circuit. Line 5 shows the input ports A, B, and the CI carry in bit. Line 6 declares the output bits S sum bit and COUT carry out bit. Whenever A, B, and CI have odd numbers of '1', S will be '1', otherwise, S will be '0'. This is modeled in line 10 by a 3 input xor. Line 11 generates COUT output. When 2 of A, B, and CI are '1', then COUT will be '1', otherwise, COUT will be '0'. Figure 5-2 shows one possible synthesized schematic for entity FA, architecture RTL.

```
1     library IEEE;
2     use IEEE.std_logic_1164.all;
3     entity FA is
4       port (
5         A, B, CI : in  std_logic;
6         S, COUT  : out std_logic);
7     end FA;
8     architecture RTL of FA is
9     begin
10      S    <= A xor B xor CI;
11      COUT <= (A and B) or (A and CI) or (B and CI);
12    end RTL;
```

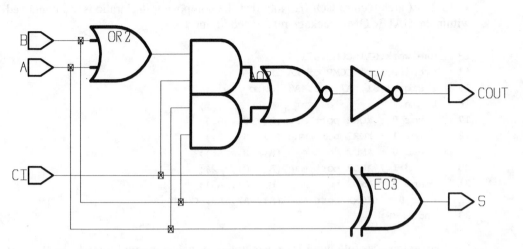

Figure 5-2 Synthesized schematic for FA RTL.

Figure 5-2 implements the FA function correctly, but suppose that we like the exact schematic implementation as shown in Figure 5-3. How do we achieve this so that the synthesis tools will not generate other schematics even though they may be faster and smaller?

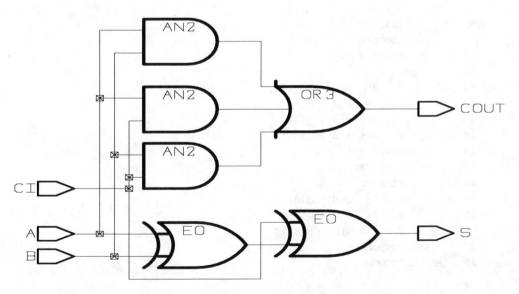

Figure 5-3 Synthesized schematic for FA COMP.

Let us begin by using **component instantiation statements** as shown in lines 17 to 22 of architecture COMP. Note that the component declarations are contained within the GATECOMP package referenced in line 13.

```
13    use work.GATECOMP.all;
14    architecture COMP of FA is
15      signal S1, A1, A2, A3 : std_logic;
16    begin
17      xor_0 : XOR2 port map (A,  B,  S1);
18      xor_1 : XOR2 port map (CI, S1, S);
19      and_0 : AND2 port map (A,  B,  A1);
20      and_1 : AND2 port map (A,  CI, A2);
21      and_2 : AND2 port map (B,  CI, A3);
22      or_0  : KOR3 port map (A1, A2, A3, COUT);
23    end COMP;
```

For reference, following is an excerpt from gatecomp.vhd that shows the port declarations for the instanced components.

```
1     library IEEE;
2     use IEEE.std_logic_1164.all;
3     package GATECOMP is
4       component AND2
5         port (
6           A : in  std_logic;
7           B : in  std_logic;
8           Y : out std_logic );
9       end component;
10      component KOR3
11        port (
12          A : in  std_logic;
13          B : in  std_logic;
14          C : in  std_logic;
15          Y : out std_logic );
16      end component;
17      component XOR2
18        port (
19          A : in  std_logic;
20          B : in  std_logic;
21          Y : out std_logic );
22      end component;
23    end GATECOMP;
```

To achieve the Figure 5-3 schematic, the synthesis script as shown in Figure 5-4 can be used. Line 2 analyzes gates.vhd, which has the functional description of the components used in GATECOMP package. Line 4 analyzes gatecomp.vhd. Line 5 analyzes the FA VHDL code. Line 6 elaborates FA architecture COMP. Line 6 makes the components unique since the same component is instantiated more than once as shown in lines 17 to 18 in FA COMP VHDL code. Line 7 synthesizes the VHDL code. Line 8 flattens the hierarchy.

```
1    design = fa
2    analyze -f vhdl VHDLSRC + "ch5/" + "gates.vhd"
3    analyze -f vhdl VHDLSRC + "ch5/" + "gatecomp.vhd"
4    analyze -f vhdl VHDLSRC + "ch5/" + design + ".vhd"
5    elaborate FA -architecture COMP
6    uniquify
7    compile
8    ungroup -all
```

Figure 5-4 Synthesis script file compile.fa.comp.

Figure 5-5 shows another possible synthesis script. Lines 1 to 5 are identical in Figure 5-4. Line 6 sets up the current design to be the subcomponent AND2. Line 7 synthesizes AND2. Lines 8 to 11 repeat the same process for subcomponents KOR2 and XOR2. Line 12 sets up the current design to be FA. Line 13 tells the synthesis tools that subcomponents should not be touched. Line 14 synthesized the FA circuit. Line 15 resets the design so that subcomponents can be flattened in line 16.

```
1    design = fa
2    analyze -f vhdl VHDLSRC + "ch5/" + "gates.vhd"
3    analyze -f vhdl VHDLSRC + "ch5/" + "gatecomp.vhd"
4    analyze -f vhdl VHDLSRC + "ch5/" + design + ".vhd"
5    elaborate FA -architecture COMP
6    current_design AND2
7    compile
8    current_design KOR3
9    compile
10   current_design XOR2
11   compile
12   current_design FA
13   set_dont_touch {AND2 KOR3 XOR2}
14   compile
15   reset_design
16   ungroup -all
```

Figure 5-5 Synthesis script file compile.fa.comp.dt.

Lines 13 to 15 in Figure 5-5 are used to illustrate the synthesis process design subcomponents that have already been synthesized. In contrast, Figure 5-4 processes the synthesis such that subcomponents are synthesized hierarchically. Since architecture COMP of FA uses all component instantiation statements, there is no need to synthesize. Removal of lines 13 to 15 generates the exact same schematic. This is left for the reader to verify in Exercise 5.2. Note that in line 2 gates.vhd is analyzed but the code is not shown here. The verification and configuration of subcomponents are left as Exercise 5.3.

Another way to replicate the schematic is illustrated in the following VHDL code. Here, the synthesis library components are referenced directly. Note that lines 9 to 17 declare library components EO, AN2, and OR3, and they are later instantiated in lines 20 to 25. By reading in the VHDL code, the exact same schematic is generated. The component declaration can also be placed in a component package provided along with the synthesis library. It is important to match the exact component and port names with the library. The same code cannot be used for another synthesis library since the component name and port names may not be exactly the same. This limits the portability of the VHDL code. Instantiation of library components directly can be used for specially designed components such as FIFO (first in, first out), clock buffers, RAM, and read-only memory (ROM) provided by silicon fabrication vendors.

```
1    library IEEE;
2    use IEEE.std_logic_1164.all;
3    entity FA is
4      port(
5        A, B, CI : in  std_logic;
6        S, COUT  : out std_logic);
7    end FA;
8    architecture LIB of FA is
9      component EO
10       port( A, B : in std_logic;  Z : out std_logic);
11     end component;
12     component AN2
13       port( A, B : in std_logic;  Z : out std_logic);
14     end component;
15     component OR3
16       port( A, B, C : in std_logic;  Z : out std_logic);
17     end component;
18     signal S1, A1, A3, A2 : std_logic;
19   begin
20     u0 : EO port map( A => A, B => B, Z => S1);
21     u1 : EO port map( A => CI, B => S1, Z => S);
22     u2 : AN2 port map( A => A, B => B, Z => A1);
23     u3 : AN2 port map( A => A, B => CI, Z => A2);
24     u4 : AN2 port map( A => B, B => CI, Z => A3);
25     u5 : OR3 port map( A => A1, B => A2, C => A3, Z => COUT);
26   end LIB;
```

5.2 CARRY RIPPLE ADDER

A carry ripple adder is an adder with its carry ripple through every bit position from right to left. It is similar to addition with pencil and paper. The following RPAD-DER VHDL shows an example of a carry ripple adder. It has input A and B with N bits defined as a generic. CI is the carry input. COUT is the carry output. S is the sum output. Architecture RTL1 uses component FA, full adder, as described in the last section. **Generate statements** are used to instantiate FA components.

```
1    ------------------ file rpadder.vhd
2    library IEEE;
3    use IEEE.std_logic_1164.all;
4    entity RPADDER is
5      generic (N : in integer := 8);
6      port (
7        A, B  : in  std_logic_vector(N-1 downto 0);
8        CI    : in  std_logic;
9        S     : out std_logic_vector(N-1 downto 0);
10       COUT  : out std_logic);
11   end RPADDER;
12   architecture RTL1 of RPADDER is
13     component FA
14     port (
15       A, B, CI : in  std_logic;
16       S, COUT  : out std_logic);
17     end component;
18     signal C : std_logic_vector(A'length-1 downto 1);
19   begin
20     gen : for j in A'range generate
21       genlsb : if j = 0 generate
22         fa0 : FA port map (A => A(0), B => B(0),
23           CI => CI, S => S(0), COUT => C(1));
24       end generate;
25       genmid : if (j > 0) and (j < A'length-1) generate
26         fa0 : FA port map (A => A(j), B => B(j),
27           CI => C(j), S => S(j), COUT => C(j+1));
28       end generate;
29       genmsb : if j = A'length-1 generate
30         fa0 : FA port map (A => A(j), B => B(j),
31           CI => C(j), S => S(j), COUT => COUT);
32       end generate;
33     end generate;
34   end RTL1;
```

The following architecture RTL2 is similar to RTL1 except that the FA function is described as a procedure in line 36 to 42. Same generate statements are used to call the FADDER procedures.

```
35    architecture RTL2 of RPADDER is
36      procedure FADDER (
37        signal A, B, CI : in  std_logic;
38        signal S, COUT  : out std_logic) is
39      begin
40        S    <= A xor B xor CI;
41        COUT <= (A and B) or (A and CI) or (B and CI);
42      end FADDER;
43      signal C : std_logic_vector(A'length-1 downto 1);
44    begin
45      gen : for j in A'range generate
46        genlab : if j = 0 generate
47          FADDER (A => A(0), B => B(0),
48            CI => CI, S => S(0), COUT => C(1));
49        end generate;
50        genmid : if (j > 0) and (j < A'length-1) generate
51          FADDER (A => A(j), B => B(j),
52            CI => C(j), S => S(j), COUT => C(j+1));
53        end generate;
54        genmsb : if j = A'length-1 generate
55          FADDER (A => A(j), B => B(j),
56            CI => C(j), S => S(j), COUT => COUT);
57        end generate;
58      end generate;
59    end RTL2;
```

To simulate RPADDER architecture RTL1, the component FA needs to be configured. The following VHDL code can be used to configure RPADDER architecture RTL1. Note that FA components are inside the generate labels and there are two levels of **generate statements**. The corresponding configuration labels are shown in lines 4 and 20 for the outer generate statement. The three inner generate statements are configured in lines 5 to 9, lines 10 to 14, and lines 15 to 19. The verification of RPADDER VHDL code is left as Exercise 5.6. RPADDER architecture RTL2 can be simulated directly without configuration, since there are no subcomponents.

```
1     ---------------------- file rpadder_cfg.vhd
2     configuration CFG_RTL1_RPADDER of RPADDER is
3       for RTL1
4         for gen
5           for genlsb
6             for all : FA
7               use entity work.FA(RTL);
```

```
8              end for;
9            end for;
10           for genmid
11             for all : FA
12               use entity work.FA(RTL);
13             end for;
14           end for;
15           for genmsb
16             for all : FA
17               use entity work.FA(RTL);
18             end for;
19           end for;
20         end for;
21       end for;
22     end CFG_RTL1_RPADDER;
```

Assume that the FA architecture COMP (Figure 5-3) has been saved in the ../db directory. Figure 5-6 synthesis script file can be used to generate the schematic shown in Figure 5-7. Line 2 reads the synthesized results of FA component. There is no compile command since architecture RTL1 does not have anything other than FA to be synthesized. All it needs is to connect all FA components together. Note that the FA components form a chain with carry ripple through in Figure 5-7.

```
1     design = rpadder
2     read -f db ../db/fa.db
3     analyze -f vhdl VHDLSRC + "ch5/" + design + ".vhd"
4     elaborate RPADDER -architecture RTL1 -param "N => 8"
5     include compile.common
6     create_clock -name clk -period CLKPERIOD -waveform {0.0
HALFCLK}
7     set_input_delay  INPDELAY -add_delay -clock clk all_inputs()
8     set_output_delay CLKPER6  -add_delay -clock clk all_outputs()
```

Figure 5-6 Synthesis script file compile.rpadder.rtl1.

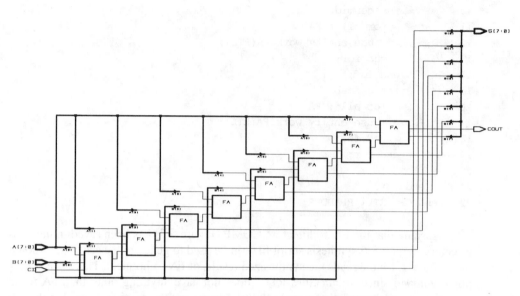

Figure 5-7 Synthesized schematic for RPADDER RTL1.

Figure 5-7 schematic has a total area of 426 units with a data arrival time of 15.31 ns. The schematic can be ungrouped with "ungroup -all" command and resynthesized with the compile command. It has a total area of 384 units with a data arrival time of 11.47 ns. The schematic is shown in Figure 5-8. The synthesis tools can further optimize the circuit when the subcomponent boundaries are removed.

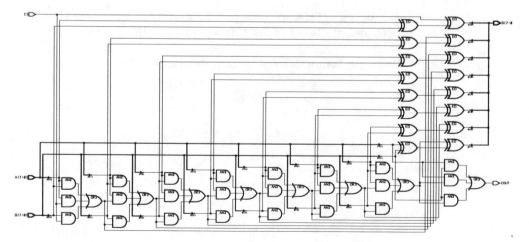

Figure 5-8 Schematic for RPADDER RTL1, ungrouped, recompiled.

RPADDER architecture RTL2 can be synthesized with the synthesis script file as shown in Figure 5-9. It has a total area of 397 units with a data arrival time of 10.61 ns. Note that both architectures RTL1 and RTL2 describe the carry ripple behavior. Architecture RTL2 does not have any subcomponents. The synthesis tools can optimize better without the boundary of the FA components. More synthesis exercises are described in Exercises 5.7 and 5.8.

```
1    design = rpadder
2    analyze -f vhdl VHDLSRC + "ch5/" + design + ".vhd"
3    elaborate RPADDER -architecture RTL2 -param "N => 8"
4    include compile.common
5    create_clock -name clk -period CLKPERIOD -waveform {0.0
HALFCLK}
6    set_clock_skew -uncertainty CLKSKEW clk
7    set_input_delay  INPDELAY -add_delay -clock clk all_inputs()
8    set_output_delay CLKPER6  -add_delay -clock clk all_outputs()
9    compile -map_effort medium
10   include compile.report
```

Figure 5-9 Synthesis script file compile.rpadder.rtl2.

5.3 CARRY LOOK AHEAD ADDER

It takes a long time for the carry to ripple through each stage of the FA in the carry ripple adder. To make the circuit faster, we should try to pass the carry to the left as fast as we can. The concept of the carry look ahead adder (CLA) was developed a long time ago. To achieve this, the carry is split into two portions. One is called *propagate carry*. The other is called *generate carry*. For each stage of the FA, a propagate carry bit P and a generate carry G are added. The following FAPG VHDL code shows output ports P and G declared in line 8. Note that P and G have simpler functions compared to the COUT of the FA circuit. It is fair to say that P and G can be generated faster than the COUT in the FA circuit. Figure 5-11 shows the synthesized schematic.

```
1    --------------------- file fapg.vhd
2    -- full adder with carry propagate and generate
3    library IEEE;
4    use IEEE.std_logic_1164.all;
5    entity FAPG is
6      port (
7        A, B, CI : in  std_logic;
8        S, P, G  : out std_logic);
9    end FAPG;
10   architecture RTL of FAPG is
11     signal Gsig, Psig : std_logic;
12   begin
```

```
13     Gsig  <= A and B;
14     Psig  <= A xor B;
15     S     <= Psig xor CI;
16     G     <= Gsig;
17     P     <= Psig;
18     end RTL;
```

FAPG VHDL code can be synthesized with the synthesis script as shown in Figure 5-10. Figure 5-11 shows the synthesized schematic.

```
1      sh date
2      design = fapg
3      analyze -f vhdl VHDLSRC + "ch5/" + design + ".vhd"
4      elaborate FAPG -architecture RTL
5      include compile.common
6      create_clock -name clk -period CLKPERIOD -waveform {0.0
HALFCLK}
7      set_clock_skew -uncertainty CLKSKEW clk
8      set_input_delay  INPDELAY -add_delay -clock clk all_inputs()
9      set_input_delay  CLKPER2  -add_delay -clock clk {CI}
10     set_output_delay CLKPER8  -add_delay -clock clk all_outputs()
11     compile -map_effort medium
12     include compile.report
13     sh date
```

Figure 5-10 Synthesis script file compile.fapg.

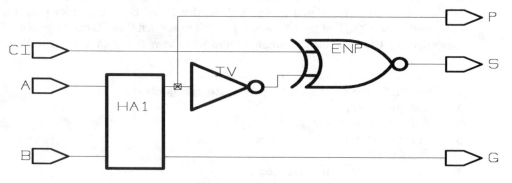

Figure 5-11 Synthesized schematic for FAPG.

Now we want to group 4 bits of propagate carries and generate carries into one carry look ahead unit (CLAU). The idea is again to produce the propagate carry and the generate carry for the group as soon as possible. The CLAU VHDL code is shown below. The CLAU has 4 bits of propagate carries and generate carries. It also has a carry input CI from the last stage. It produces 1 bit of group propagate carry GP and 1 bit of group generate carry GG as shown in line 10. Output CO is 4 bits wide to pass back to each stage of FAPG. As an example, CO(1) is either from the carry generated (line 16) or the carry input from the last stage is propagated (line 15). CO(2) is either from the carry input CI propagating through both stages 0 and 1 (line 17), or the stage 0 generates a carry and the carry propagates through stage 1 as shown in line 18, or the stage 1 generates a carry (line 19). The concept of CO(3) is similar to CO(2). The group generate GG is either from G(0) which propagates through P(1), P(2), and P(3), or G(1) which propagates through P(2) and P(3), or G(2) which propagates through P(3), or G(3) as shown in lines 24 to 27, respectively.

```
1     ---------------------- file clau.vhd
2     -- group carry look ahead unit
3     library IEEE;
4     use IEEE.std_logic_1164.all;
5     entity CLAU is
6       port (
7          P, G   : in  std_logic_vector(3 downto 0);
8          CI     : in  std_logic;
9          CO     : out std_logic_vector(3 downto 0);
10         GP, GG : out std_logic);
11    end CLAU;
12    architecture RTL of CLAU is
13    begin
14       CO(0) <= CI;
15       CO(1) <= (CI and P(0)) or
16               G(0);
17       CO(2) <= (CI and P(0) and P(1)) or
18               (G(0) and P(1)) or
19               G(1);
20       CO(3) <= (CI and P(0) and P(1) and P(2)) or
21               (G(0) and P(1) and P(2)) or
22               (G(1) and P(2)) or
23               G(2);
24       GG    <= (G(0) and P(1) and P(2) and P(3)) or
25               (G(1) and P(2) and P(3)) or
26               (G(2) and P(3)) or
27               G(3);
28       GP    <= P(3) and P(2) and P(1) and P(0);
29    end RTL;
```

CLAU VHDL code can be synthesized with the synthesis script file as shown in Figure 5-12. The synthesis script is simple. Line 9 sets input CI with more input delay than the rest of the inputs, since we want to make the carry to pass on faster. Figure 5-13 shows the synthesized schematic.

```
1    sh date
2    design = clau
3    analyze -f vhdl VHDLSRC + "ch5/" + design + ".vhd"
4    elaborate CLAU -architecture RTL
5    include compile.common
6    create_clock -name clk -period CLKPERIOD -waveform {0.0
HALFCLK}
7    set_clock_skew -uncertainty CLKSKEW clk
8    set_input_delay  INPDELAY -add_delay -clock clk all_inputs()
9    set_input_delay  CLKPER2  -add_delay -clock clk {CI}
10   set_output_delay CLKPER8  -add_delay -clock clk all_outputs()
11   compile -map_effort medium
12   include compile.report
13   sh date
```

Figure 5-12 Synthesis script file compile.clau.

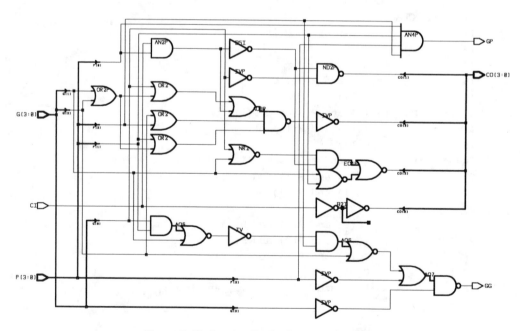

Figure 5-13 Synthesized schematic for CLAU.

Now we have the building blocks of FAPG and CLAU. Figure 5-14 shows a block diagram for the 16-bit carry look ahead adder. Inputs A and B come from the top to 16 FAPG blocks. Four FAPG blocks are grouped as a set in the diagram so that the signals between FAPG and CLAU can be seen easier. The sum output S comes out from FAPG blocks. The carry input CI connects to both levels of the rightmost CLAU units. Refer to the first level of four CLAU units. Each CLAU unit gets the P and G inputs from the four FAPG units P and G output ports. Each CLAU unit's CO (four bits) outputs connect to each carry input CI of the FAPG unit. The GP and GG outputs of the CLAU connect to the next level of the CLAU unit. Referring to the second level of the CLAU (only one unit), its CO output (4 bits) is connected to the CI input of the first level of CLAU units through signal GC.

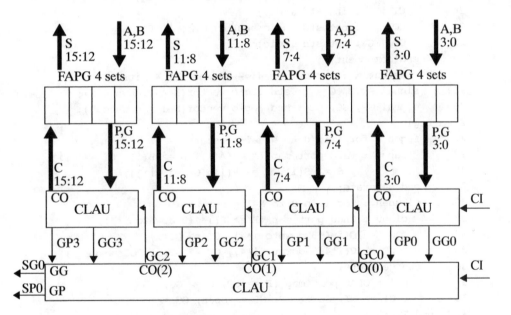

Figure 5-14 CLA16 block diagram.

CLA16 VHDL code is shown below. Lines 12 to 16 and lines 17 to 23 declare FAPG and CLAU components. Line 24 declares constant N as the number of a 4-bit set. For a 16-bit adder, N is set to 4. Lines 25 and 26 declare signals to connect FAPG and CLAU together. Note that P, G; and C are 16-bit signals. GP, GG, GC are 4-bit signals.

```
1    --------------------- file cla16.vhd
2    library IEEE;
3    use IEEE.std_logic_1164.all;
4    entity CLA16 is
```

```
 5        port (
 6          A, B : in   std_logic_vector(15 downto 0);
 7          CI   : in   std_logic;
 8          S    : out  std_logic_vector(15 downto 0);
 9          CO   : out  std_logic);
10      end CLA16;
11      architecture RTL of CLA16 is
12        component FAPG
13        port (
14          A, B, CI : in   std_logic;
15          S, P, G  : out  std_logic);
16        end component;
17        component CLAU
18        port (
19          P, G  : in   std_logic_vector(3 downto 0);
20          CI    : in   std_logic;
21          CO    : out  std_logic_vector(3 downto 0);
22          GP, GG : out std_logic);
23        end component;
24        constant N : integer := 4; -- number of 4 bits
25        signal P, G, C     : std_logic_vector(N*4-1 downto 0);
26        signal GP, GG, GC : std_logic_vector(N-1    downto 0);
27      begin
28        fagen : for i in 0 to N*4-1 generate
29          fapg0 : FAPG port map (A => A(i), B => B(i), CI => C(i),
30                       S => S(i), P => P(i), G  => G(i));
31        end generate;
32        claugen : for i in 0 to N-1 generate
33          clau0 : CLAU port map (P => P(i*4+3 downto i*4),
34            G  => G(i*4+3 downto i*4), CI => GC(i),
35            CO => C(i*4+3 downto i*4), GP => GP(i), GG => GG(i));
36        end generate;
37        clau1 : CLAU port map (P => GP, G => GG,
38            CI => CI, CO => GC, GP => open, GG => CO);
39      end RTL;
```

Lines 28 to 31 instantiate 16 FAPG blocks with a **generate statement**. Lines 32 to 36 instantiate four CLAU blocks with a **generate statement**. Lines 37 and 38 instantiate the second level of the only CLAU block.

To verify the correctness of the CLA16 VHDL code, the following ADDER VHDL code is used as shown below. Note that ADDER also has the carry input CI and carry output COUT as shown in lines 9 and 11. Line 16 extends input signals A and B by 1 bit so that the carry out can be obtained in line 18.

```
1       -------------------- file adder.vhd
2       library IEEE;
3       use IEEE.std_logic_1164.all;
4       use IEEE.std_logic_unsigned.all;
5       entity ADDER is
6         generic (N : in integer := 16);
7         port (
8           A, B : in  std_logic_vector(N-1 downto 0);
9           CI   : in  std_logic;
10          S    : out std_logic_vector(N-1 downto 0);
11          COUT : out std_logic);
12      end ADDER;
13      architecture RTL of ADDER is
14        signal RESULT : std_logic_vector(A'length downto 0);
15      begin
16        RESULT <= ('0' & A) + ('0' & B) + CI;
17        S      <= RESULT(A'length-1 downto 0);
18        COUT   <= RESULT(A'length);
19      end RTL;
```

The CLA16 can be configured with the following VHDL code.

```
1       -------------------- file cla16_cfg.vhd
2       configuration CFG_CLA16 of CLA16 is
3          for RTL
4             for fagen
5                for all : FAPG
6                   use entity work.FAPG(RTL);
7                end for;
8             end for;
9             for claugen
10               for all : CLAU
11                  use entity work.CLAU(RTL);
12               end for;
13            end for;
14            for clau1 : CLAU
15               use entity work.CLAU(RTL);
16            end for;
17         end for;
18      end CFG_CLA16;
```

The following TBCLA16 VHDL code is used to verify CLA16 VHDL code. Components CLA16 and ADDER are declared in lines 8 to 22. They are instantiated in lines 58 to 62. Note that line 25 declares constant N, which is mapped to ADDER generic N in line 61. Both components are configured in lines 64 to 73. Note that line 67 configures CLA16 with the CFG_CLA16 configuration. Inputs A, B, and CI are

generated with a linear feedback array in line 45. The test bench has run for a while
without any single miscompare so we assume that CLA16 VHDL code is correct.

```
1          -------------------- file tbcla16.vhd
2     library IEEE;
3     use IEEE.std_logic_1164.all;
4     use IEEE.std_logic_arith.all;
5     entity TBCLA16 is
6     end TBCLA16;
7     architecture BEH of TBCLA16 is
8       component CLA16
9       port (
10        A, B : in  std_logic_vector(15 downto 0);
11        CI   : in  std_logic;
12        S    : out std_logic_vector(15 downto 0);
13        CO   : out std_logic);
14      end component;
15      component ADDER
16      generic (N : in integer := 16);
17      port (
18        A, B  : in  std_logic_vector(N-1 downto 0);
19        CI    : in  std_logic;
20        S     : out std_logic_vector(N-1 downto 0);
21        COUT  : out std_logic);
22      end component;
23      constant PERIOD : time := 50 ns;
24      constant STROBE : time := PERIOD - 5 ns;
25      constant N      : integer := 16;
26      signal   A      : std_logic_vector(N-1 downto 0);
27      signal   B      : std_logic_vector(N-1 downto 0);
28      signal   CI     : std_logic;
29      signal   CO, CO1 : std_logic;
30      signal   S, S1  : std_logic_vector(N-1 downto 0);
31    begin
32      p0 : process
33        variable Q : std_logic_vector(N-1 downto 0);
34      begin
35        A <= (A'range => '1');
36        B <= (B'range => '1');
37        CI <= '1';
38        wait for PERIOD;
39        Q := (Q'range => '0');
40        loop
41          A  <= Q;
42          B  <= (Q(N-1) xor (not Q(2))) & Q(N-1 downto 1);
43          CI <= Q(0) and Q(1);
44          wait for PERIOD;
45          Q := Q(N-2 downto 0) & (Q(N-1) xor (not Q(2)));
46        end loop;
47      end process;
```

```
48      check : process
49      begin
50        wait for STROBE;
51        loop
52          if (S /= S1) or (CO /= CO1) then
53            assert FALSE report "not compared" severity WARNING;
54          end if;
55          wait for PERIOD;
56        end loop;
57      end process;
58      cla0 : CLA16
59        port map (A => A, B => B, CI => CI, S => S, CO => CO);
60      adder0 : ADDER
61        generic map (N => N)
62        port map (A => A, B => B, CI => CI, S => S1, COUT => CO1);
63    end BEH;
64    configuration CFG_TBCLA16 of TBCLA16 is
65      for BEH
66        for all : CLA16
67          use configuration work.CFG_CLA16;
68        end for;
69        for all : ADDER
70          use entity work.ADDER(RTL);
71        end for;
72      end for;
73    end CFG_TBCLA16;
```

The following simple synthesis script in Figure 5-15 can be used to synthesize CLA16 VHDL code. Note that FAPG and CLAU are read in as presynthesized blocks in lines 3 to 4. There is no compile command. Figure 5-16 shows the CLA16 schematic.

```
1     sh date
2     design = cla16
3     read ../db/clau.db
4     read ../db/fapg.db
5     analyze -f vhdl VHDLSRC + "ch5/" + design + ".vhd"
6     elaborate CLA16 -architecture RTL
7     include compile.common
8     create_clock -name clk -period CLKPERIOD -waveform {0.0
HALFCLK}
9     set_clock_skew -uncertainty CLKSKEW clk
10    set_input_delay  INPDELAY -add_delay -clock clk all_inputs()
11    set_output_delay CLKPER8  -add_delay -clock clk all_outputs()
12    sh date
```

Figure 5-15 Synthesis script file compile.cla16.

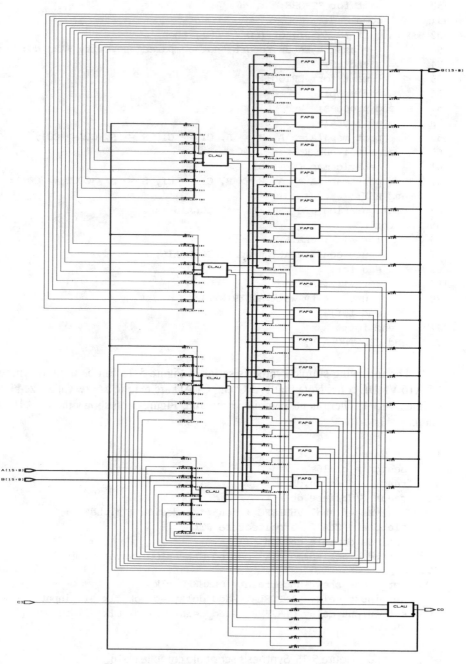

Figure 5-16 Synthesized schematic for CLA16.

The schematic in Figure 5-16 has the following area and timing statistic. Note that the timing path starts at the FAPG (line 21), goes through the first level of the CLAU unit (lines 22 to 25), goes through the second level of the CLAU unit (lines 26 to 29), back to the first level of the CLAU unit (lines 30 to 32), then to the FAPG unit to generate S(15).

```
1    Number of ports:           50
2    Number of nets:            197
3    Number of cells:           148
4    Number of references:      16
5
6    Combinational area:        355.000000
7    Noncombinational area:       0.000000
8    Net Interconnect area:     1012.431335
9
10   Total cell area:           355.000000
11   Total area:                1367.431396
12
13   Operating Conditions: WCCOM   Library: lca300k
14   CLA16                  B9X9              lca300k
15     Point                        Incr      Path
16     -------------------------------------------------
17     clock clk (rise edge)        0.00      0.00
18     clock network delay (ideal)  0.00      0.00
19     input external delay         2.50      2.50 f
20     A[5] (in)                    0.00      2.50 f
21     fapg0_5/U4/S (HA1)           1.97      4.47 r
22     clau0_1/U14/Z (AO6)          0.70      5.17 f
23     clau0_1/U15/Z (IV)           0.24      5.41 r
24     clau0_1/U16/Z (AO6)          0.39      5.80 f
25     clau0_1/U19/Z (AO7)          1.11      6.91 r
26     clau1/U9/Z (OR2P)            0.73      7.65 r
27     clau1/U21/Z (OR2)            0.46      8.11 r
28     clau1/U23/Z (AO3P)           0.53      8.64 f
29     clau1/U6/Z (IVP)             0.22      8.86 r
30     clau0_3/U10/Z (AN2P)         0.63      9.50 r
31     clau0_3/U23/Z (AO3P)         0.54     10.03 f
32     clau0_3/U6/Z (IVP)           0.15     10.18 r
33     fapg0_15/U5/Z (ENP)          0.99     11.17 f
34     S[15] (out)                  0.00     11.18 f
35     data arrival time                     11.18
```

The schematic in Figure 5-16 can be ungrouped with the "ungroup -all" command. It can be synthesized with the compile command. Figure 5-17 shows the schematic.

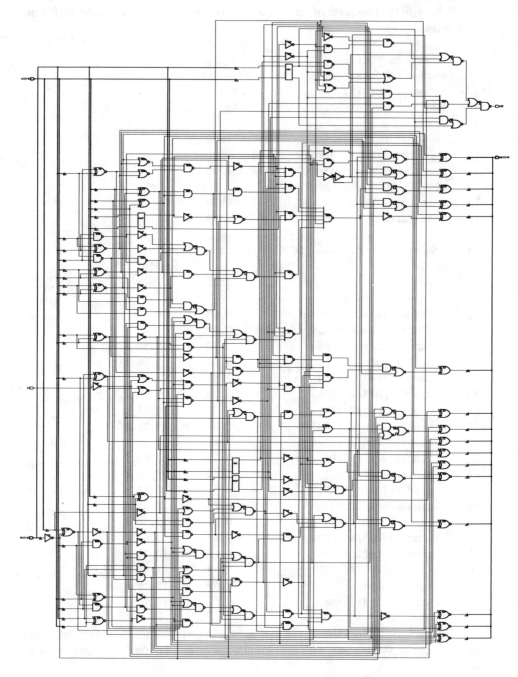

Figure 5-17 Synthesized schematic for CLA16 after ungroup.

The following shows the area and timing statistic for the schematic in Figure 5-17.

```
1    Number of ports:              50
2    Number of nets:               191
3    Number of cells:              153
4    Number of references:         35
5
6    Combinational area:         347.000000
7    Net Interconnect area:     1137.748535
8
9    Total cell area:            347.000000
10   Total area:                1484.748535
11
12   Operating Conditions: WCCOM   Library: lca300k
13   CLA16                    B9X9           lca300k
14     Point                      Incr       Path
15     ------------------------------------------------
16     clock clk (rise edge)      0.00       0.00
17     clock network delay (ideal) 0.00      0.00
18     input external delay       2.50       2.50 f
19     A[3] (in)                  0.00       2.50 f
20     U31/Z (ENP)                1.34       3.84 r
21     U64/Z (OR2P)               0.62       4.46 r
22     U116/Z (AO3)               0.61       5.07 f
23     U25/Z (ND4)                0.38       5.45 r
24     U121/Z (ND6P)              1.17       6.62 f
25     U126/Z (AO6P)              0.44       7.06 r
26     U127/Z (EOP)               0.75       7.82 f
27     S[14] (out)                0.00       7.82 f
28     data arrival time                     7.82
```

Let's compare the result with the ripple carry adder. RPADDER is synthesized with N mapped to 16 so that the same 16-bit adder can be compared. Under the same condition, it has a total cell area of 224 units, a total area of 966 units, and the data arrival time of 28.25 ns. After ungroup and compile commands, the total cell area is 274 units, total area is 1,112 units, and the data arrival time is 15.62 ns.

The carry look ahead adder can be expanded. For example, another level of CLAU can be used. This allows the 64-bit carry look ahead adder to be constructed. CLA64 is shown below.

```
1    ---------------------- file cla64.vhd
2    library IEEE;
3    use IEEE.std_logic_1164.all;
4    entity CLA64 is
5      port (
```

```
6            A, B : in  std_logic_vector(63 downto 0);
7            CI   : in  std_logic;
8            S    : out std_logic_vector(63 downto 0);
9            CO   : out std_logic);
10   end CLA64;
11   architecture RTL of CLA64 is
12     component FAPG
13     port (
14       A, B, CI : in  std_logic;
15       S, P, G  : out std_logic);
16     end component;
17     component CLAU
18     port (
19       P, G  : in  std_logic_vector(3 downto 0);
20       CI    : in  std_logic;
21       CO    : out std_logic_vector(3 downto 0);
22       GP, GG : out std_logic);
23     end component;
24     constant N : integer := 16; -- number of 4 bits
25     signal P, G, C   : std_logic_vector(N*4-1 downto 0);
26     signal GP, GG, GC : std_logic_vector(N-1   downto 0);
27     signal SP, SG, SC : std_logic_vector(N/4-1 downto 0);
28     signal AA, BB, SS : std_logic_vector(N*4-1 downto 0);
29   begin
30     AA <= A; BB <= B; S  <= SS;
31     fagen : for i in 0 to N*4-1 generate
32       fapg0 : FAPG port map (A => AA(i), B => BB(i), CI => C(i),
33                 S => SS(i), P => P(i), G  => G(i));
34     end generate;
35     claugen : for i in 0 to N-1 generate
36       clau0 : CLAU port map (P => P(i*4+3 downto i*4),
37         G  => G(i*4+3 downto i*4), CI => GC(i),
38          CO => C(i*4+3 downto i*4), GP => GP(i), GG => GG(i));
39     end generate;
40     gclaugen : for i in 0 to N/4-1 generate
41       clau0 : CLAU port map (P => GP(i*4+3 downto i*4),
42         G  => GG(i*4+3 downto i*4), CI => SC(i),
43          CO => GC(i*4+3 downto i*4), GP => SP(i), GG => SG(i));
44     end generate;
45     clau1 : CLAU port map (P => SP, G => SG,
46         CI => CI, CO => SC, GP => open, GG => CO);
47   end RTL;
```

Note that lines 31 to 34 instantiate FAPG 64 times. Lines 35 to 39 instantiate CLAU 16 times as the first level. Lines 40 to 44 instantiate CLAU four times as the second level. Lines 45 and 46 instantiate CLAU unit 1 time as the third level. The verification of CLA64 is left as Exercise 5.9. CLA64 is synthesized with the area and timing summary shown below. Again the timing path goes down from FAPG through two CLAU levels to the third level, then goes up two levels of CLAU, and finally goes to the FAPG unit.

```
1     Number of ports:              194
2     Number of nets:               446
3     Number of cells:               85
4     Number of references:           2
5
6     Combinational area:        1459.000000
7     Noncombinational area:        0.000000
8     Net Interconnect area:     4178.340332
9
10    Total cell area:           1459.000000
11 ˙  Total area:                5637.340332
12    Operating Conditions: WCCOM    Library: lca300k
13
14    Design              Wire Loading Model        Library
15    ------------------------------------------------------
16    CLA64                    B9X9                  lca300k
17    FAPG                     B9X9                  lca300k
18    CLAU                     B9X9                  lca300k
19     Point                             Incr        Path
20    ------------------------------------------------------
21     clock clk (rise edge)            0.00        0.00
22     clock network delay (ideal)      0.00        0.00
23     input external delay             2.50        2.50 f
24     A[17] (in)                       0.00        2.50 f
25     fapg0_17/A (FAPG)                0.20        2.70 f
26     fapg0_17/U4/S (HA1)              1.77        4.47 r
27     fapg0_17/P (FAPG)                0.00        4.47 r
28     clau0_4/P[1] (CLAU)              0.00        4.47 r
29     clau0_4/U14/Z (AO6)              0.70        5.17 f
30     clau0_4/U15/Z (IV)               0.24        5.41 r
31     clau0_4/U16/Z (AO6)              0.39        5.80 f
32     clau0_4/U19/Z (AO7)              1.22        7.02 r
33     clau0_4/GG (CLAU)                0.00        7.02 r
34     clau0_1_2/G[0] (CLAU)            0.00        7.02 r
35     clau0_1_2/U14/Z (AO6)            0.84        7.87 f
36     clau0_1_2/U15/Z (IV)             0.26        8.13 r
37     clau0_1_2/U16/Z (AO6)            0.40        8.53 f
38     clau0_1_2/U19/Z (AO7)            1.12        9.65 r
39     clau0_1_2/GG (CLAU)              0.00        9.65 r
40     clau1/G[1] (CLAU)                0.00        9.65 r
41     clau1/U9/Z (OR2P)                0.73       10.38 r
```

42	clau1/U21/Z (OR2)	0.46	10.85 r
43	clau1/U23/Z (AO3P)	0.53	11.38 f
44	clau1/U6/Z (IVP)	0.22	11.60 r
45	clau1/CO[3] (CLAU)	0.00	11.60 r
46	clau0_3_2/CI (CLAU)	0.00	11.60 r
47	clau0_3_2/U10/Z (AN2P)	0.63	12.23 r
48	clau0_3_2/U11/Z (B5I)	0.28	12.51 f
49	clau0_3_2/U25/Z (EO1P)	0.41	12.92 r
50	clau0_3_2/CO[2] (CLAU)	0.00	12.92 r
51	clau0_14/CI (CLAU)	0.00	12.92 r
52	clau0_14/U10/Z (AN2P)	0.71	13.64 r
53	clau0_14/U23/Z (AO3P)	0.54	14.17 f
54	clau0_14/U6/Z (IVP)	0.15	14.32 r
55	clau0_14/CO[3] (CLAU)	0.00	14.32 r
56	fapg0_59/CI (FAPG)	0.00	14.32 r
57	fapg0_59/U5/Z (ENP)	0.99	15.31 f
58	fapg0_59/S (FAPG)	0.00	15.31 f
59	S[59] (out)	0.00	15.32 f
60	data arrival time		15.32

The same process of "ungroup -all" and then compile is used on CLA64. The following shows the area and timing summary. Note that it is faster and smaller. This is because the lack of block boundary constraints allows greater optimization.

```
1    Number of ports:          194
2    Number of nets:           727
3    Number of cells:          579
4    Number of references:      51
5
6    Combinational area:       1272.000000
7    Net Interconnect area:    4234.403320
8
9    Total cell area:          1272.000000
10   Total area:               5506.403320
11
12   Operating Conditions: WCCOM   Library: lca300k
13   CLA64                  B9X9            lca300k
14     Point                       Incr      Path
15     ---------------------------------------------
16     clock clk (rise edge)       0.00      0.00
17     clock network delay (ideal) 0.00      0.00
18     input external delay        2.50      2.50 f
19     A[42] (in)                  0.00      2.50 f
20     U113/Z (ENP)                1.28      3.78 f
21     U116/Z (OR4P)               1.33      5.12 f
22     U134/Z (NR2P)               0.29      5.41 r
```

23	U399/Z (AN2P)	0.62	6.03 r
24	U248/Z (AN2P)	0.57	6.59 r
25	U449/Z (ND3)	0.56	7.16 f
26	U451/Z (ND4P)	0.42	7.58 r
27	U140/Z (IVP)	0.41	7.99 f
28	U460/Z (AO7P)	0.41	8.40 r
29	U12/Z (IVAP)	0.40	8.80 f
30	U154/Z (AO7P)	0.36	9.16 r
31	U461/Z (AO6P)	0.40	9.56 f
32	U462/Z (EOP)	0.71	10.27 f
33	S[63] (out)	0.00	10.28 f
34	data arrival time		10.28

Note that ADDER VHDL code can be synthesized with N mapped to 64 as shown in the following synthesis script file compile.adder.64. Figure 5-18 shows the schematic which uses a DesignWare component.

```
1    sh date
2    design = adder
3    analyze -f vhdl VHDLSRC + "ch5/" + design + ".vhd"
4    elaborate ADDER -architecture RTL -param "N => 64"
5    include compile.common
6    create_clock -name clk -period CLKPERIOD -waveform {0.0
HALFCLK}
7    set_clock_skew -uncertainty CLKSKEW clk
8    set_input_delay  INPDELAY -add_delay -clock clk all_inputs()
9    set_output_delay CLKPER7  -add_delay -clock clk all_outputs()
10   compile -map_effort medium
11   include compile.report
```

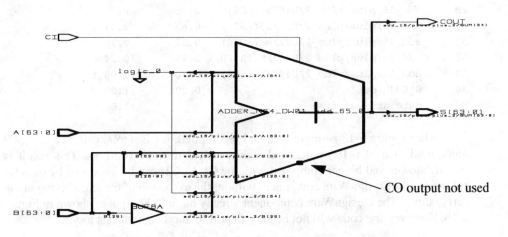

Figure 5-18 Synthesized schematic for ADDER, N=>64.

The schematic in Figure 5-18 has the following area and timing statistics.

```
1    Number of ports:              194
2    Number of nets:               789
3    Number of cells:              647
4    Number of references:          42
5
6    Combinational area:      1173.000000
7    Net Interconnect area:   4633.439941
8
9    Total cell area:         1173.000000
10   Total area:              5806.439941
11   Operating Conditions: WCCOM   Library: lca300k
12
13   Design            Wire Loading Model        Library
14   ADDER_N64               B9X9                 lca300k
15     Point                            Incr       Path
16     ----------------------------------------------------
17     clock clk (rise edge)            0.00       0.00
18     clock network delay (ideal)      0.00       0.00
19     input external delay             2.50       2.50 f
20     B[53] (in)                       0.00       2.50 f
21     add_16/plus/plus_2/U130/Z  (NR2) 1.43       3.93 r
22     add_16/plus/plus_2/U335/Z  (AO7) 0.85       4.78 f
23     add_16/plus/plus_2/U336/Z  (IV)  0.26       5.04 r
24     add_16/plus/plus_2/U338/Z  (AO7) 0.44       5.48 f
25     add_16/plus/plus_2/U339/Z  (EO1) 0.46       5.95 r
26     add_16/plus/plus_2/U40/Z   (AO7) 1.08       7.03 f
27     add_16/plus/plus_2/U646/Z  (ND2) 0.39       7.42 r
28     add_16/plus/plus_2/U436/Z  (ND2) 0.57       7.99 f
29     add_16/plus/plus_2/U46/Z   (AO6) 0.78       8.77 r
30     add_16/plus/plus_2/U56/Z   (AN2P)0.63       9.41 r
31     add_16/plus/plus_2/U67/Z   (AO7P)0.37       9.78 f
32     add_16/plus/plus_2/U442/Z  (AO6P)0.46      10.24 r
33     add_16/plus/plus_2/U443/Z  (EOP) 0.76      11.00 f
34     S[63] (out)                      0.00      11.00 f
35     data arrival time                          11.00
```

After Figure 5-18 is ungrouped and recompiled, it has a total cell area of 1,309 units, total area of 6,166 units, and a data arrival time of 10.61 ns. This result is slightly slower and bigger compared with the CLA64. It can be improved by directly calling out the DesignWare component without the extension of the extra bit to get the carry output. The DesignWare component already has a CO output as shown in Figure 5-18. However, the code will not be easy to port to another synthesis tool.

5.4 COUNTONE CIRCUIT

A countone circuit counts the number of '1's in an input data. For example, the following COUNTONE VHDL code has an input DIN of 32 bits. The output NUM has 6 bits to represent numbers 0 to 32 inclusive. In a microprocessor design, a bit pattern may be set in an instruction to indicate which CPU registers are to be stored in a stack. Each bit position indicates a particular register. When the bit is set to '1', the register value is to be pushed to the stack. The countone circuit can be used to calculate the number of clock cycles required to complete the data transfer or to control the data transfer count.

```
1      library IEEE;
2      use IEEE.std_logic_1164.all;
3      use IEEE.std_logic_unsigned.all;
4      entity COUNTONE is
5        port (
6           DIN : in  std_logic_vector(31 downto 0);
7           NUM : out std_logic_vector(5  downto 0));
8      end COUNTONE;
```

The following architecture RTL1 simply uses a **loop statement** to scan through the input data. Increase the count when '1' is seen.

```
9      architecture RTL1 of COUNTONE is
10     begin
11       comb : process ( DIN )
12         variable TEMP : std_logic_vector(5 downto 0);
13       begin
14         TEMP := "000000";
15         for j in DIN'range loop
16           if (DIN(j) = '1') then
17              TEMP := TEMP + 1;
18            end if;
19         end loop;
20         NUM <= TEMP;
21       end process;
22     end RTL1;
```

The following architecture RTL2 is similar to RTL1 except that the increasing of the counter is broken into eight groups of 4 bits. Lines 33 to 37 describe an inner loop which checks a group of 4 bits using an incrementer. Lines 31 to 39 use an outer loop with an adder inferred in line 38.

```
23    architecture RTL2 of COUNTONE is
24    begin
25      comb : process ( DIN )
26        constant PARTS  : integer := 8;
27        variable TEMP   : std_logic_vector(5 downto 0);
28        variable TEMP1  : std_logic_vector(5 downto 0);
29      begin
30        TEMP := "000000";
31        for j in 0 to PARTS-1 loop
32          TEMP1 := "000000";
33          for k in 32/PARTS-1 downto 0 loop
34            if (DIN(32/PARTS * j + k) = '1') then
35              TEMP1 := TEMP1 + 1;
36            end if;
37          end loop;
38          TEMP := TEMP + TEMP1;
39        end loop;
40        NUM <= TEMP;
41      end process;
42    end RTL2;
```

The following architecture RTL3 is similar to RTL2 except that it uses four groups of 8 bits.

```
43    architecture RTL3 of COUNTONE is
44    begin
45      comb : process ( DIN )
46        constant PARTS  : integer := 4;
47        variable TEMP   : std_logic_vector(5 downto 0);
48        variable TEMP1  : std_logic_vector(5 downto 0);
49      begin
50        TEMP := "000000";
51        for j in 0 to PARTS-1 loop
52          TEMP1 := "000000";
53          for k in 32/PARTS-1 downto 0 loop
54            if (DIN(32/PARTS * j + k) = '1') then
55              TEMP1 := TEMP1 + 1;
56            end if;
57          end loop;
58          TEMP := TEMP + TEMP1;
59        end loop;
60        NUM <= TEMP;
61      end process;
62    end RTL3;
```

The following architecture RTL4 implements the incrementing with logical operators.

```
63    architecture RTL4 of COUNTONE is
64    begin
65      comb : process ( DIN )
66        variable TEMP    : std_logic_vector(5 downto 0);
67        variable Cin, Cout : std_logic;
68      begin
69        TEMP := "000000";
70        for j in DIN'reverse_range loop
71          if (DIN(j) = '1') then
72            Cin := '1';
73            for k in TEMP'reverse_range loop
74              Cout     := TEMP(k) and Cin;
75              TEMP(k) := TEMP(k) xor Cin;
76              Cin      := Cout;
77            end loop;
78          end if;
79        end loop;
80        NUM <= TEMP;
81      end process;
82    end RTL4;
```

The following VHDL code shows a test bench example to verify the COUNT-ONE VHDL code. The **component declaration**, **component instantiation**, and **configuration** are about the same as before. In process *p0* (lines 16 to 27), line 19 sets DIN to all '1's. Line 21 initializes variable Q to be all '0's. Lines 22 to 26 are an infinite **loop statement**. Line 25 implements a simple linear feedback shifting operation to generate somewhat random patterns for DIN. The **check** process (lines 28 to 37) is also an infinite **loop statement**.

```
1     library IEEE;
2     use IEEE.std_logic_1164.all;
3     entity TBCOUNTONE is
4     end TBCOUNTONE;
5     architecture BEH of TBCOUNTONE is
6       component COUNTONE
7       port (
8         DIN : in  std_logic_vector(31 downto 0);
9         NUM : out std_logic_vector(5  downto 0));
10      end component;
11      signal  DIN : std_logic_vector(31 downto 0);
12      signal Y1, Y2, Y3, Y4 : std_logic_vector(5  downto 0);
13      constant PERIOD : time := 50 ns;
14      constant STROBE : time := PERIOD - 5 ns;
```

```
15    begin
16    p0 : process
17      variable Q : std_logic_vector(31 downto 0);
18    begin
19      DIN <= (DIN'range => '1');
20      wait for PERIOD;
21      Q := (Q'range => '0');
22      loop
23        DIN <= Q;
24        wait for PERIOD;
25        Q := Q(30 downto 0) & (Q(31) xor (not Q(2)));
26      end loop;
27    end process;
28    check : process
29    begin
30      wait for STROBE;
31      loop
32        if (Y1 /= Y2) or (Y2 /= Y3) or (Y3 /= Y4) then
33          assert FALSE report "not compared" severity WARNING;
34        end if;
35        wait for PERIOD;
36      end loop;
37    end process;
38    countone1 : COUNTONE port map ( DIN => DIN, NUM => Y1);
39    countone2 : COUNTONE port map ( DIN => DIN, NUM => Y2);
40    countone3 : COUNTONE port map ( DIN => DIN, NUM => Y3);
41    countone4 : COUNTONE port map ( DIN => DIN, NUM => Y4);
42   end BEH;
43   configuration CFG_TBCOUNTONE of TBCOUNTONE is
44     for BEH
45       for countone1 : COUNTONE
46         use entity work.COUNTONE(RTL1);
47       end for;
48       for countone2 : COUNTONE
49         use entity work.COUNTONE(RTL2);
50       end for;
51       for countone3 : COUNTONE
52         use entity work.COUNTONE(RTL3);
53       end for;
54       for countone4 : COUNTONE
55         use entity work.COUNTONE(RTL4);
56       end for;
57     end for;
58   end CFG_TBCOUNTONE;
```

The TBCOUNTONE VHDL code first checks the input patterns of all '1's and all '0'. The infinite loop can run forever if the simulation time is not controlled. The linear feedback shift operation is simple to implement and efficient to simulate. However, how do we know that the test bench covers enough input pattern variations? One possible improvement to the test bench is to generate a histogram of the counter output value (such as Y1). If the histogram shows that Y1 does not have a count of 5 after running a certain amount of time, the test bench can be improved based on the histogram results. This is left as Exercises 5.15 and 5.16.

Architecture RTL1 of COUNTONE VHDL is synthesized with the synthesis script file shown in Figure 5-19. It has a combinational area of 1,884 units, an interconnect area of 4,072 units, and a data arrival time of 52.79 ns.

```
1     sh date
2     design = countone
3     analyze -f vhdl VHDLSRC + "ch5/" + design + ".vhd"
4     elaborate COUNTONE -architecture RTL1
5     include compile.common
6     create_clock -name clk -period CLKPERIOD -waveform {0.0
HALFCLK}
7     set_clock_skew -uncertainty CLKSKEW clk
8     set_input_delay  INPDELAY -add_delay -clock clk all_inputs()
9     set_output_delay CLKPER6  -add_delay -clock clk all_outputs()
10    compile -map_effort medium
11    include compile.report
12    sh date
```

Figure 5-19 Synthesis script file compile.countone.rtl1.

Architecture RTL4 can be synthesized by changing line 4 of Figure 5-19 from RTL1 to RTL4. It has a combinational area of 724 units, an interconnect area of 1958 units, and a data arrival time of 25.93 ns. It also takes a lot less central processing unit (CPU) time than the synthesis of architecture RTL1. Architecture RTL1 infers 32 incrementers. It takes a longer time to optimize. In general, using operators ">", "<", "+", "-" should be done carefully. The rest of the synthesis is left as Exercise 5.17.

Certainly, there are other ways to design the COUNTONE circuit much more efficiently. As we will show in Chapter 13, this circuit is a special case in a multiplier design when all the partial products at a certain bit position are to be added together. This is left as Exercise 5.37 for the reader.

5.5 LEADING ZERO CIRCUIT

A leading zero circuit calculates the number of leading zeroes in a given data. This is commonly used for the floating-point number operation to normalize the data.

For example, the following LEADZERO VHDL code shows input DIN of 32 bits. The output ZERO has 5 bits. It is possible for all 32 bits to be '0'. The output ZERO has only 5 bits. Since shifting left 32 bits of all '0', the data is the same as the original data without any shift. The LEADZERO circuit generates ZERO output as "00000" when the input data DIN has all '0's.

```
1     library IEEE;
2     use IEEE.std_logic_1164.all;
3     use IEEE.std_logic_unsigned.all;
4     entity LEADZERO is
5       port (
6          DIN  : in  std_logic_vector(31 downto 0);
7          ZERO : out std_logic_vector(4 downto 0));
8     end LEADZERO;
```

The following architecture RTL1 uses a **loop statement** to scan from left to right. Whenever a '0' is seen, variable **num** is increased by 1 in line 17. When the first '1' is seen, the loop is exited in line 19. The output ZERO is assigned in line 22.

```
9     architecture RTL1 of LEADZERO is
10    begin
11      p0 : process (DIN)
12        variable num : std_logic_vector(4 downto 0);
13      begin
14        num := (others => '0');
15        for j in 31 downto 0 loop
16          if (DIN(j) = '0') then
17             num := num + 1;
18          else
19             exit;
20          end if;
21        end loop;
22        ZERO <= num;
23      end process;
24    end RTL1;
```

The following architecture RTL2 models the incrementer to generate the carry Cout in line 36 and the 5-bit value in line 37.

```
25    architecture RTL2 of LEADZERO is
26    begin
27      p0 : process (DIN)
28        variable NUM   : std_logic_vector(4 downto 0);
29        variable Cin, Cout : std_logic;
30      begin
31        NUM := (others => '0');
```

```
32        for j in 31 downto 0 loop
33          if (DIN(j) = '0') then
34            Cin := '1';
35            for k in 0 to 4 loop
36              Cout   := NUM(k) and Cin;
37              NUM(k) := NUM(k) xor Cin;
38              Cin    := Cout;
39            end loop;
40          else
41            exit;
42          end if;
43        end loop;
44        ZERO <= num;
45      end process;
46    end RTL2;
```

The following architecture RTL3 is similar to RTL2 except that lines 61 to 63
exit the inner loop sooner.

```
47    architecture RTL3 of LEADZERO is
48    begin
49      p0 : process (DIN)
50        variable NUM   : std_logic_vector(4 downto 0);
51        variable Cin, Cout : std_logic;
52      begin
53        NUM := (others => '0');
54        for j in 31 downto 0 loop
55          if (DIN(j) = '0') then
56            Cin := '1';
57            for k in 0 to 4 loop
58              Cout   := NUM(k) and Cin;
59              NUM(k) := NUM(k) xor Cin;
60              Cin    := Cout;
61              if (Cout = '0') then
62                exit;
63              end if;
64            end loop;
65          else
66            exit;
67          end if;
68        end loop;
69        ZERO <= num;
70      end process;
71    end RTL3;
```

The following architecture RTL4 uses a **loop statement** to compare whether the data is 32-bit '0' or 31-bit '0', etc. When line 80 is compared true, line 81 assigns the output and exits the loop in line 82.

```
72    architecture RTL4 of LEADZERO is
73    begin
74      p0 : process (DIN)
75        variable ZVEC : std_logic_vector(31 downto 0);
76      begin
77        ZVEC := (ZVEC'range => '0');
78        ZERO <= "00000";
79        for j in 0 to 31 loop
80          if (DIN(31 downto j) = ZVEC(31 downto j)) then
81            ZERO <= "00000" - j;
82            exit;
83          end if;
84        end loop;
85      end process;
86    end RTL4;
```

The following architecture RTL5 simply uses a long chain of an **if statement**.

```
87    architecture RTL5 of LEADZERO is
88    begin
89      p0 : process (DIN)
90        variable ZVEC : std_logic_vector(31 downto 0);
91      begin
92        ZVEC := (ZVEC'range => '0');
93        if (DIN(31 downto 0) = ZVEC(31 downto 0)) then
94          ZERO <= "00000";
95        elsif (DIN(31 downto 1) = ZVEC(31 downto 1)) then
96          ZERO <= "11111";
97        elsif (DIN(31 downto 2) = ZVEC(31 downto 2)) then
98          ZERO <= "11110";
99        elsif (DIN(31 downto 3) = ZVEC(31 downto 3)) then
100         ZERO <= "11101";
101       elsif (DIN(31 downto 4) = ZVEC(31 downto 4)) then
102         ZERO <= "11100";
103       elsif (DIN(31 downto 5) = ZVEC(31 downto 5)) then
104         ZERO <= "11011";
105       elsif (DIN(31 downto 6) = ZVEC(31 downto 6)) then
106         ZERO <= "11010";
107       elsif (DIN(31 downto 7) = ZVEC(31 downto 7)) then
108         ZERO <= "11001";
109       elsif (DIN(31 downto 8) = ZVEC(31 downto 8)) then
110         ZERO <= "11000";
```

```
111        elsif (DIN(31 downto 9) = ZVEC(31 downto 9)) then
112           ZERO <= "10111";
113        elsif (DIN(31 downto 10) = ZVEC(31 downto 10)) then
114           ZERO <= "10110";
115        elsif (DIN(31 downto 11) = ZVEC(31 downto 11)) then
116           ZERO <= "10101";
117        elsif (DIN(31 downto 12) = ZVEC(31 downto 12)) then
118           ZERO <= "10100";
119        elsif (DIN(31 downto 13) = ZVEC(31 downto 13)) then
120           ZERO <= "10011";
121        elsif (DIN(31 downto 14) = ZVEC(31 downto 14)) then
122           ZERO <= "10010";
123        elsif (DIN(31 downto 15) = ZVEC(31 downto 15)) then
124           ZERO <= "10001";
125        elsif (DIN(31 downto 16) = ZVEC(31 downto 16)) then
126           ZERO <= "10000";
127        elsif (DIN(31 downto 17) = ZVEC(31 downto 17)) then
128           ZERO <= "01111";
129        elsif (DIN(31 downto 18) = ZVEC(31 downto 18)) then
130           ZERO <= "01110";
131        elsif (DIN(31 downto 19) = ZVEC(31 downto 19)) then
132           ZERO <= "01101";
133        elsif (DIN(31 downto 20) = ZVEC(31 downto 20)) then
134           ZERO <= "01100";
135        elsif (DIN(31 downto 21) = ZVEC(31 downto 21)) then
136           ZERO <= "01011";
137        elsif (DIN(31 downto 22) = ZVEC(31 downto 22)) then
138           ZERO <= "01010";
139        elsif (DIN(31 downto 23) = ZVEC(31 downto 23)) then
140           ZERO <= "01001";
141        elsif (DIN(31 downto 24) = ZVEC(31 downto 24)) then
142           ZERO <= "01000";
143        elsif (DIN(31 downto 25) = ZVEC(31 downto 25)) then
144           ZERO <= "00111";
145        elsif (DIN(31 downto 26) = ZVEC(31 downto 26)) then
146           ZERO <= "00110";
147        elsif (DIN(31 downto 27) = ZVEC(31 downto 27)) then
148           ZERO <= "00101";
149        elsif (DIN(31 downto 28) = ZVEC(31 downto 28)) then
150           ZERO <= "00100";
151        elsif (DIN(31 downto 29) = ZVEC(31 downto 29)) then
152           ZERO <= "00011";
153        elsif (DIN(31 downto 30) = ZVEC(31 downto 30)) then
154           ZERO <= "00010";
155        elsif (DIN(31 downto 31) = ZVEC(31 downto 31)) then
156           ZERO <= "00001";
```

```
157        else
158           ZERO <= "00000";
159        end if;
160     end process;
161   end RTL5;
```

All the above architectures implement the LEADZERO function. The following test bench verifies the correctness of all architectures. In this test bench, the compared results are written into a text file, rather than asserting messages in the simulation transcript. Lines 43, 44, 45, 51, 53, and 55 call textio procedures. These procedures are referenced in lines 4 and 5. The generation of DIN data is almost the same as before. The **component declaration** (lines 13 to 17), **component instantiation** (lines 58 to 67), and **configurations** (lines 69 to 87) are also similar. Note also that the constant PERIOD and STROBE are now declared as generics in lines 8 and 9. Line 10 declares a FILENAME generic with a default value of "tb.out". The file FOUT is declared in line 36. Line 37 declares a variable L as in **line** type. The idea is to write messages into a text file so that the results can be checked later without looking at the messages in the simulation transcript. This allows many simulations to be run as a batch (maybe with other Unix commands). The results can be checked either manually or automatically (such as comparing results between versions or between RTL and structural simulations). Most VHDL simulators also support simulation command input files. For example, we can specify the amount of time to run the simulation. When the specified simulation time elapses, the VHDL simulator terminates. Unix commands can be used to save simulation result files and to copy another set of input files for preparing the next simulation. The simulation of this test bench is left as Exercise 5.19.

```
1     library IEEE;
2     use IEEE.std_logic_1164.all;
3     use IEEE.std_logic_arith.all;
4     use IEEE.std_logic_textio.all;
5     use STD.TEXTIO.all;
6     entity TBLEADZERO is
7       generic (
8         PERIOD   : time   := 50 ns;
9         STROBE   : time   := 45 ns;
10        FILENAME : string := "tb.out");
11    end TBLEADZERO;
12    architecture BEH of TBLEADZERO is
13      component LEADZERO
14      port (
15        DIN  : in  std_logic_vector(31 downto 0);
16        ZERO : out std_logic_vector(4 downto 0));
17      end component;
18      signal  DIN            : std_logic_vector(31 downto 0);
19      signal  Y, Y1, Y2, Y3 : std_logic_vector( 4 downto 0);
```

```
20      signal  Y4, Y5        : std_logic_vector( 4 downto 0);
21   begin
22     p0 : process
23     begin
24       for j in 31 downto 0 loop
25         Y    <= conv_std_logic_vector(j,5);
26         DIN  <= (DIN'range => '0');
27         DIN(31-j) <= '1';
28         wait for PERIOD;
29       end loop;
30       DIN <= (DIN'range => '0');
31       Y   <= "00000";
32       wait;
33     end process;
34     check : process
35       variable err_cnt : integer := 0;
36       file FOUT : text is out FILENAME;
37       variable L : line;
38     begin
39       wait for STROBE;
40       for j in 0 to 32 loop
41         if (Y /= Y1) or (Y /= Y2) or (Y /= Y3) or
42            (Y /= Y4) or (Y /= Y5) then
43           write(L,string'(" Not compared at time "));
44           write(L,NOW);
45           writeline(FOUT, L);
46           err_cnt := err_cnt + 1;
47         end if;
48         wait for PERIOD;
49       end loop;
50       if (err_cnt = 0) then
51         write(L,string'(" Test passed "));
52       else
53         write(L,string'(" Test failed "));
54       end if;
55       writeline(FOUT, L);
56       wait;
57     end process;
58     leadzero1 : LEADZERO
59       port map ( DIN => DIN, ZERO => Y1);
60     leadzero2 : LEADZERO
61       port map ( DIN => DIN, ZERO => Y2);
62     leadzero3 : LEADZERO
63       port map ( DIN => DIN, ZERO => Y3);
64     leadzero4 : LEADZERO
65       port map ( DIN => DIN, ZERO => Y4);
```

```
66    leadzero5 : LEADZERO
67       port map ( DIN => DIN, ZERO => Y5);
68    end BEH;
69    configuration CFG_TBLEADZERO of TBLEADZERO is
70      for BEH
71        for leadzero1 : LEADZERO
72          use entity work.LEADZERO(RTL1);
73        end for;
74        for leadzero2 : LEADZERO
75          use entity work.LEADZERO(RTL2);
76        end for;
77        for leadzero3 : LEADZERO
78          use entity work.LEADZERO(RTL3);
79        end for;
80        for leadzero4 : LEADZERO
81          use entity work.LEADZERO(RTL4);
82        end for;
83        for leadzero5 : LEADZERO
84          use entity work.LEADZERO(RTL5);
85        end for;
86      end for;
87    end CFG_TBLEADZERO;
```

The above test bench verifies all five architectures of the LEADZERO model. Figure 5-20 shows a sample synthesis script file.

```
1    sh date
2    design = leadzero
3    analyze -f vhdl VHDLSRC + "ch5/" + design + ".vhd"
4    elaborate LEADZERO -architecture RTL1
5    include compile.common
6    create_clock -name clk -period CLKPERIOD -waveform {0.0
HALFCLK}
7    set_clock_skew -uncertainty CLKSKEW clk
8    set_input_delay  INPDELAY -add_delay -clock clk all_inputs()
9    set_output_delay CLKPER6  -add_delay -clock clk all_outputs()
10   compile -map_effort medium
11   include compile.report
12   sh date
```

Figure 5-20 Synthesis script file compile.leadzero.rtl1.

The LEADZERO architecture RTL1 is synthesized with the following area and timing statistics. Note that it takes about 4 minutes (lines 1 and 37) to synthesize the VHDL code. The synthesis script file in Figure 5-20 can be changed (line 4) so that other architectures can be synthesized. Architectures RTL2 and RTL3 arrive at area and timing results that are very close to the results of architecture RTL1 in only a little more than 1 minute. Architecture RTL5 takes less than 1 minute to get approximately the same result. Architecture RTL4 takes about 1 hour to arrive at approximately the same timing and area results. Note all the run time is on a Hewlett-Packard (HP) 755 workstation. The synthesis of all architectures is left as Exercise 5.17.

```
1    Fri Dec 26 15:40:12 PST 1997
2    Design : LEADZERO
3    Number of ports:             37
4    Number of nets:             162
5    Number of cells:            130
6    Number of references:        21
7    Combinational area:      189.000000
8    Net Interconnect area:   860.731567
9    Total cell area:         189.000000
10   Total area:             1049.731567
11
12   Operating Conditions: WCCOM   Library: lca300k
13   LEADZERO              B9X9              lca300k
14     Point                      Incr        Path
15     --------------------------------------------------
16     clock clk (rise edge)      0.00        0.00
17     clock network delay (ideal) 0.00       0.00
18     input external delay       2.50        2.50 f
19     DIN[29] (in)               0.00        2.50 f
20     U203/Z (NR2P)              0.85        3.35 r
21     U234/Z (AN2P)              0.62        3.98 r
22     U206/Z (ND2P)              0.42        4.40 f
23     U236/Z (NR2P)              0.46        4.87 r
24     U239/Z (ND2P)              0.50        5.37 f
25     U241/Z (NR2P)              0.41        5.78 r
26     U243/Z (AN2P)              0.72        6.51 r
27     U246/Z (AN2P)              0.72        7.23 r
28     U248/Z (AN2P)              0.58        7.81 r
29     U250/Z (AN2P)              0.54        8.36 r
30     U231/Z (IV)                0.36        8.72 f
31     U255/Z (EON1)          ·   0.75        9.47 r
32     U265/Z (OR2P)              0.58       10.06 r
33     U274/Z (NR2)               0.34       10.40 f
34     U275/Z (ND2P)              0.41       10.81 r
35     ZERO[4] (out)              0.00       10.81 r
36     data arrival time                     10.81
37   Fri Dec 26 15:44:07 PST 1997
```

All the VHDL descriptions for the LEADZERO circuit are based on the function. There are other ways to design the LEADZERO circuit. This is left as Exercise 5.36.

5.6 BARREL SHIFTER

A barrel shifter shifts an N-bit bus any number of bits between 0 and $N-1$ inclusive. The following BARRELL VHDL code shows an example of a barrel shifter to the left direction with 16-bits input data IN0 in line 5. Line 6 specifies the number of bits to be shifted. Line 10 declares a constant N to be the width of the data. Line 11 declares the width of the shift signal. Lines 12 and 13 define an array type with $M+1$ stages for signals defined in line 14. Line 15 declares a ZEROS signals with all bits set to '0' in line 17. Figure 5-21 shows a schematic to implement the BARRELL.

```
1    library IEEE;
2    use IEEE.std_logic_1164.all;
3    entity BARRELL is
4      port (
5        IN0     : in  std_logic_vector(15 downto 0);
6        S       : in  std_logic_vector( 3 downto 0);
7        Y       : out std_logic_vector(15 downto 0));
8    end BARRELL;
9    architecture RTL1 of BARRELL is
10     constant   N   : integer := 16;
11     constant   M   : integer := 4;
12     type arytype is array(M downto 0) of
13       std_logic_vector(N-1 downto 0);
14     signal INTSIG, LEFT, PASS : arytype;
15     signal ZEROS   : std_logic_vector(N-1 downto 0);
16   begin
17     ZEROS <= (ZEROS'range => '0');
18     INTSIG(0) <= IN0;
19     muxgen : for j in 1 to M generate
20       PASS(j) <= INTSIG(j-1);
21       LEFT(j) <= INTSIG(j-1)(N-2**(j-1)-1 downto 0) &
22                  ZEROS(2**(j-1)-1 downto 0);
23       INTSIG(j) <= PASS(j) when S(j-1) = '0' else LEFT(j);
24     end generate;
25     Y <= INTSIG(M);
26   end RTL1;
```

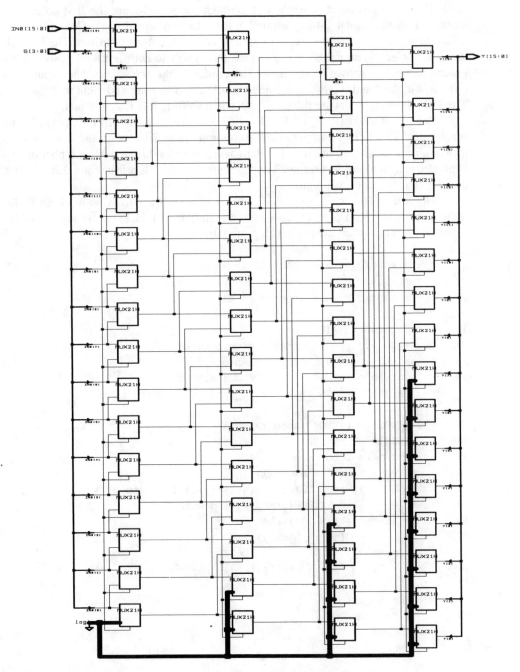

Figure 5-21 BARRELL schematic with all multiplexers.

Observe the Figure 5-21 schematic. There are four columns of 16 two-input multiplexers. The leftmost column multiplexer either shifts left of the input 1-bit position or passes through the input. The selection signal for the two-input multiplexer is $S(0)$. The second column either passes through the signal from the first column or shifts left of the signal 2-bit positions, with selection signal $S(1)$. Column 3 uses selection signal $S(2)$ to either pass through or to shift left 4-bit positions. The last column uses selection signal S3 to either pass through or to shift left 8 bits. Note that a ground symbol is shown in the lower left corner of the schematic. The ground net (which is highlighted) connects to one input of one, two, four, and eight multiplexers. The least significant bits are filled with '0' when shifting left 1, 2, 4, 8 bits.

Refer to BARRELL RTL1 VHDL code. Line 18 initializes $INTSIG(0)$ to be the input signal IN0 before the **generate statement**. Line 20 sets up the data for pass through. Line 21 sets up the shift left data based on the value of j. The number of '0's appended is also based on j. Line 23 implements the multiplexer function. Line 25 sets the output as the last stage signal.

The following architecture RTL2 uses a **process statement** with a **loop statement** inside. The VHDL statements are very similar to RTL1, except that variables (lines 34 to 35) are used here rather than signals.

```
27    architecture RTL2 of BARRELL is
28       constant  N   : integer := 16;
29       constant  M   : integer := 4;
30    begin
31      p0 : process (IN0, S)
32        type arytype is array(M downto 0) of
33           std_logic_vector(N-1 downto 0);
34        variable INTSIG, LEFT, PASS : arytype;
35        variable ZEROS : std_logic_vector(N-1 downto 0);
36      begin
37        ZEROS := (ZEROS'range => '0');
38        INTSIG(0) := IN0;
39        for j in 1 to M loop
40          PASS(j) := INTSIG(j-1);
41          LEFT(j) := INTSIG(j-1)(N-2**(j-1)-1 downto 0) &
42                     ZEROS(2**(j-1)-1 downto 0);
43          if (S(j-1) = '0') then
44             INTSIG(j) := PASS(j);
45          else
46             INTSIG(j) := LEFT(j);
47          end if;
48        end loop;
49        Y <= INTSIG(M);
50      end process;
51    end RTL2;
```

The verification of BARRELL architectures RTL1 and RTL2 is left as Exercise 5.21. Architecture RTL1 can be synthesized with the synthesis script file as shown in Figure 5-22. Figure 5-23 shows the synthesized schematic.

```
1    sh date
2    design = barrell
3    analyze -f vhdl VHDLSRC + "ch5/" + design + ".vhd"
4    elaborate BARRELL -architecture RTL1
5    include compile.common
6    create_clock -name clk -period CLKPERIOD -waveform {0.0
HALFCLK}
7    set_clock_skew -uncertainty CLKSKEW clk
8    set_input_delay  INPDELAY -add_delay -clock clk all_inputs()
9    set_output_delay CLKPER6  -add_delay -clock clk all_outputs()
10   compile -map_effort medium
11   include compile.report
```

Figure 5-22 Synthesis script file compile.barrell.rtl1.

Figure 5-23 schematic has the following area and timing statistics. There are many variations of barrel shifters. They are left as exercises (5.23 to 5.29). The schematic of Figure 5-21 is used as an illustration. It is left for the reader in Exercise 5.30 to generate the schematic with VHDL code and synthesis commands. Figure 5-21 has a combinational area of 256 units, a net interconnect area of 683 units, a total area of 939 units, and the data arrival time is 9.77 ns under the same operating conditions and wire load model.

```
1    Number of ports:            36
2    Number of nets:             98
3    Number of cells:            76
4    Number of references:       19
5    Combinational area:     178.000000
6    Net Interconnect area:  768.392517
7    Total area:             946.392517
8    *****************************************
9    Operating Conditions: WCCOM   Library: lca300k
10   BARRELL                 B9X9            lca300k
11   -----------------------------------------------------
12   clock clk (rise edge)            0.00     0.00
13   clock network delay (ideal)      0.00     0.00
14   input external delay             2.50     2.50 r
15   S[0] (in)                        0.00     2.50 r
16   U37/Z1 (B2I)                     1.31     3.81 f
17   U37/Z2 (B2I)                     1.39     5.20 r
18   U46/Z (ND2)                      1.24     6.44 f
19   U62/Z (EON1)                     1.30     7.74 r
20   U100/Z (AO11P)                   1.07     8.82 f
21   U102/Z (ND2)                     0.66     9.48 r
22   Y[13] (out)                      0.00     9.48 r
23   data arrival time                         9.48
```

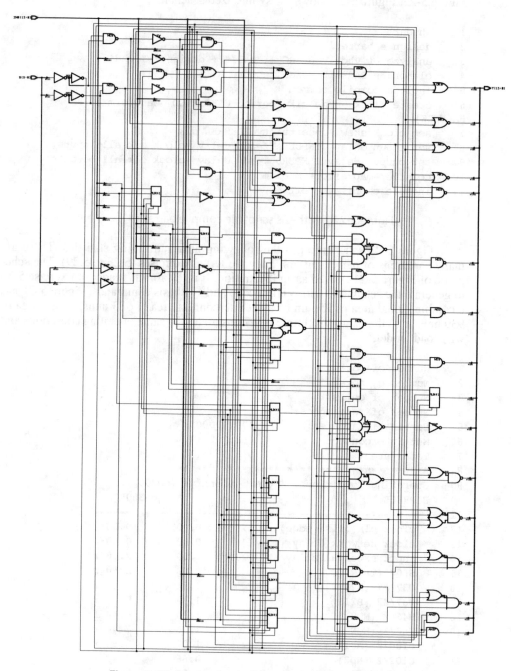

Figure 5-23 Synthesized schematic for BARRELL RTL1.

There are many other basic arithmetic circuits such as incrementer and 2's complementer. They are left as Exercises 5.31 and 5.35.

There are many other types of binary arithmetic algorithms. Readers are referred to E. E. Swartzlander, Jr. (ed.), *Computer Arithmetic,* IEEE Computer Society Press, 1990. It is a collection of many technical papers.

5.7 EXERCISES

5.1 Refer to FA COMP VHDL code. Use the synthesis script file compile.fa.comp to synthesize the circuit step by step. You can cut and paste the command one by one to the Synopsys dc_shell command window. Observe the number of components created by the Synopsys design_analyzer. Repeat the same process for the synthesis script file compile.fa.comp.dt. Compare the number of components created in both synthesis script files.

5.2 Refer to FA COMP VHDL code. Remove lines 13 to 15 in synthesis script file compile.fa.comp.dt. Run the synthesis command "include compile.fa.comp.dt" in the Synopsys dc_shell command window. Verify that you get the same schematic as shown in Figure 5-3.

5.3 Refer to FA COMP VHDL code. The simulation was not done. Write the VHDL code for gates.vhd. Write a test bench to configure the VHDL hierarchy for all components. Verify the test bench.

5.4 Refer to FA LIB VHDL code. Find out the exact component names with their port names used in your synthesis library. Change the FA LIB VHDL code so that the exact same schematic can be generated. Verify your VHDL code by reading in the VHDL code in Synopsys design_analyzer to view the schematic.

5.5 Use the techniques discussed in FA COMP VHDL code. Write VHDL code to generate a schematic with six inverters in series. Note that this circuit can be used to generate a delay.

5.6 Refer to RPADDER VHDL code. Develop a test bench to verify both architectures.

5.7 Refer to RPADDER VHDL code. Both architectures are synthesized with different approaches. Change the synthesis script to synthesize both architectures with N mapped to 64. Compare the synthesis area and timing results with CPU run time. What are advantages and disadvantages of both architectures?

5.8 Refer to RPADDER VHDL code. Architecture RTL1 uses FA architecture COMP for synthesis. Use FA architecture RTL instead and synthesize RPADDER architecture RTL1. Compare the synthesis results with using FA architecture COMP.

5.9 Refer to CLA64 VHDL code. Develop a test bench to verify its functional correctness.

5.10 Refer to CLA64 VHDL code. Develop synthesis scripts to synthesize CLA64.

5.11 Refer to CLA64 VHDL code and ADDER VHDL code. Synthesize ADDER VHDL code with N mapped to 64. Compare the synthesis results with the CLA64 results.

5.12 Refer to CLA16 VHDL code. Develop a 16-bit borrow look ahead subtracter. Develop a test bench to verify the VHDL code. Develop synthesis script files to synthesize the VHDL code.

5.13 Refer to CLA64 VHDL code. Develop a 64-bit borrow look ahead subtracter. Develop a test bench to verify the VHDL code. Develop synthesis script files to synthesize the VHDL code.

5.14 Refer to CLA64 VHDL code. What would you do if you need exactly a 42-bit CLA? Develop VHDL code for CLA42 42-bit carry look ahead adder. Modify the test bench to verify your VHDL code. Develop synthesis script files to synthesize the VHDL code.

5.15 Refer to TBCOUNTONE VHDL code. Improve the test bench to include a histogram of Y1 values. Does the test bench cover all possible values of Y1? If not, how do you improve the test bench?

5.16 Refer to TBCOUNTONE VHDL code. An infinite loop for verification may not be the best option because there is no clear indication how long the simulation should run. Develop a test bench with a fixed number of patterns to cover all possible values of Y1. State your assumptions and reasons why your test bench accommodates all possible variables.

5.17 Refer to COUNTONE VHDL code. Develop synthesis scripts to synthesize all four architectures. Compare the results of area, speed, and run time.

5.18 Refer to COUNTONE VHDL code. Develop an architecture that is faster than the four architectures. Synthesize your architecture to verify that it is indeed faster.

5.19 Refer to TBLEADZERO VHDL code. How long does the simulation need to run before all the intended inputs are verified? Is there a text file "tb.out" generated in your simulation directory? Does it say " Test passed "?

5.20 Refer to LEADZERO VHDL code. Synthesize all five architectures with your synthesis library. Compare the area and timing of the synthesis results. Also compare the amount of CPU time for synthesis. What conclusions can you make? Which architecture would you recommend and why?

5.21 Refer to BARRELL VHDL code. Write a test bench to verify both RTL1 and RTL2 architectures.

5.22 Refer to BARRELL VHDL code. Write a synthesis script file to compile architecture RTL2. Compare the results with RTL1. Are they about the same?

5.23 Refer to BARRELL VHDL code. Change the code for the BARRELL with 32-bit input data. Synthesize the VHDL code.

5.24 Refer to BARRELL VHDL code. It is a logical shift left barrel shifter. Change the VHDL code to do a logical shift right barrel shifter.

5.25 Refer to BARRELL VHDL code. It is a logical shift left barrel shifter. Change the VHDL code to do a logical rotate right barrel shifter.

5.26 Refer to BARRELL VHDL code. It is a logical shift left barrel shifter. Change the VHDL code to do a logical rotate left barrel shifter.

5.27 Refer to BARRELL VHDL code. It is a logical shift left barrel shifter. Change the VHDL code to do an arithmetic shift left barrel shifter, assuming 2's complement with the leftmost bit as the sign bit.

5.28 Refer to BARRELL VHDL code. It is a logical shift left barrel shifter. Change the VHDL code to do an arithmetic shift right barrel shifter, assuming 2's complement with the left most bit as the sign bit.

5.29 Refer to BARRELL VHDL code. Modify the VHDL code with another input FUNC of 3 bits. When FUNC(0) = '0', it refers to the left direction, while '1' designates the right direction. When FUNC(1) = '0', it is a logical shift, '1' for arithmetic shift. When FUNC(2) = '0', it is a shift operation, '1' is for the rotate operation. Write a test bench to verify all combinations of FUNC. Develop synthesis script file to synthesize the VHDL code.

5.30 Refer to Figure 5-21 schematic. Write VHDL code and synthesis script to generate the same schematic as shown in Figure 5-21. Compare the area and timing with BARRELL RTL1 results.

5.31 An incrementer is similar to an adder with one addend as 1. The following VHDL code can be used to model an incrementer. There are four architectures for entity INC. UNIT4 VHDL code is also shown, which is used as a subcomponent in architecture RTL3. Develop a test bench to verify all these four architectures. Develop synthesis scripts to synthesize all architectures. Compare synthesis results in area and timing. What is the purpose of using the UNIT4 component?

```
1      ------------------------file inc.vhd
2      library IEEE;
3      use IEEE.std_logic_1164.all;
4      use IEEE.std_logic_unsigned.all;
5      entity INC is
6        port (
7          A : in  std_logic_vector(31 downto 0);
8          Y : out std_logic_vector(31 downto 0));
9      end INC;
10     architecture RTL1 of INC is
11     begin
12       Y <= A + 1;
13     end RTL1;
14     architecture RTL2 of INC is
15     begin
16       p0 : process (A)
17         variable CARRY : std_logic;
18       begin
19         CARRY := '1';
20         for j in A'reverse_range loop
21           Y(j)  <= A(j) xor CARRY;
22           CARRY := A(j) and CARRY;
23         end loop;
24       end process;
25     end RTL2;
26     architecture RTL3 of INC is
27       component UNIT4
28       port (
29         D  : in  std_logic_vector(3 downto 0);
30         Ci : in  std_logic;
31         Co : out std_logic;
32         Y  : out std_logic_vector(3 downto 0));
33       end component;
34       signal C : std_logic_vector(8 downto 0);
35     begin
36       C(0) <= '1';
37       g0 : for j in 0 to 7 generate
38         u0 : UNIT4
```

```
39          port map (A(j*4+3 downto j*4), C(j),
40              C(j+1), Y(j*4+3 downto j*4));
41       end generate;
42     end RTL3;
43     architecture RTL4 of INC is
44       procedure UNIT4 (
45         signal D  : in  std_logic_vector(3 downto 0);
46         signal Ci : in  std_logic;
47         signal Co : out std_logic;
48         signal Y  : out std_logic_vector(3 downto 0)) is
49       begin
50         Co   <= Ci and D(0) and D(1) and D(2) and D(3);
51         Y(0) <= D(0) xor Ci;
52         Y(1) <= D(1) xor (Ci and D(0));
53         Y(2) <= D(2) xor (Ci and D(0) and D(1));
54         Y(3) <= D(3) xor (Ci and D(0) and D(1) and D(2));
55       end UNIT4;
56       signal C : std_logic_vector(8 downto 0);
57     begin
58       C(0) <= '1';
59       g0 : for j in 0 to 7 generate
60         u0 : UNIT4 (A(j*4+3 downto j*4), C(j),
61                   C(j+1), Y(j*4+3 downto j*4));
62       end generate;
63     end RTL4;


1      --------------------file unit4.vhd
2      library IEEE;
3      use IEEE.std_logic_1164.all;
4      entity UNIT4 is
5        port (
6          D  : in  std_logic_vector(3 downto 0);
7          Ci : in  std_logic;
8          Co : out std_logic;
9          Y  : out std_logic_vector(3 downto 0));
10     end UNIT4;
11     architecture RTL of UNIT4 is
12     begin
13       Co   <= Ci and D(0) and D(1) and D(2) and D(3);
14       Y(0) <= D(0) xor Ci;
15       Y(1) <= D(1) xor (Ci and D(0));
16       Y(2) <= D(2) xor (Ci and D(0) and D(1));
17       Y(3) <= D(3) xor (Ci and D(0) and D(1) and D(2));
18     end RTL;
```

5.32 Refer to the INC VHDL code above. Develop another architecture with the similar idea of the carry look ahead features. Verify and synthesize your VHDL code. Compare the results of the synthesis with the last exercise.

5.33 Refer to INC VHDL code. Modify the VHDL code to design a decrementer. Develop a test bench for verification. Write synthesis scripts to synthesize the VHDL code. Analyze the synthesis results.

5.34 A 2's complementer is often used in digital designs. The following COMP2 VHDL code shows three possible architectures. Develop a test bench to verify all three architectures. Develop synthesis script files to synthesize all three architectures. Compare the synthesis results in regard to area and timing.

```
1    ---------------------- file comp2.vhd
2    library IEEE;
3    use IEEE.std_logic_1164.all;
4    use IEEE.std_logic_unsigned.all;
5    entity COMP2 is
6      port (
7        A : in  std_logic_vector(31 downto 0);
8        Z : out std_logic_vector(31 downto 0));
9    end COMP2;
10   architecture RTL1 of COMP2 is
11   begin
12     p0 : process (A)
13       variable found1 : std_logic;
14     begin
15       found1 := '0';
16       for i in A'reverse_range loop
17         if (found1 = '1') then
18           Z(i) <= not A(i);
19         elsif (A(i) = '1') then
20           found1 := '1';
21           Z(i) <= '1';
22         else
23           Z(i) <= '0';
24         end if;
25       end loop;
26     end process;
27   end RTL1;
28   architecture RTL2 of COMP2 is
29   begin
30     Z <= (not A) + 1;
31   end RTL2;
32   architecture RTL3 of COMP2 is
33   begin
34     Z <= 0 - A;
35   end RTL3;
```

5.35 Refer to the above COMP2 VHDL code. Write another architecture with the idea of carry look ahead. Verify and synthesize your VHDL code. Compare the synthesis results with the above three architectures.

5.36 Refer to the LEADZERO VHDL code. V. G. Oklobdzija published a paper entitled "An Implementation Algorithm and Design of a Novel Leading Zero Detector Circuit" in *IEEE Asilomar Conference on Signals, Systems and Computers*, Vol. 1, 1992, pp. 391–395. Implement his algorithm and compare with the results obtained by the VHDL code presented in this chapter.

5.37 The COUNTONE circuit can be implemented with array of full adders, the same way to add a column of bits at a certain bit position for a multiplier design. Each column of full adders will generate a sum bit. The carry bits from of a column of full adders will be passed to the next column of full adders. Develop VHDL code to capture this design concept. Synthesize the VHDL code. Compare the synthesis results with the results obtained in Exercise 5.17.

5.38 Refer to COMP2 architecture RTL1 VHDL code in Exercise 5.34. It goes through the bits from the least significant bit to the most significant bit (right to left). The output bit is the same as the input bit until the first '1' is found. After that, the output bit is the inverse of the input bit. Use the same idea but traverse the bits from right to left. Given an input pattern, after the first '1' is found, the output bit is set to '1'. For example, an input pattern of 000001XXXXXXXXXX will result an output of 0000011111111111 where 'X' indicates a don't care bit. Write VHDL code for such a design. Synthesize the VHDL code. Develop a test bench to verify the correctness of the design.

Chapter 6

Basic Sequential Circuits

A combinational circuit produces its output based solely on the current inputs. On the other hand, a sequential circuit generates its outputs based on the current inputs and its previous states. In other words, a sequential circuit "remembers" its previous states. Other than the memory, latches and flip-flops are the common sequential circuits. In this chapter, we will concentrate on circuits based on flip-flops, since flip-flops are the backbones of the synchronous digital circuits. We will discuss signal manipulators to chop, extend, or generate a certain signal based on existing signals. Shift registers, serial to parallel converters, parallel to serial converters, and synchronous counters are also presented. Finite state machines are also important sequential circuits which have been discussed in K.C. Chang, *Digital Design and Modeling with VHDL and Synthesis*, IEEE Computer Society Press, Los Alamitos, California, 1997. Readers are referred to that book; the information therein will not be repeated here.

6.1 SIGNAL MANIPULATOR

Digital designs are often required to manipulate signals. The most common signal manipulation is to delay a signal by a full clock cycle. A simple D-type flip-flop can do just that. A **signal chopper** is a circuit that cuts the length of a signal into a signal that is a full or half-cycle long. The following CHOP VHDL code shows an example of a signal chopper. In line 5, CLK is the clock signal. In line 6, CLKn is the inverted clock input. In line 7, RSTn is the asynchronous reset input to the flip-flops. In line 8, SOURCE is the input signal that is long and will be chopped. In line 9, TARGET_H is an output signal that is a chopped SOURCE signal at the first falling edge of the clock CLK. In line 10, TARGET_FH is an output signal that is a chopped signal of SOURCE with a full half of the clock period starting at the falling edge of the clock CLK. In line 11, TARGET_F is an output signal that is a chopped signal of SOURCE at the first rising edge of the clock CLK. In line 12, TARGET_FF is an output signal that is a chopped signal having the full length of a clock cycle, starting at the first rising edge of the clock CLK.

```
1       library IEEE;
2       use IEEE.std_logic_1164.all;
3       entity CHOP is
4         port (
5           CLK            : in  std_logic;
6           CLKn           : in  std_logic;
7           RSTn           : in  std_logic;
8           SOURCE         : in  std_logic;
9           TARGET_H       : out std_logic;
10          TARGET_FH      : out std_logic;
11          TARGET_F       : out std_logic;
12          TARGET_FF      : out std_logic);
13      end CHOP;
```

Figure 6-1 shows a simulation waveform. CLK and CLKn are reversed to each other. Signal SOURCE is active high and is longer than one clock cycle. When SOURCE is going from '0' to '1', signals TARGET_H and TARGET_F are asserted right away. Signal TARGET_H (TARGET_F) is deasserted at the first falling edge (rising edge). Depending on when signal SOURCE is asserted '1', the output signals may have different widths at different starting times. For example, if signal SOURCE is asserted before the falling edge of the clock CLK, it will cause TARGET_H to be shorter if signal SOURCE is asserted after the falling edge of the clock CLK. In both situations, TARGET_FH will have the same width. However, TARGET_FH signal will be asserted one clock sooner when signal SOURCE is asserted before the falling edge of the clock CLK. The same analysis applies to signals TARGET_F and TARGET_FF.

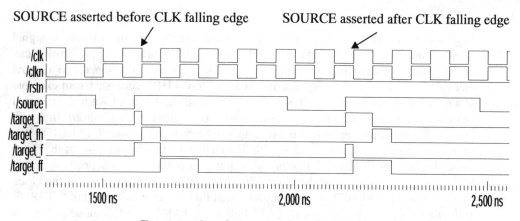

Figure 6-1 Simulation waveform for TBCHOP.

The following VHDL code shows an example of implementing the signal chopper. Lines 17 to 28 infer flip-flops with the rising edge of clock CLK. Lines 31 to 38 infer flip-flops with the falling edge of clock CLK. Rather than using (CLK'event and CLK= '0'), an inverted clock signal CLKn is used. We assume that signal CLKn is inverted outside of this block. All flip-flops are asynchronously reset, since all signals are assumed active '1'. Lines 29, 30, 39, and 40 generate output signals.

```
14    architecture RTL of CHOP is
15      signal FFH0, FFH1, FFF0, FFF1 : std_logic;
16    begin
17      posedge : process (RSTn, CLK)
18      begin
19        if (RSTn = '0') then
20          FFF0  <= '0';
21          FFF1  <= '0';
22          FFH1  <= '0';
23        elsif (CLK'event and CLK = '1') then
24          FFF0  <= SOURCE;
25          FFF1  <= FFF0;
26          FFH1  <= FFH0;
27        end if;
28      end process;
29      TARGET_F  <= (not SOURCE) nor FFF0;
30      TARGET_FF <= (not FFF0  ) nor FFF1;
31      negedge : process (RSTn, CLKn)
32      begin
33        if (RSTn = '0') then
34          FFH0  <= '0';
35        elsif (CLKn'event and CLKn = '1') then
36          FFH0  <= SOURCE;
37        end if;
38      end process;
39      TARGET_H  <= (not SOURCE) nor FFH0;
40      TARGET_FH <= (not FFH0  ) nor FFH1;
41    end RTL;
```

To verify the function of CHOP VHDL code, the following test bench is used. Lines 23 and 24 generate a RSTn signal. Lines 25 to 29 generate a clock CLK signal. Line 30 inverts the CLK signal to become the signal CLKn. Lines 31 to 44 generate signal SOURCE. Depending on the value of vec(6) in line 37, SOURCE is asserted after the falling edge of clock CLK. Lines 45 to 50 instantiate component CHOP. Lines 52 to 55 declare empty configuration. The default architecture (there is only one) of CHOP will be used for simulation. Part of the simulation waveform is shown in Figure 6-1.

```
1     library IEEE;
2     use IEEE.std_logic_1164.all;
3     entity TBCHOP is
4     end TBCHOP;
5     architecture BEH of TBCHOP is
6       component CHOP
```

```vhdl
7        port (
8          CLK             : in  std_logic;
9          CLKn            : in  std_logic;
10         RSTn            : in  std_logic;
11         SOURCE          : in  std_logic;
12         TARGET_H        : out std_logic;
13         TARGET_FH       : out std_logic;
14         TARGET_F        : out std_logic;
15         TARGET_FF       : out std_logic);
16       end component;
17       constant HALFCLK : time := 50 ns;
18       constant DELAY   : time := 30 ns;
19       signal CLK, CLKn, RSTn, SOURCE  : std_logic;
20       signal TARGET_H, TARGET_FH : std_logic;
21       signal TARGET_F, TARGET_FF : std_logic;
22     begin
23       RSTn <= '1', '0' after HALFCLK*3 + DELAY,
24                     '1' after HALFCLK*7 + DELAY;
25       clkp : process
26       begin
27         CLK <= '0'; wait for HALFCLK;
28         CLK <= '1'; wait for HALFCLK;
29       end process;
30       CLKn <= not CLK;
31       datap : process
32         variable vec : std_logic_vector(7 downto 0) := "01100011";
33       begin
34         vec := vec(6 downto 0) & (vec(7) xor (not vec(2)));
35         SOURCE <= '0' after DELAY;
36         wait until (CLK'event and CLK = '1') ;
37         if (vec(6) = '1') then
38           wait until (CLK'event and CLK = '0') ;
39         end if;
40         SOURCE <= '1' after DELAY;
41         for j in 1 to 4 loop
42           wait until (CLK'event and CLK = '1') ;
43         end loop;
44       end process;
45       chop0 : CHOP
46       port map (
47         CLK       => CLK,      CLKn      => CLKn,
48         RSTn      => RSTn,     SOURCE    => SOURCE,
49         TARGET_H  => TARGET_H, TARGET_FH => TARGET_FH,
50         TARGET_F  => TARGET_F, TARGET_FF => TARGET_FF);
51     end BEH;
52     configuration CFG_TBCHOP of TBCHOP is
53       for BEH
54       end for;
55     end CFG_TBCHOP;
```

To synthesize the CHOP VHDL code, the following synthesis script file can be used. Line 6 declares port CLK as a clock. Line 7 declares port CLKn also as a clock, but the waveform is reversed. Line 12 specifies that the interconnects to CLK, CLKn, and RSTn are not to be touched so that the ports are connected directly to flip-flops. Figure 6-2 shows the synthesized schematic.

```
1    sh date
2    design = chop
3    analyze -f vhdl VHDLSRC + "ch6/" + design + ".vhd"
4    elaborate CHOP -architecture RTL
5    include compile.common
6    create_clock "CLK" -name clk -period CLKPERIOD -waveform {0.0
HALFCLK}
7    create_clock "CLKn" -name clkn -period CLKPERIOD -waveform
{HALFCLK CLKPERIOD}
8    set_clock_skew -uncertainty CLKSKEW clk
9    set_input_delay  INPDELAY -add_delay -clock clk all_inputs()
10   set_output_delay CLKPER4  -add_delay -clock clk all_outputs()
11   set_input_delay  0 -add_delay -clock clk {CLK}
12   set_dont_touch_network {CLK CLKn RSTn}
13   compile -map_effort medium
14   include compile.report
15   sh date
```

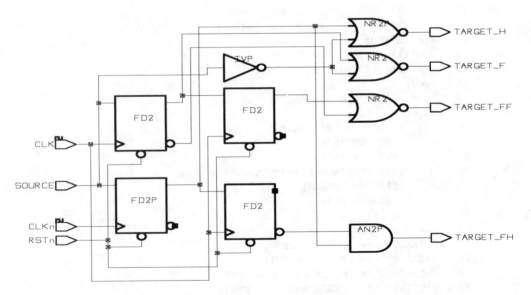

Figure 6-2 Synthesized schematic for CHOP VHDL code.

Note the differences among TARGET_H, TARGET_FH, TARGET_F, and TARGET_FF when signal SOURCE is asserted before and after the falling edge of the clock. In designing circuits, caution should be taken to know (and to control) when SOURCE is asserted. This may take careful gate level simulation with exact timing, since VHDL simulation is zero delay. The static timing analyzer in the synthesis tools can be very helpful. However, the timing delay after the layout is often different. A good practice may be to use the rising edge for all signals. This is easier for timing analysis in the synthesis process. Use the falling edge with caution and ensure careful timing analysis.

The reverse of the signal chopper is the *signal extender*. A signal extender extends the length of a signal. The following EXTEND VHDL code is such an example. Lines 17 to 26 delay signal SOURCE two times. Lines 27 to 31 generate output signals. Figure 6-3 shows the synthesized schematic.

```
1     library IEEE;
2     use IEEE.std_logic_1164.all;
3     entity EXTEND is
4       port (
5         CLK           : in  std_logic;
6         RSTn          : in  std_logic;
7         SOURCE        : in  std_logic;
8         TARGET1       : out std_logic;
9         TARGET2       : out std_logic;
10        TARGET3       : out std_logic;
11        TAIL1         : out std_logic;
12        TAIL2         : out std_logic);
13    end EXTEND;
14    architecture RTL of EXTEND is
15      signal FF0, FF1 : std_logic;
16    begin
17      posedge : process (RSTn, CLK)
18      begin
19        if (RSTn = '0') then
20          FF0  <= '0';
21          FF1  <= '0';
22        elsif (CLK'event and CLK = '1') then
23          FF0  <= SOURCE;
24          FF1  <= FF0;
25        end if;
26      end process;
27      TARGET1 <= SOURCE or FF0;
28      TARGET2 <= FF0    or FF1;
29      TARGET3 <= SOURCE or FF0 or FF1;
30      TAIL1   <= SOURCE nor (not FF0);
31      TAIL2   <= FF0    nor (not FF1);
32    end RTL;
```

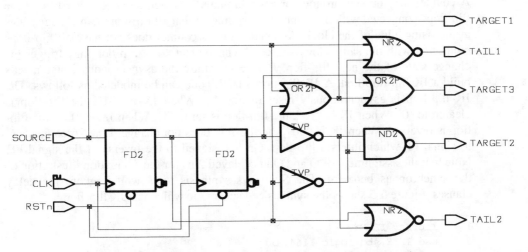

Figure 6-3 Synthesized schematic for EXTEND VHDL code.

Figure 6-4 shows a portion of the simulation waveform. Signal TARGET1 is asserted at the same time as SOURCE and deasserted at the first rising edge of the clock CLK when SOURCE is deasserted. TARGET2 is asserted at the rising edge of the clock CLK and deasserted two clock rising edges after SOURCE is deasserted. TARGET3 is asserted at the same time as SOURCE and deasserted two clock rising edges after SOURCE is deasserted. TAIL1 is asserted when SOURCE is deasserted. TAIL1 is deasserted at the first rising edge of the clock when SOURCE is deasserted. TAIL2 is exactly one full clock cycle wide. The verification of the EXTEND VHDL code is left as Exercise 6.1.

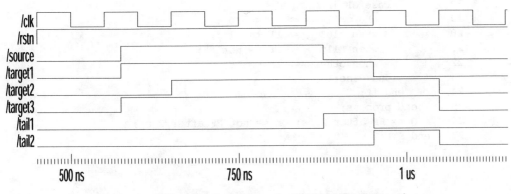

Figure 6-4 Simulation waveform for TBEXTEND VHDL code.

6.2 COUNTER

A counter may be synchronous or asynchronous. A synchronous counter has all the flip-flops connected with the same clock source so that all flip-flops can change states at the rising edge of the clock. An asynchronous counter does not have all flip-flops connected with the same clock source. A change state of a flip-flop may trigger the change state of the next flip-flop. Before we study the asynchronous counter, let's build a JK flip-flop first. A JK flip-flop VHDL code can be modeled as follows. The JK flip-flop is asynchronously reset in line 15. When JK is "01", the flip-flop is cleared to '0'. When JK is "10", the flip-flop is set to '1'. When JK is "11", the flip-flop is reversed. When JK is "00", the flip-flop does not change. Line 26 connects Q to signal FF which infers a flip-flop. Qn is assigned to the inverse of the signal FF. Note that the assignment of Q and Qn is delayed 2 ns for the simulation illustration of the synchronous behavior. The Synopsys synthesis tools would ignore the delay clauses. Figure 6-5 shows the synthesized schematic which is a JK flip-flop.

```
1     library IEEE;
2     use IEEE.std_logic_1164.all;
3     entity JKFF is
4       port (
5         CLK, RSTn, J, K : in  std_logic;
6         Q, Qn              : out std_logic);
7     end JKFF;
8     architecture RTL of JKFF is
9       signal FF : std_logic;
10    begin
11      process (CLK, RSTn)
12        variable JK : std_logic_vector(1 downto 0);
13      begin
14        if (RSTn = '0') then
15          FF <= '0';
16        elsif (CLK'event and CLK = '1') then
17          JK := J & K;
18          case JK is
19            when "01"   => FF <= '0';
20            when "10"   => FF <= '1';
21            when "11"   => FF <= not FF;
22            when others => FF <= FF;
23          end case;
24        end if;
25      end process;
26      Q <= FF after 2 ns; Qn <= not FF after 2 ns;
27    end RTL;
```

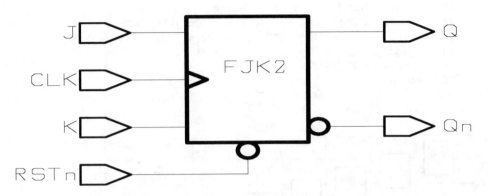

Figure 6-5 Synthesized schematic for JKFF.

An asynchronous counter can be constructed using JK flip-flop by using the special property that reverses the flip-flop state when both J and K are '1'. The following ACNT VHDL code shows an example of 8-bit asynchronous counter using JK flip-flops. Lines 10 to 14 declare the JKFF component. Lines 20 to 23 instantiate JKFF eight times. Note that the CLK input of the JKFF is connected to the previous stage of the JK flip-flop Qn output. For the first stage, which is the least significant bit, the CLK input should connect to the enable input EN as shown in line 19. Signal VDD is tie to high in line 19, which connects to J and K inputs of all JK flip-flops. Line 24 connects the output CNTR to signal FFQ(8 downto 1). Figure 6-6 shows the synthesized schematic of the ACNT VHDL code. From the schematic, it is easy to see that all J and K inputs are connected to "logic_1". The CLK input of the JK flip-flop connects to the previous stage of the JK flip-flop Qn output, except for the first stage which connects to input EN.

```
1    ---------------------- file acnt.vhd
2    library IEEE;
3    use IEEE.std_logic_1164.all;
4    entity ACNT is
5      port (
6        RSTn, EN : in  std_logic;
7        CNTR     : out std_logic_vector(7 downto 0));
8    end ACNT;
9    architecture RTL of ACNT is
10     component JKFF
11     port (
12       CLK, RSTn, J, K : in  std_logic;
13       Q, Qn           : out std_logic);
14     end component;
15     signal FFQ  : std_logic_vector(8 downto 0);
16     signal FFQn : std_logic_vector(8 downto 0);
```

```
17      signal VDD  : std_logic;
18    begin
19      VDD <= '1'; FFQn(0) <= EN;
20      jk0 : for j in 1 to 8 generate
21        b17 : JKFF port map (CLK => FFQn(j-1), RSTn => RSTn,
22          J => VDD, K => VDD, Q => FFQ(j), Qn => FFQn(j));
23      end generate;
24      CNTR <= FFQ(8 downto 1);
25    end RTL;
```

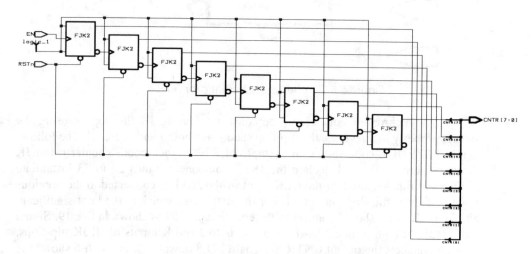

Figure 6-6 Synthesized schematic for ACNT.

To illustrate the asynchronous behavior, the following simple test bench is used.

```
1     library IEEE;
2     use IEEE.std_logic_1164.all;
3     entity TBACNT is
4     end TBACNT;
5     architecture BEH of TBACNT is
6       component ACNT
7       port (
8         RSTn, EN : in  std_logic;
9         CNTR     : out std_logic_vector(7 downto 0));
10      end component;
11      constant HALFCLK : time := 50 ns;
12      constant DELAY   : time := 30 ns;
13      signal RSTn, EN  : std_logic;
14      signal CNTR      : std_logic_vector(7 downto 0);
15    begin
16      RSTn <= '1', '0' after HALFCLK*3 + DELAY,
17                  '1' after HALFCLK*7 + DELAY;
```

```
18        datap : process
19          variable vec : std_logic_vector(7 downto 0) := "01100011";
20        begin
21          vec := vec(6 downto 0) & (vec(7) xor (not vec(2)));
22          EN <= vec(7) after DELAY;
23          wait for 2 * HALFCLK;
24        end process;
25        acnt0 : ACNT
26        port map (
27          RSTn  => RSTn, EN => EN, CNTR => CNTR);
28      end BEH;
29      configuration CFG_TBACNT of TBACNT is
30        for BEH
31        end for;
32      end CFG_TBACNT;
```

Figure 6-7 shows part of the simulation waveform. Note that when the counter value is "01111110", EN changes from '0' to '1', the counter value is changed to "01111111" faster than the counter changes value from "01111111" to "10000000". Changing from "01111110" to "01111111" requires only a change to the least significant bit of the JK flip-flop which has a delay of 2 ns. Changing from "01111111" to "10000000" requires all eight JK flip-flops to change state. Each stage requires a 2-ns delay. Figure 6-8 shows the enlarged simulation waveform by zooming in the place when the counter changes to "10000000".

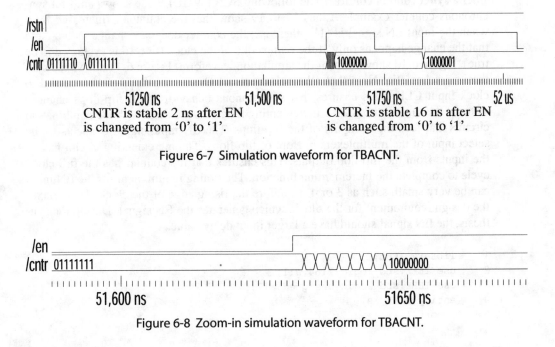

Figure 6-7 Simulation waveform for TBACNT.

Figure 6-8 Zoom-in simulation waveform for TBACNT.

We have used JK flip-flops to illustrate an 8-bit, asynchronous counter. Some libraries provide another type of flip-flop called T (toggle) flip-flop. When T is '1', the T-type flip-flop toggles (reverses) its state, otherwise, it remains state. Figure 6-6 schematic can also be synthesized to Figure 6-9 schematic with using only D-type flip-flops. The synthesis needed to produce Figure 6-9 as a schematic is left as Exercise 6.3.

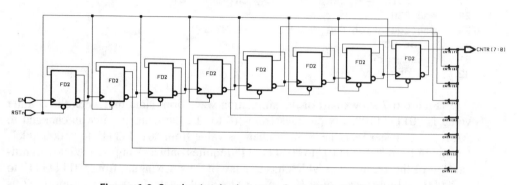

Figure 6-9 Synthesized schematic for ACNT with D flip-flops.

We have shown that an asynchronous counter has a timing delay propagating through each flip-flop. If we have a delay of 2 ns for each stage of the flip-flop for a 32-bit asynchronous counter, it would equal a 64 ns delay. This is too slow. Let's consider a **synchronous counter**. The following SCNT VHDL code shows an 8-bit synchronous counter example. Lines 17 to 19 show that the counter is increased by 1 when the input EN signal is '1', otherwise, the counter does not change value. Note that the change happens only at the rising edge of the clock (line 16 is evaluated to be true). Figure 6-10 shows part of the simulation waveform. Figure 6-11 shows the synthesized schematic. Note that all clock inputs of D-type flip-flops are connected to clock input CLK. This ensures that synchronous behavior of all flip-flops changes states at the same time. All the incrementing functions are done with the combination circuits that feed the D inputs of the flip-flops. The EN input signal connects to the select input of the multiplexers in front of flip-flops. The incrementing circuit has all the inputs from the flip-flop outputs. This ensures that the counter has the full clock cycle to complete the incrementing function. The timing requirement for the EN input can be very small, such as 3 or 4 ns, before the rising edge of the clock. This relaxes the design requirement for the block, which generates the EN signal. During the synthesis, the EN signal should have a larger input delay value.

```
1    library IEEE;
2    use IEEE.std_logic_1164.all;
3    use IEEE.std_logic_unsigned.all;
4    entity SCNT is
5      port (
```

```
 6         CLK, RSTn, EN : in  std_logic;
 7         CNTR          : out std_logic_vector(7 downto 0));
 8     end SCNT;
 9     architecture RTL of SCNT is
10       signal FF : std_logic_vector(7 downto 0);
11     begin
12       process (CLK, RSTn)
13       begin
14         if (RSTn = '0') then
15           FF <= (FF'range => '0');
16         elsif (CLK'event and CLK = '1') then
17           if (EN = '1') then
18             FF <= FF + 1;
19           end if;
20         end if;
21       end process;
22       CNTR <= FF;
23     end RTL;
```

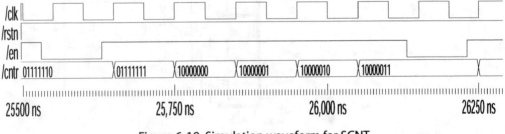

Figure 6-10 Simulation waveform for SCNT.

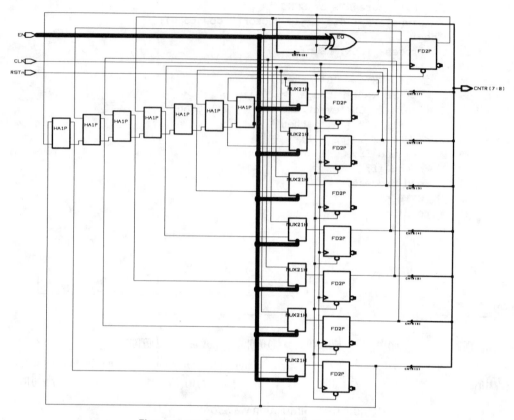

Figure 6-11 Synthesized schematic for SCNT.

The SCNT VHDL code can be easily extended for a bigger counter. In a popular 32-bit design, a 32-bit synchronous counter is often used. The following SCNT32 VHDL code shows a synchronous counter with parallel load PL input. An output signal EC to indicate the counter reaching the end of the count is added as shown in line 8.

```
1    library IEEE;
2    use IEEE.std_logic_1164.all;
3    use IEEE.std_logic_unsigned.all;
4    entity SCNT32 is
5      port (
6        CLK, RSTn, PL, EN : in  std_logic;
7        DIN      : in  std_logic_vector(31 downto 0);
8        EC       : out std_logic;
9        CNTR     : out std_logic_vector(31 downto 0));
10   end SCNT32;
```

The following architecture RTL1 parallel loads the counter with input data DIN when PL is asserted '1' as shown in lines 19 and 20. In many applications, counters will not count to the largest binary value the counter can represent. Lines 22 to 23 show that the counter only counts up to a particular value. When that value is reached, the counter wraps around to start with 0. The counter only counts when it is enabled by the input signal EN, as shown in line 21. Note also the parallel load has higher priority than the enable signal since line 21 is executed only when line 19 is evaluated false. Line 30 assigns the flip-flops output to the CNTR output port. The end of count is generated in lines 31 and 32. Note that the-end-of-count can only be asserted when the counter is enabled to ensure that the EC signal is only one clock cycle wide.

```
11    architecture RTL1 of SCNT32 is
12      signal FF : std_logic_vector(31 downto 0);
13    begin
14      process (CLK, RSTn)
15      begin
16        if (RSTn = '0') then
17          FF <= (FF'range => '0');
18        elsif (CLK'event and CLK = '1') then
19          if (PL = '1') then
20            FF <= DIN;
21          elsif (EN = '1') then
22            if (FF < "10001100111011111000110011101111") then
23              FF <= FF + 1;
24            else
25              FF <= (FF'range => '0');
26            end if;
27          end if;
28        end if;
29      end process;
30      CNTR <= FF;
31      EC   <= '1' when (FF = "10001100111011111000110011101111")
32                   and (EN = '1') else '0';
33    end RTL1;
```

The following architecture RTL2 is similar to RTL1, except for EC output which is registered (see lines 48 and 51).

```
34    architecture RTL2 of SCNT32 is
35      signal FF : std_logic_vector(31 downto 0);
36    begin
37      process (CLK, RSTn)
38      begin
39        if (RSTn = '0') then
40          FF <= (FF'range => '0');
41          EC <= '0';
42        elsif (CLK'event and CLK = '1') then
43          if (PL = '1') then
44            FF <= DIN;
```

```
45              EC <= '0';
46            elsif (EN = '1') then
47              if (FF = "10001100111011111000110011101111") then
48                EC <= '1';
49                FF <= (FF'range => '0');
50              else
51                EC <= '0';
52                FF <= FF + 1;
53              end if;
54            end if;
55          end if;
56        end process;
57        CNTR <= FF;
58      end RTL2;
```

To show the differences between architectures RTL1 and RTL2, the following test benches are used. Lines 29 to 30 initialize PL, EN, and DIN values. It then waits for 16 clock cycles. During this time, the counter is not enabled, so it should not change value. Lines 34 to 37 enable the counter by setting EN to '1' to count 16 times. Line 38 sets the PL to '1' to check that the parallel load has a higher priority than the EN input. Line 44 sets the DIN input value to be a little less than the end of count value. Line 52 sets the DIN input value to be a little larger than the end of count value. Lines 59 to 60 generate a signal DIFF to be used in the **assert statement** in line 66.

```
1     library IEEE;
2     use IEEE.std_logic_1164.all;
3     use IEEE.std_logic_unsigned.all;
4     entity TBSCNT32 is
5     end TBSCNT32;
6     architecture BEH of TBSCNT32 is
7       component SCNT32
8       port(
9         CLK, RSTn, PL, EN : in  std_logic;
10        DIN       : in  std_logic_vector(31 downto 0);
11        EC        : out std_logic;
12        CNTR      : out std_logic_vector(31 downto 0));
13      end component;
14      constant PERIOD2 : time := 50 ns;
15      constant DELAY   : time := 80 ns;
16      constant STROBE  : time := 2 * PERIOD2 - 5 ns;
17      signal CLK, RSTn, PL, EN : std_logic;
18      signal DIN, CNTR1, CNTR2 : std_logic_vector(31 downto 0);
19      signal EC1, EC2, DIFF    : std_logic;
20    begin
21      RSTn <= '0', '1' after 10 * PERIOD2;
22      clkgen : process
23      begin
24        CLK <= '0'; wait for PERIOD2;
25        CLK <= '1'; wait for PERIOD2;
```

```
26       end process;
27       dgen : process
28       begin
29         PL  <= '0'; EN <= '0';
30         DIN <= (DIN'range => '0'); DIN(1) <= '1';
31         for k in 0 to 15 loop
32           wait until (CLK'event) and (CLK = '1');
33         end loop;
34         EN  <= '1' after DELAY;
35         for k in 0 to 15 loop
36           wait until (CLK'event) and (CLK = '1');
37         end loop;
38         PL  <= '1' after DELAY;
39         wait until (CLK'event) and (CLK = '1');
40         PL  <= '0' after DELAY;
41         for k in 0 to 15 loop
42           wait until (CLK'event) and (CLK = '1');
43         end loop;
44         DIN <= "1000110011101111000110011101101" after DELAY;
45         PL  <= '1' after DELAY;
46         wait until (CLK'event) and (CLK = '1');
47         PL  <= '0' after DELAY;
48         for k in 0 to 15 loop
49           wait until (CLK'event) and (CLK = '1');
50         end loop;
51         PL  <= '1' after DELAY;
52         DIN <= "1000110011101111000110011111111" after DELAY;
53         wait until (CLK'event) and (CLK = '1');
54         PL  <= '0' after DELAY;
55         for k in 0 to 15 loop
56           wait until (CLK'event) and (CLK = '1');
57         end loop;
58       end process;
59       DIFF <= '0' when (EC1 = EC2) and (CNTR1 = CNTR2)
60                   else '1';
61       check : process
62       begin
63         loop
64           wait until (CLK'event) and (CLK = '1');
65           wait for STROBE;
66           assert (DIFF = '0') report "Not equal" severity NOTE;
67         end loop;
68       end process;
69       scnt320 : SCNT32
70       port map (
71         CLK => CLK, RSTn => RSTn, PL => PL, EN => EN,
72         DIN => DIN, EC => EC1, CNTR => CNTR1);
73       scnt321 : SCNT32
74       port map (
75         CLK => CLK, RSTn => RSTn, PL => PL, EN => EN,
```

```
76          DIN => DIN, EC => EC2, CNTR => CNTR2);
77      end BEH;
78      configuration CFG_TBSCNT32 of TBSCNT32 is
79        for BEH
80          for scnt320 : SCNT32
81            use entity work.SCNT32(RTL1);
82          end for;
83          for scnt321 : SCNT32
84            use entity work.SCNT32(RTL2);
85          end for;
86        end for;
87      end CFG_TBSCNT32;
```

Figures 6-12 and 6-13 show part of the simulation waveform. In Figure 6-12, after the counter is loaded with a value a little smaller than the end-of-count value, the counter reaches its end-of-count value in just another two clock cycles. Note that EC1 is generated one clock cycle before EC2. In Figure 6-13, when the counter is parallel loaded with a value larger than the end of count value, RTL1 wraps around to 0 and counts. RTL2 does not wrap around to 0 until every bit of the counter is '1'. In practice, caution should be taken to design the counter precisely as required.

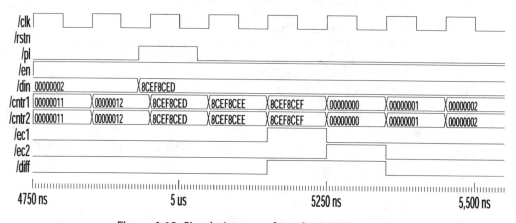

Figure 6-12 Simulation waveform for SCNT32, part 1.

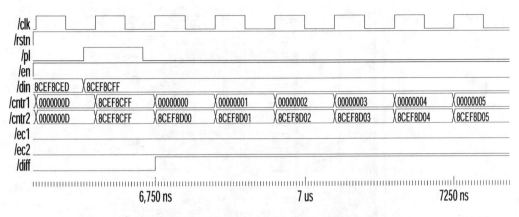

Figure 6-13 Simulation waveform for SCNT32, part 2.

Both architectures RTL1 and RTL2 can be synthesized. The following is a simple synthesis script to synthesize architecture RTL1. Figure 6-14 shows the synthesized schematic. Note that there are two DesignWare components. When DesignWare components are used, Synopsys timing analyzer automatically selects a wire load model for the DesignWare component based on its size. The wire load model selected is usually smaller than when we have specified. For example, the design uses B9X9. The wire load model selected for the DesignWare component is B1X1.

```
1     design = scnt32
2     analyze -f vhdl VHDLSRC + "ch6/" + design + ".vhd"
3     elaborate SCNT32 -architecture RTL1
4     include compile.common
5     create_clock "CLK" -name clk -period CLKPERIOD -waveform {0.0
HALFCLK}
6     set_clock_skew -uncertainty CLKSKEW clk
7     set_input_delay  INPDELAY -add_delay -clock clk all_inputs()
8     set_input_delay  CLKPER3  -add_delay -clock clk {PL EN}
9     set_output_delay CLKPER6  -add_delay -clock clk all_outputs()
10    set_dont_touch_network {CLK RSTn}
11    compile -map_effort medium
12    include compile.report
```

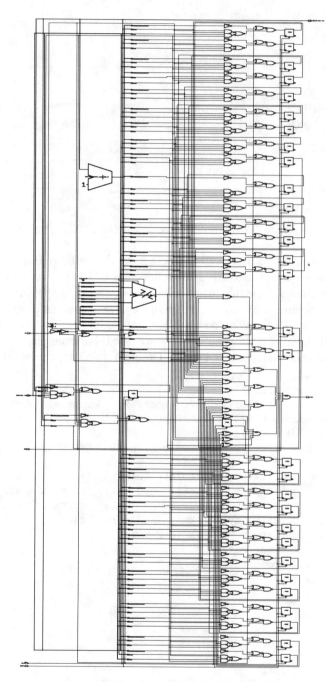

Figure 6-14 Synthesized schematic for SCNT32 RTL1.

The above schematic satisfies the timing constraints we specified. However, if we ungroup the DesignWare components so that the same wire load model is used, the schematic violates the timing constraints required. The clock period of 30 ns is not enough. The following shows the timing and area summary.

```
1    Number of ports:              69
2    Number of nets:              223
3    Number of cells:             150
4    Combinational area:       402.000000
5    Noncombinational area:    256.000000
6    Net Interconnect area:   1837.715454
7    Total cell area:          658.000000
8    Total area:              2495.715332
9
10   Operating Conditions: WCCOM   Library: lca300k
11   ----------------------------------------------------
12   SCNT32                  B9X9          lca300k
13     Point                        Incr      Path
14   ----------------------------------------------------
15     clock clk (rise edge)        0.00      0.00
16     clock network delay (ideal)  0.00      0.00
17     FF_reg[0]/CP (FD2)           0.00      0.00 r
18     FF_reg[0]/Q (FD2)            3.00      3.00 r
19     add_23/plus/plus/U16/CO (HA1)  1.03    4.03 r
20     add_23/plus/plus/U17/Z (IVP)   0.35    4.38 f
21     add_23/plus/plus/U76/Z (NR2)   0.54    4.92 r
22     add_23/plus/plus/U15/CO (HA1)  0.98    5.90 r
23     add_23/plus/plus/U18/Z (IVP)   0.30    6.20 f
24     add_23/plus/plus/U38/Z (NR2)   0.53    6.73 r
25     add_23/plus/plus/U14/CO (HA1)  0.98    7.72 r
26     add_23/plus/plus/U19/Z (IVP)   0.30    8.02 f
27     add_23/plus/plus/U40/Z (NR2)   0.53    8.55 r
28     add_23/plus/plus/U13/CO (HA1)  0.98    9.53 r
29     add_23/plus/plus/U20/Z (IVP)   0.30    9.83 f
30     add_23/plus/plus/U41/Z (NR2)   0.53   10.37 r
31     add_23/plus/plus/U12/CO (HA1)  0.98   11.35 r
32     add_23/plus/plus/U21/Z (IVP)   0.30   11.65 f
33     add_23/plus/plus/U43/Z (NR2)   0.53   12.18 r
34     add_23/plus/plus/U11/CO (HA1)  0.98   13.16 r
35     add_23/plus/plus/U22/Z (IVP)   0.30   13.47 f
36     add_23/plus/plus/U45/Z (NR2)   0.53   14.00 r
37     add_23/plus/plus/U10/CO (HA1)  0.98   14.98 r
38     add_23/plus/plus/U23/Z (IVP)   0.30   15.28 f
39     add_23/plus/plus/U47/Z (NR2)   0.53   15.82 r
40     add_23/plus/plus/U9/CO (HA1)   0.98   16.80 r
41     add_23/plus/plus/U24/Z (IVP)   0.30   17.10 f
42     add_23/plus/plus/U49/Z (NR2)   0.53   17.63 r
43     add_23/plus/plus/U8/CO (HA1)   0.98   18.61 r
44     add_23/plus/plus/U25/Z (IVP)   0.30   18.91 f
```

```
45        add_23/plus/plus/U51/Z  (NR2)          0.53          19.45 r
46        add_23/plus/plus/U7/CO  (HA1)          0.98          20.43 r
47        add_23/plus/plus/U26/Z  (IVP)          0.30          20.73 f
48        add_23/plus/plus/U53/Z  (NR2)          0.53          21.27 r
49        add_23/plus/plus/U6/CO  (HA1)          0.98          22.25 r
50        add_23/plus/plus/U27/Z  (IVP)          0.30          22.55 f
51        add_23/plus/plus/U31/Z  (NR2)          0.53          23.08 r
52        add_23/plus/plus/U5/CO  (HA1)          0.98          24.06 r
53        add_23/plus/plus/U28/Z  (IVP)          0.30          24.36 f
54        add_23/plus/plus/U32/Z  (NR2)          0.53          24.90 r
55        add_23/plus/plus/U4/CO  (HA1)          0.98          25.88 r
56        add_23/plus/plus/U29/Z  (IVP)          0.30          26.18 f
57        add_23/plus/plus/U33/Z  (NR2)          0.53          26.71 r
58        add_23/plus/plus/U3/CO  (HA1)          0.98          27.70 r
59        add_23/plus/plus/U30/Z  (IVP)          0.30          28.00 f
60        add_23/plus/plus/U58/Z  (OR2)          0.69          28.69 f
61        add_23/plus/plus/U34/Z  (NR2)          0.52          29.21 r
62        add_23/plus/plus/U80/Z  (EO)           0.58          29.80 r
63        U141/Z  (IVP)                          0.19          29.99 f
64        U142/Z  (AO7)                          0.41          30.40 r
65        FF_reg[31]/D  (FD2)                    0.00          30.40 r
66        data arrival time                                    30.40
67
68        clock clk (rise edge)                 30.00          30.00
69        clock network delay (ideal)            0.00          30.00
70        clock uncertainty                     -1.00          29.00
71        FF_reg[31]/CP  (FD2)                    0.00          29.00 r
72        library setup time                    -0.52          28.48
73        data required time                                   28.48
74        ----------------------------------------------------------
75        data required time                                   28.48
76        data arrival time                                   -30.40
77        ----------------------------------------------------------
78        slack (VIOLATED)                                     -1.92
```

After the Figure 6-14 schematic is ungrouped, another iteration of synthesis can be performed by just using the "compile" command. The following data shows the area and timing summary. It satisfies the 30 ns clock period requirement.

```
1     Number of ports:              69
2     Number of nets:              308
3     Number of cells:             257
4     Combinational area:       390.000000
5     Noncombinational area:    288.000000
6     Net Interconnect area:   2196.349365
7     Total cell area:          678.000000
8     Total area:              2874.349365
9
10    Operating Conditions: WCCOM    Library: lca300k
11    SCNT32                  B9X9             lca300k
```

```
12       -----------------------------------------------------
13       clock clk (rise edge)           0.00      0.00
14       clock network delay (ideal)     0.00      0.00
15       FF_reg[2]/CP (FD2P)             0.00      0.00 r
16       FF_reg[2]/Q (FD2P)              2.23      2.23 r
17       U237/Z (ND4)                    2.23      4.46 f
18       U293/Z (NR2)                    1.39      5.85 r
19       U304/Z (ND3)                    1.59      7.44 f
20       U316/Z (OR3)                    1.32      8.76 f
21       U328/Z (NR2)                    1.06      9.83 r
22       U339/Z (ND3)                    1.62     11.45 f
23       U351/Z (IVP)                    0.28     11.73 r
24       U352/Z (ND3)                    1.14     12.87 f
25       U365/Z (NR2)                    1.30     14.17 r
26       U377/Z (AN3)                    1.31     15.49 r
27       U389/Z (ND3)                    1.21     16.70 f
28       U245/Z (IVP)                    0.34     17.04 r
29       U402/Z (ND3)                    0.81     17.85 f
30       U244/Z (IVP)                    0.36     18.22 r
31       U413/Z (ND3)                    1.06     19.27 f
32       U243/Z (IVP)                    0.34     19.61 r
33       U426/Z (ND3)                    0.81     20.42 f
34       U242/Z (IVP)                    0.36     20.79 r
35       U437/Z (ND3)                    1.06     21.85 f
36       U241/Z (AO6)                    1.18     23.02 r
37       U440/Z (MUX21L)                 0.57     23.59 f
38       U441/Z (OR2P)                   0.57     24.16 f
39       FF_reg[30]/D (FD2P)             0.00     24.16 f
40       data arrival time                        24.16
41
42       clock clk (rise edge)          30.00     30.00
43       clock network delay (ideal)     0.00     30.00
44       clock uncertainty              -1.00     29.00
45       FF_reg[30]/CP (FD2P)            0.00     29.00 r
46       library setup time             -0.29     28.71
47       data required time                       28.71
48       -----------------------------------------------------
49       data required time                       28.71
50       data arrival time                       -24.16
51       slack (MET)                               4.55
```

By going through the same synthesis process, RTL2 is synthesized. It has about the same timing as RTL1 in terms of the clock period timing requirements. The area report is shown below.

```
1       Number of ports:             69
2       Number of nets:             296
3       Number of cells:            250
4       Number of references:        19
5       Combinational area:     378.000000
```

```
6    Noncombinational area:       297.000000
7    Net Interconnect area:      2136.988525
8    Total cell area:             675.000000
9    Total area:                 2811.988525
```

Another major difference between RTL1 and RTL2 is the output timing delay of the EC outputs. In architecture RTL1, EC output comes from the combinational circuit and is affected by the EN input delay. In architecture RTL2, EC output is registered from the flip-flop output. However, it comes one clock cycle later than the EC output in architecture RTL1. In practice, there are usually two types of timing requirements: one is the timing to the D-input of the D-type flip-flops. The other is the output timing. The following are some command examples to analyze the timing requirements. There are many variations of the report_timing command. You are encouraged to explore the capabilities of the command in Exercise 6.6. After all, ensuring the design to satisfy timing requirements is a major task for practical designs.

```
1    report_timing -from PL
2    report_timing -from EN -to *reg*/D
3    report_timing -from *reg*/Q -to *reg*/D
4    report_timing -from EN -to EC
5    report_timing -to   *reg*/D
6    report_timing -to   all_outputs()
7    report_timing -from all_inputs()
```

A synchronous counter is an essential element for many digital design blocks. For example, a program counter register in a microprocessor is used to address the instruction in a memory so that the instruction can be fetched and executed. The program counter can be increased to point to the next instruction. For a branch instruction, the program counter can be parallel loaded with the subroutine address. A timer can also be based on a synchronous counter. The timing can periodically generate an interrupt. The EC output can serve this purpose. Also, the counter may be a down counter or an up/down counter. This is left as Exercises 6.9 and 6.10. A stack pointer is an address pointer for the last-in first-out data structure. The stack pointer can be implemented as an up/down counter when the data is pushed into or pulled from the stack.

6.3 SHIFT REGISTER

A *shift register* shifts data either left or right through a number of cascaded flip-flops. The following SHIFTR VHDL code shows a shift register with the shift right function through eight flip-flops. Line 13 and 14 asynchronously reset the flip-flops. Lines 15 and 16 shift right the flip-flops when the rising edge of the clock CLK occurs. Line 19 ties the SO output directly to FF8(0), the rightmost bit of the eight flip-flops. Figure 6-15 shows the schematic. Figure 6-16 shows the simulation waveform. It is clear to see that the SI waveform is delayed eight clock cycles to generate the SO waveform.

```
1     library IEEE;
2     use IEEE.std_logic_1164.all;
3     entity SHIFTR is
4       port (
5         CLK, RSTn, SI : in  std_logic;
6         SO            : out std_logic);
7     end SHIFTR;
8     architecture RTL of SHIFTR is
9       signal FF8 : std_logic_vector(7 downto 0);
10    begin
11      posedge : process (RSTn, CLK)
12      begin
13        if (RSTn = '0') then
14          FF8   <= (FF8'range => '0');
15        elsif (CLK'event and CLK = '1') then
16          FF8 <= SI & FF8(FF8'length-1 downto 1);
17        end if;
18      end process;
19      SO <= FF8(0);
20    end RTL;
```

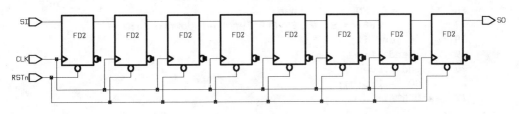

Figure 6-15 Synthesized schematic for SHIFTR.

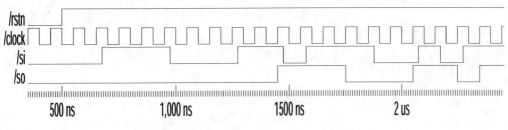

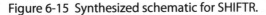

Figure 6-16 Simulation waveform for SHIFTR.

The SHIFTR VHDL code always shifts when the rising edge of the clock occurs. In practice, a shift register is usually controlled so that it shifts when it is enabled, otherwise, the shift register retains its states. The following SHIFTREN VHDL shows such an example. Input port EN is added in line 5. Lines 16 to 18 implement the enable function. Figure 6-17 shows the synthesized schematic. Note that there is a multiplexer in front of each flip-flop. The multiplexer uses signal EN to select either the previous stage of the data or the output of the current stage of the flip-flop. The verification of the SHIFTREN VHDL code is left as Exercise 6.11.

```
1      library IEEE;
2      use IEEE.std_logic_1164.all;
3      entity SHIFTREN is
4        port (
5          CLK, RSTn, SI, EN : in  std_logic;
6          SO                : out std_logic);
7      end SHIFTREN;
8      architecture RTL of SHIFTREN is
9        signal FF8 : std_logic_vector(7 downto 0);
10     begin
11       posedge : process (RSTn, CLK)
12       begin
13         if (RSTn = '0') then
14           FF8  <= (FF8'range => '0');
15         elsif (CLK'event and CLK = '1') then
16           if (EN = '1') then
17             FF8 <= SI & FF8(FF8'length-1 downto 1);
18           end if;
19         end if;
20       end process;
21       SO <= FF8(0);
22     end RTL;
```

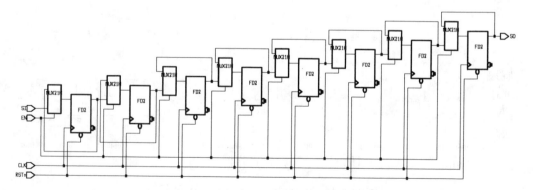

Figure 6-17 Synthesized schematic for SHIFTREN.

The above SHIFTR and SHIFTREN VHDL code shift the data input from SI signal. In practice, the shift register may be loaded with the data to be shifted before the shift operations are performed. This is referred to as the parallel load function. The following SHIFT VHDL code shows such an example. Note that input data is coming from DIN port. Input ports PL and LRn are added in line 5.

```
1    library IEEE;
2    use IEEE.std_logic_1164.all;
3    entity SHIFT is
4      port (
5        CLK, RSTn, SI, PL, EN, LRn : in  std_logic;
6        DIN     : in  std_logic_vector(7 downto 0);
7        SO      : out std_logic);
8    end SHIFT;
```

The following architecture RTL1 is similar to SHIFTREN. Lines 17 and 18 first check whether input PL is asserted. If PL is asserted, the flip-flops are parallel loaded with the DIN value. In line 19, if PL is not asserted, the EN signal is checked to see whether a shift operation should be performed. If EN is asserted, the shift register is either shift right (when LRn = '0') in line 21 or shift left (when LRn = '1') in line 23. Line 28 generates SO output. Figure 6-18 shows the synthesized schematic.

```
9    architecture RTL1 of SHIFT is
10     signal FF8 : std_logic_vector(DIN'length-1 downto 0);
11   begin
12     posedge : process (RSTn, CLK)
13     begin
14       if (RSTn = '0') then
15         FF8   <= (FF8'range => '0');
16       elsif (CLK'event and CLK = '1') then
17         if (PL = '1') then
18           FF8 <= DIN;
19         elsif (EN = '1') then
20           if (LRn = '0') then
21             FF8 <= SI & FF8(FF8'length-1 downto 1);
22           else
23             FF8 <= FF8(FF8'length-2 downto 0) & SI;
24           end if;
25         end if;
26       end if;
27     end process;
28     SO <= FF8(0) when LRn = '0' else FF8(FF8'length-1);
29   end RTL1;
```

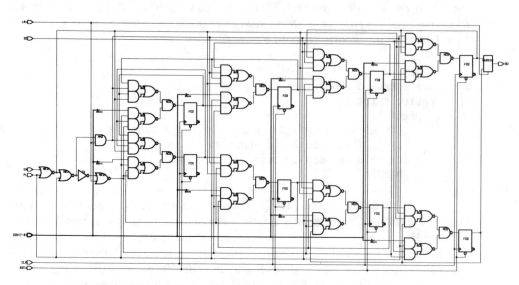

Figure 6-18 Synthesized schematic for SHIFT RTL1.

The following architecture RTL2 is similar to RTL1 except that output SO is registered, which directly connects to a flip-flop output. Figure 6-19 shows the synthesized schematic. Note that there are nine flip-flops in the Figure 6-19 schematic, while there are only eight flip-flops in the Figure 6-18 schematic.

```
30    architecture RTL2 of SHIFT is
31      signal FF8 : std_logic_vector(DIN'length-1 downto 0);
32    begin
33      posedge : process (RSTn, CLK)
34      begin
35        if (RSTn = '0') then
36          FF8  <= (FF8'range => '0');
37          SO   <= '0';
38        elsif (CLK'event and CLK = '1') then
39          if (PL = '1') then
40            FF8 <= DIN;
41            SO  <= '0';
42          elsif (EN = '1') then
43            if (LRn = '0') then
44              FF8 <= SI & FF8(FF8'length-1 downto 1);
45              SO  <= FF8(0);
46            else
47              FF8 <= FF8(FF8'length-2 downto 0) & SI;
48              SO  <= FF8(FF8'length-1);
```

```
49              end if;
50            end if;
51          end if;
52       end process;
53    end RTL2;
```

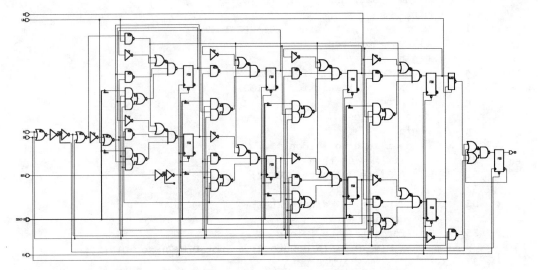

Figure 6-19 Synthesized schematic for SHIFT RTL2.

To show the functional differences between RTL1 and RTL2, the following test bench TBSHIFT is developed. Line 13 declares constant PERIOD2 as half of the clock cycle. Input signals are delayed with a DELAY constant of 80 ns. The data is strobed at 5 ns before the rising edge of the clock as STROBE is declared in line 15. The process **dgen** in lines 27 to 49 sets up values for SI, PL, LRn, and DIN. Note that EN is not assigned a value here. The EN signal can be set up with a simulator command. For example, in Mentor QuickVHDL, "force EN 1" can be used. In Synopsys VSS, "assign '1' EN" can be used. Lines 50 to 57 compare both architecture outputs SO1 and SO2, exactly 5 ns before the rising edge of the clock.

```
1     library IEEE;
2     use IEEE.std_logic_1164.all;
3     use IEEE.std_logic_unsigned.all;
4     entity TBSHIFT is
5     end TBSHIFT;
6     architecture BEH of TBSHIFT is
7       component SHIFT
8       port(
9         CLK, RSTn, SI, PL, EN, LRn : in  std_logic;
```

```
10          DIN    : in  std_logic_vector(7 downto 0);
11          SO     : out std_logic);
12      end component;
13      constant PERIOD2 : time := 50 ns;
14      constant DELAY   : time := 80 ns;
15      constant STROBE  : time := 2 * PERIOD2 - 5 ns;
16      signal CLK, RSTn, SI, PL, EN, LRn : std_logic;
17      signal DIN : std_logic_vector(7 downto 0) :=
18                      (others => '0');
19      signal  SO1, SO2    : std_logic;
20    begin
21      RSTn <= '0', '1' after 10 * PERIOD2;
22      clkgen : process
23      begin
24        CLK <= '0'; wait for PERIOD2;
25        CLK <= '1'; wait for PERIOD2;
26      end process;
27      dgen : process
28      begin
29        for i in 0 to 1 loop
30          if (i = 0) then
31            SI <= '0' after DELAY;
32          else
33            SI <= '1' after DELAY;
34          end if;
35          for j in 0 to 1 loop
36            DIN <= DIN(6 downto 0) & (not DIN(7) xor DIN(2))
37                    AFTER DELAY;
38            LRn <= DIN(7) after DELAY;
39            PL  <= '1' after DELAY;
40            wait until (CLK'event) and (CLK = '1');
41            LRn <= DIN(4) after DELAY;
42            PL  <= '0' after DELAY;
43            wait until (CLK'event) and (CLK = '1');
44            for k in 0 to 15 loop
45              wait until (CLK'event) and (CLK = '1');
46            end loop;
47          end loop;
48        end loop;
49      end process;
50      check : process
51      begin
52        loop
53          wait until (CLK'event) and (CLK = '1');
54          wait for STROBE;
55          assert (SO1 = SO2) report "Not compared" severity NOTE;
56        end loop;
57      end process;
58      shift0 : SHIFT
59      port map (
```

```
60        CLK  => CLK, RSTn => RSTn, SI => SI, PL => PL,
61         EN => EN, LRn => LRn, DIn => DIN, SO => SO1);
62      shift1 : SHIFT
63      port map (
64         CLK  => CLK, RSTn => RSTn, SI => SI, PL => PL,
65         EN => EN, LRn => LRn, DIn => DIN, SO => SO2);
66    end BEH;
67    configuration CFG_TBSHIFT of TBSHIFT is
68      for BEH
69        for shift0 : SHIFT
70          use entity work.SHIFT(RTL1);
71        end for;
72        for shift1 : SHIFT
73          use entity work.SHIFT(RTL2);
74        end for;
75      end for;
76    end CFG_TBSHIFT;
```

Figure 6-20 shows the simulation waveform for TBSHIFT. Note that SO2 is one clock cycle behind SO1. Also, SO1 has short pulses in two places that are referred to as signal glitches. The first glitch is enlarged in Figure 6-21. Why does signal SO1 have a glitch while signal SO2 has none?

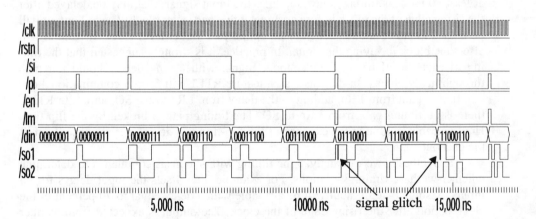

Figure 6-20 Simulation waveform for TBSHIFT.

Refer back to SHIFT RTL1 VHDL code. Output SO is a combinational output generated in line 28. This is confirmed by the Figure 6-18 schematic such that SO is connected to the output of a multiplexer. SO output is affected by the LRn input. In the test bench, LRn input has an 80-ns delay. SO may have a glitch of a 20-ns pulse. On the other hand, in RTL2, SO is a registered output such that SO is connected to a flip-flop output. Signal LRn will have some combinational circuits to reach the D input of the flip-flop. The SO output timing is not dependent on the timing of LRn.

The timing of LRn affects the timing to the D-input of the flip-flop. Since SO is registered output in RTL2, this is the reason that SO2 is one clock cycle behind SO1.

Figure 6-21 Simulation waveform for TBSHIFT, enlarged.

The timing delay of 80 ns for all input signals reflects a short pulse of 20 ns wide in signal SO1. If the delay is changed to 20 ns, the SO1 pulse will have pulses of at least 80 ns. This timing delay simulates the input signals that may be delayed after the rising edge of the clock before arriving at this particular block. The SO output will be passed to either another block or an output to a higher-level block. Its timing will affect the block receiving the signal. In practice, it is strongly suggested that the timing of each signal be well understood. Signal with a long delay should not pass through many blocks. In this case, architecture RTL1 will form a combinational circuit timing path from LRn, adding to the delay from LRn to the SO output. In RTL2, there is no timing path from LRn to SO. The timing path is broken by the flip-flop. However, this is done at the expense of the flip-flop and an extra cycle delay to shift the data out.

In synthesis, we can analyze the timing with synthesis commands to determine whether we have timing violations. For example, line 4 sets the input delays for all inputs. Assume that all input will be available with a delay equal to 70 percent of the clock period after the rising edge of the clock. The output is expected from another block equal to 40 percent of the clock period before the rising edge of the clock as shown in line 6.

```
1    sh date
2    include compile.common
3    create_clock "CLK" -name clk -period CLKPERIOD -waveform {0.0
HALFCLK}
4    set_clock_skew -uncertainty CLKSKEW clk
5    set_input_delay  CLKPER7 -add_delay -clock clk all_inputs()
6    set_output_delay CLKPER4 -add_delay -clock clk all_outputs()
```

```
7      set_input_delay  0 -add_delay -clock clk {CLK}
```

For the RTL2 schematic, the following timing report is generated when the "report_timing" command is used. It shows that the timing from input PL, with an input delay of 21 ns, reaching the D-input of a flip-flop in 29.66 ns in line 9. The clock skew is 1 ns as shown in line 14. The flip-flop requires a set up time of 0.29 ns as shown in line 16. The timing is 0.95 ns slower than what we expected as shown in line 22.

```
1      input external delay        21.00      21.00 f
2      PL (in)                      0.00       21.00 f
3      U83/Z (NR2)                  2.39       23.39 r
4      U73/Y (IVDA)                 1.02       24.41 f
5      U73/Z (IVDA)                 1.77       26.18 r
6      U84/Z (NR2)                  1.12       27.29 f
7      U86/Z (ND2)                  1.33       28.63 r
8      U75/Z (AO3)                  1.03       29.66 f
9      FF8_reg[0]/D (FD2)           0.00       29.66 f
10     data arrival time                       29.66
11
12     clock clk (rise edge)       30.00      30.00
13     clock network delay (ideal)  0.00       30.00
14     clock uncertainty           -1.00      29.00
15     FF8_reg[0]/CP (FD2)          0.00       29.00 r
16     library setup time          -0.29      28.71
17     data required time                      28.71
18     --------------------------------------------------
19     data required time                      28.71
20     data arrival time                      -29.66
21     --------------------------------------------------
22     slack (VIOLATED)                        -0.95
```

The same timing constraints are used for the RTL1 schematic. The following timing report is generated. It shows the delay of LRn is 21 ns in line 1. It arrives SO at time 22.70 ns in line 5. The clock skew is 1 ns as shown in line 9. The output delay is expected 12 ns before the rising edge of the clock as shown in line 10. The timing is 5.7 ns slower than what we wanted as shown in line 16.

```
1      input external delay        21.00      21.00 f
2      LRn (in)                     0.00       21.00 f
3      U76/Z (MUX21H)               1.70       22.70 f
4      SO (out)                     0.00       22.70 f
5      data arrival time                       22.70
6
7      clock clk (rise edge)       30.00      30.00
8      clock network delay (ideal)  0.00       30.00
9      clock uncertainty           -1.00      29.00
10     output external delay      -12.00      17.00
11     data required time                      17.00
```

```
12     --------------------------------------------------
13     data required time                        17.00
14     data arrival time                        -22.70
15     --------------------------------------------------
16     slack (VIOLATED)                          -5.70
```

Another form of the shift register is a linear feedback shift register (LFSR). The following VHDL code shows a 4-bit LFSR. The least significant bit is taken by exclusive NOR of the least and the most significant bit. Figure 6-22 shows the simulation waveform. Figure 6-23 shows the synthesized schematic.

```
1      library IEEE;
2      use IEEE.std_logic_1164.all;
3      entity LFSR is
4        port (
5          CLK, RSTn : in  std_logic;
6          CNT       : out std_logic_vector(3 downto 0));
7      end LFSR;
8      architecture RTL of LFSR is
9        signal FF : std_logic_vector(CNT'length-1 downto 0);
10     begin
11       posedge : process (RSTn, CLK)
12       begin
13         if (RSTn = '0') then
14           FF  <= (FF'range => '0');
15         elsif (CLK'event and CLK = '1') then
16           FF  <= FF(FF'length-2 downto 0) &
17                  not (FF(0) xor FF(FF'length-1));
18         end if;
19       end process;
20       CNT <= FF;
21     end RTL;
```

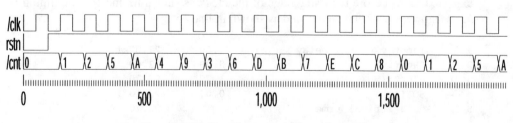

Figure 6-22 Simulation waveform for LFSR.

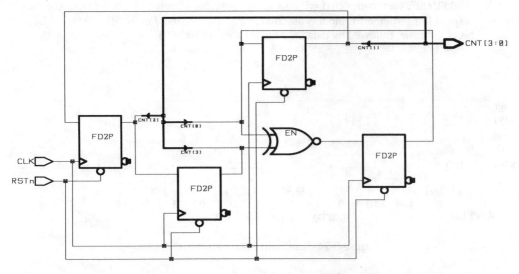

Figure 6-23 Synthesized schematic for LFSR.

Note that the LFSR output value is reset to 0. Its value sequence is 0, 1, 2, 5, A, 4, 9, 3, 6, D, B, 7, E, C, 8, and then the sequence repeats. The sequence is a pseudo-random order. Compared with the binary counter, the LFSR requires less circuit. A LFSR can be used as a counter in applications where the nonsequential nature of the output values is of no consequence. These include circular buffer address generators, FIFO read write pointers, and programmable frequency dividers. The LFSR can be run much faster compared to the same size of the binary counter. However, a LFSR does not traverse all possible binary values. Depending how the values feedback, the value sequence will not be the same. For example, if we feedback the exclusive NOR of the most significant bit and bit 1 to the least significant bit, the value sequence is 0, 1, 3, 6, C, 8, and then the sequence repeats. Only six values are traversed.

6.4 PARALLEL TO SERIAL CONVERTER

A *parallel to serial converter* is a special function of a shift register. The data is parallel loaded to the shift register, the data is then shifted out bit by bit and is bounded by a start bit and a stop bit. In digital designs, a parallel to serial converter is the essential circuit to transfer a data word out through a serial interface such as RS-232 or Universal Asynchronous Receiver and Transmitter (UART). The following P2S VHDL code shows a parallel to serial converter. The data to be transferred is first parallel loaded then transmitted by a start bit of value '1'. This is followed by the 8-bit data with the leftmost bit first; the even parity bit of the 8-bit data is used as the stop bit. The converter holds the output low when the transmission is completed. Figure 6-24 shows two examples of data transmission. When input signal PL is asserted '1', the data DIN "00111000" is parallel-loaded into the P2S circuit. In the next clock cycle, a start bit of '1' is output, followed by the data "00111000", then completed with an even parity bit of value '1'. After that, the output stays at low until the PL input is asserted again. The second data is "01110001". The start bit is the same, followed by data, and the even parity bit is '0'.

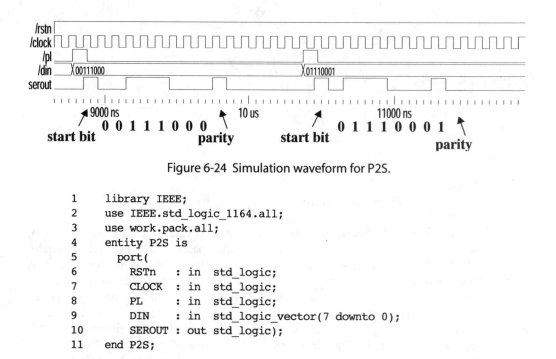

Figure 6-24 Simulation waveform for P2S.

```
1     library IEEE;
2     use IEEE.std_logic_1164.all;
3     use work.pack.all;
4     entity P2S is
5       port(
6         RSTn   : in  std_logic;
7         CLOCK  : in  std_logic;
8         PL     : in  std_logic;
9         DIN    : in  std_logic_vector(7 downto 0);
10        SEROUT : out std_logic);
11    end P2S;
```

The following architecture RTL declares signals DFF, START, and PARBIT in lines 13 and 14. Signal DFF is used to store the 8-bit input data. Signal START is used to generate the start bit. Signal PARBIT is used for the even parity bit. All DFF, START, and PARBIT signals infer flip-flops and form a 10-bit shift left shift register. These flip-flops are synchronously reset in lines 19 to 21. In lines 23 to 26, when PL is asserted '1', START is assigned to '1' and DFF is assigned to data input DIN. PAR-BIT is assigned the even parity of data input DIN using the REDUCE_XOR function in package pack. When PL is not asserted in lines 27 to 30, a left shift operation is performed. The rightmost bit of the 10-bit shift register PARBIT, is assigned '0' so that the output will stay at '0' when it is completed. Line 34 connects the output SEROUT to the output of the START flip-flop. Figure 6-25 shows the synthesized schematic. Note that there are 10 flip-flops. The START flip-flop is in the right end of the schematic. The PARBIT flip-flop is in the left end of the schematic with exclusive OR gates for the even parity generation.

```
12    architecture RTL of P2S is
13      signal DFF   : std_logic_vector(7 downto 0);
14      signal START, PARBIT : std_logic;
15    begin
16      p0 : process (RSTn, CLOCK)
17      begin
18        if (RSTn = '0') then
19          START  <= '0';
20          PARBIT <= '0';
21          DFF    <= (DFF'range => '0');
22        elsif (CLOCK'event and CLOCK = '1') then
23          if (PL = '1') then
24            START  <= '1';
25            DFF    <= DIN;
26            PARBIT <= REDUCE_XOR(DIN);
27          else
28            START  <= DFF(7);
29            DFF    <= DFF(6 downto 0) & PARBIT;
30            PARBIT <= '0';
31          end if;
32        end if;
33      end process;
34      SEROUT <= START;
35    end RTL;
```

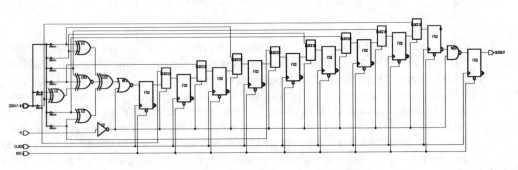

Figure 6-25 Synthesized schematic for P2S.

A more complex parallel to serial circuit may include different formats of the start bit and the stop bit. For example, there may be more than one start bit. Some designs use the "sync" sequence of bits, say "111000" 6 bits. The stop bit can be more than 1 bit. The stop bit can also be fixed at '0', '1', and even or odd parity. The P2S circuit transmits one data element at a time. A more complete design may transmit a number of words as a frame. The first data element transmitted may be the number of words to be transferred. It is followed by the data words. It may end with a check word which represents the total number of bits of value '1' in all the transmitted data.

The check word can be used to detect whether the transmission is wrong. Readers are encouraged to read other programmable communication interface designs such as Intel M8251A to develop a feeling of the flexibility and variations of the designs.

6.5 SERIAL TO PARALLEL CONVERTER

A *serial to parallel converter* is somewhat the reverse of the operation of a parallel to serial converter. The following S2P VHDL code shows a serial to parallel converter example. The data comes in serially from the input port SERIN. The parallel data is output from DOUT port. Output port DRDY is asserted '1' when the start bit, 8-bit data, and the parity bit are received. Output port PERRn is asserted '0' when the parity bit received is different from the parity bit generated inside the S2P circuit. When parity error is detected, the S2P circuit would be reset before its normal operation can be performed.

```
1     library IEEE;
2     use IEEE.std_logic_1164.all;
3     use IEEE.std_logic_unsigned.all;
4     use work.pack.all;
5     entity S2P is
6       port(
7         RSTn    : in  std_logic;
8         CLOCK   : in  std_logic;
9         SERIN   : in  std_logic;
10        PERRn   : out std_logic;
11        DRDY    : out std_logic;
12        DOUT    : out std_logic_vector(7 downto 0));
13    end S2P;
```

Figure 6-26 shows a simulation waveform for a transmission of data "10011111". The transmission begins with the start bit being asserted '1' in the SERIN input. Then, the SERIN input receives serial data '1', '0', '0', '1', '1', '1', '1', '1', followed by the even parity bit of value '0'. After the parity bit is received, output signal DRDY is asserted '1' in the next clock cycle. The DRDY signal may be used to tell another circuit block to get the parallel data from DOUT right away. Otherwise, the data may be lost when the next word comes. This allows DRDY and the start bit to be asserted at the same time as shown in Figure 6-28. Note that DOUT's value is changed right after DRDY is deasserted. The old data is shifted out bit by bit. Output PERRn is not asserted since the parity error is not detected.

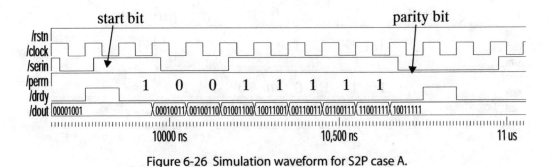

Figure 6-26 Simulation waveform for S2P case A.

Figure 6-27 shows another simulation waveform. The start bit is asserted not at the same time as the DRDY signal. Data "10111010" is received without parity error.

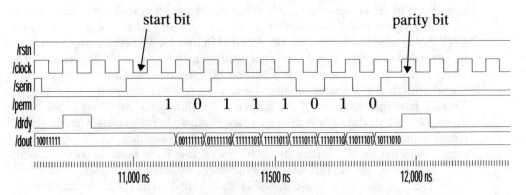

Figure 6-27 Simulation waveform for S2P case B.

Figure 6-28 shows another simulation waveform. Data "11100011" is transmitted. However, the input parity bit is not the same as the even parity bit generated inside the S2P circuit. The parity error signal PERRn is asserted '0'. The RSTn is asserted which initializes the S2P circuit. The DOUT output is synchronously reset to all '0's.

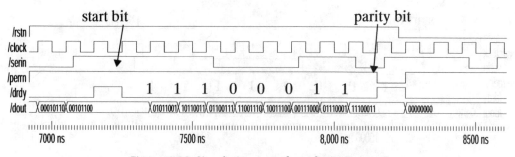

Figure 6-28 Simulation waveform for S2P case C.

To implement the S2P circuit, we need at least an 8-bit shift register to store the serial input data as shown in line 15, which declares DFF of 8 bits. We may also need a signal to tell the shift register when to shift left and when not to shift. SL_EN signal as declared in line 17 is used for this purpose. To be able to tell the shift register when to shift and when to stop shifting, a counter that can count from 0 to 7 is a reasonable selection. The serial input data needs to be shifted eight times. The counter also needs an enable signal to tell the counter when to count and when not to count. Signal CNT_EN is used for this purpose as declared in line 18. The even parity bit needs to be generated to compare with the input parity. The parity should be generated when all data are shifted in. Signal PAR_EN (line 18) is used to enable the parity checking. Signal PERRn_FF declared in line 18 is used to infer the parity error flip-flop. Another important function of the S2P circuit is to flag the condition of the circuit. Signal NORMAL (line 17) is used as a flag to remember the status whether the S2P circuit is in the normal condition or a parity error has been detected. When the parity error is detected, the S2P circuit should be reset before it can function normally again.

In this example, all flip-flops are synchronously set or reset as shown in lines 22 to 30. The 3-bit counter is implemented in lines 38 to 40. The counter is enabled when (NORMAL = '1') and (PAR_EN = '0') and (CNT7_FF = '0') and (PERRn_FF = '1') and (the start bit SERIN = '1') and the counter is "000" or the counter is not "000" as shown in lines 50 to 53. The shift register is enabled when the counter is valued from 1 to 7, and the cycle followed by the counter is valued at 7 as shown in line 49. Signal CNT7_FF is used to infer a flip-flop that is one clock delay when the counter value is seven as shown in line 45. Line 46 delays the CNT7_FF as the PAR_EN signal. Lines 35 to 37 set the NORMAL flag to '0' when the parity error is detected. Signal NORMAL will not be set to '1' unless the S2P circuit is reset. When signal NORMAL is '0', the counter will not be enabled. The circuit cannot function. Lines 54 and 55 connect the output directly to flip-flop outputs.

```
14    architecture RTL of S2P is
15      signal DFF   : std_logic_vector(7 downto 0);
16      signal CNT   : std_logic_vector(2 downto 0);
17      signal CNT7_FF, SL_EN, NORMAL : std_logic;
```

```
18       signal PERRn_FF, PAR_EN, CNT_EN : std_logic;
19     begin
20       regs : process
21       begin
22         wait until (CLOCK'event and CLOCK = '1');
23         if (RSTn = '0') then
24           DFF        <= (DFF'range => '0');
25           NORMAL     <= '1';
26           PERRn_FF   <= '1';
27           DRDY       <= '0';
28           PAR_EN     <= '0';
29           CNT7_FF    <= '0';
30           CNT        <= "000";
31         else
32           if (SL_EN = '1') then
33             DFF <= DFF(6 downto 0) & SERIN;
34           end if;
35           if (PERRn_FF = '0') then
36             NORMAL <= '0';
37           end if;
38           if (CNT_EN = '1') then
39             CNT <= CNT + 1;
40           end if;
41           if (PAR_EN = '1') then
42             PERRn_FF <= not (REDUCE_XOR(DFF) xor SERIN);
43           end if;
44           DRDY       <= PAR_EN;
45           CNT7_FF    <= CNT(2) and CNT(1) and CNT(0);
46           PAR_EN     <= CNT7_FF;
47         end if;
48       end process;
49       SL_EN  <= NORMAL and (CNT7_FF or CNT(2) or CNT(1) or CNT(0));
50       CNT_EN <= '1' when (NORMAL = '1') and (PAR_EN = '0') and
51                          (CNT7_FF = '0') and (PERRn_FF = '1') and
52                          ((CNT = "000" and SERIN = '1') or
53                           (CNT /= "000")) else '0';
54       PERRn  <= PERRn_FF;
55       DOUT   <= DFF;
56     end RTL;
```

The following test bench is used to generate the simulation waveform shown in Figures 6-26, 6-27, and 6-28. Lines 33 to 39 generate a SERIN signal with a simple linear feedback function. The randomness of the SERIN input enables the parity error being tested. Also, the test bench assigns RSTn to '0' when PERRn is asserted '0' in lines 29 to 31.

```
1      library IEEE;
2      use IEEE.std_logic_1164.all;
3      entity TBS2P is
```

```
4    end TBS2P;
5    architecture BEH of TBS2P is
6      component S2P
7      port(
8        RSTn    : in  std_logic;
9        CLOCK   : in  std_logic;
10       SERIN   : in  std_logic;
11       PERRn   : out std_logic;
12       DRDY    : out std_logic;
13       DOUT    : out std_logic_vector(7 downto 0));
14     end component;
15     constant PERIOD2 : time := 50 ns;
16     constant DELAY   : time := 25 ns;
17     signal  RSTn   : std_logic := '0';
18     signal  CLOCK  : std_logic;
19     signal  SERIN  : std_logic := '0';
20     signal  PERRn  : std_logic;
21     signal  DRDY   : std_logic;
22     signal  DOUT   : std_logic_vector(7 downto 0);
23   begin
24     clkgen : process
25     begin
26       CLOCK <= '0'; wait for PERIOD2;
27       CLOCK <= '1'; wait for PERIOD2;
28       RSTn  <= '1'  after 10 * PERIOD2;
29       if (PERRn = '0') then
30         RSTn <= '0' after DELAY;
31       end if;
32     end process;
33     dgen : process
34       variable RAND : std_logic_vector(4 downto 0) := "01001";
35     begin
36       wait until (CLOCK'event) and (CLOCK = '1');
37       RAND  := RAND(3 downto 0) & (RAND(4) xor RAND(2));
38       SERIN <= RAND(3) after DELAY;
39     end process;
40     s2p0 : S2P
41     port map (
42       RSTn    => RSTn,  CLOCK  => CLOCK, SERIN   => SERIN,
43       PERRn   => PERRn, DRDY   => DRDY,  DOUT    => DOUT);
44   end BEH;
45   configuration CFG_TBS2P of TBS2P is
46     for BEH
47     end for;
48   end CFG_TBS2P;
```

Note also that there are detailed timing relationships among signals in the S2P circuit, especially the enable signals with their relationship to the counter value. Figure 6-29 shows a simulation waveform example. The waveform shows that the counter has enabled eight clock cycles. CNT7_FF is asserted in the next cycle when the counter has a value of 7. The PAR_EN is asserted in the next cycle of CNT7_FF. The PERRn_FF is asserted in the next cycle of PAR_EN. The NORMAL signal is reset to '0' in the next cycle of PERRn_FF. In this simulation, RSTn is set to 'X' so that the signal values inside S2P are not changed since NORMAL flag is not reset to '1' to return to its normal operation. Figure 6-30 shows the synthesized schematic of S2P VHDL code.

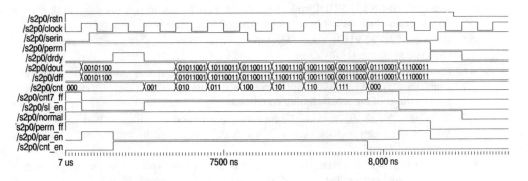

Figure 6-29 Simulation waveform for S2P internal signals.

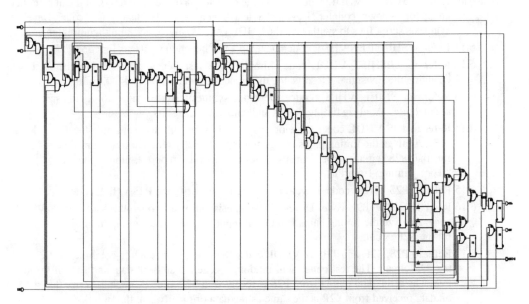

Figure 6-30 Synthesized schematic for S2P.

6.6 EXERCISES

6.1 Refer to EXTEND VHDL code. Develop a test bench to verify the EXTEND VHDL code.

6.2 Refer to EXTEND VHDL code. Develop a synthesis script file to synthesize the EXTEND VHDL code.

6.3 Refer to JKFF and ACNT VHDL code. Develop synthesis script files to synthesize VHDL codes to get the schematic shown in Figure 6-9 such that only D-type flip-flops are used. There are no extra combinational gates.

6.4 Refer to TBSCNT32 VHDL code. Change the VHDL code to perform more testing. How do you ensure that the test bench verifies SCNT32 functions completely?

6.5 Refer to SCNT32 VHDL code. Develop synthesis scripts to synthesize both architectures. Compare synthesis results.

6.6 Read the on-line help on the "report_timing" command. Develop timing analysis commands as a synthesis script to report the timing into a file. Try to use various combinations of the report_timing command.

6.7 Refer to SCNT32 VHDL code. Reduce the clock period to synthesize both architectures. What is the shortest clock period in which the design can be synthesized without any timing violations?

6.8 Refer to SCNT32 VHDL code. Use the carry look ahead concept to develop a synchronous counter that can run faster with a smaller clock period. Synthesize your VHDL code. Compare with the results you get by using the DesignWare components.

6.9 Refer to SCNT32 VHDL code. It is an up counter. Modify the code to make a down counter. Develop synthesis script to synthesize the VHDL code. Analyze the timing requirements.

6.10 Refer to SCNT32 VHDL code. Modify the code to make the design so that it can be either an up or a down counter. Develop a test bench to verify the VHDL code. Develop synthesis script files to synthesize the VHDL code. Analyze the timing requirements.

6.11 Refer to SHIFTREN VHDL code. Develop a test bench to verify its function.

6.12 Refer to TBSHIFT VHDL code. Simulate the test bench by changing the DELAY, changing the EN signal. Observe the differences between SO1 and SO2.

6.13 Refer to SHIFT VHDL code. Develop synthesis script files to synthesize both architectures. Analyze timing delays and compare the results.

6.14 Refer to P2S VHDL code. Develop a synthesis script file to synthesize the P2S VHDL code. Analyze the timing paths from each input to determine the minimum time required for the P2S circuit. You can use the commands "report_timing -from DIN", and "report_timing -from PL".

6.15 Refer to P2S VHDL code. Develop a test bench to verify the P2S VHDL code.

6.16 Refer to S2P VHDL code. Develop a synthesis script file to synthesize the S2P VHDL code. Analyze the timing paths from each input to determine the minimum time required for the S2P circuit.

6.17 Refer to P2S, S2P, and TBS2P VHDL code. Develop a test bench to combine the S2P and the P2S circuit. The test bench will randomly generate an 8-bit data and send it through P2S. The serial data output from P2S is received in S2P as an 8-bit data. Verify that the 8-bit data received from S2P is the same as the data originally sent through P2S.

Chapter **7**

Registers

Registers are sequential circuits which remember the states of the hardware. There are many types of registers for various applications. For example, registers for handling interrupts, addresses, counters, for controlling other circuits, and microprocessor registers. From users point of view, registers can be read-only, and both readable and writable. General-purpose microprocessor registers can be accessed with assembly language code. Some registers inside the microprocessor are not accessible by users. Registers can be implemented with latches, flip-flops, or memory arrays. In this chapter, we will use flip-flops to implement registers, since edge-triggered flip-flops are the backbone of synchronous digital designs. Several types of registers are illustrated, simulated, and synthesized. These register blocks are then connected as a higher-level block. We will also discuss how higher-level blocks can be synthesized and analyzed.

7.1 GENERAL FRAMEWORK FOR DESIGNING REGISTERS

Some general registers are set aside that can be written or read by users. Therefore, a register must have an associated address so that the register can be uniquely identified. As a flip-flop, latch, or memory array, the register can be written from the D input and read from the Q output at the same time. In this chapter we assume the memory map of the registers in Figure 7-1. Here, only 6 bits of address are used to identify each register. The register block has three interrupt registers, four direct memory access (DMA) related registers, and 16 configuration registers. These registers will be discussed in more detail in the following sections. Note that there are gaps between registers. In other words, not all addresses are specified. In practice, registers are grouped with gaps in between. These gaps provide the flexibility to add or to delete registers without greatly affecting other registers during the design process. For example, if all addresses are in sequence, allowing no gaps, adding a register will automatically push those that follow it down registers down, thus changing more addresses. Grouping related registers together allows related registers to be read or written as a single read or write transaction, which improves bus performance. However, gaps present a design issue. What would happen if an address in the gap is to be written? In general, this should not affect registers specified by other addresses. What would happen if the address in the gap is read? Two general approaches are usually taken. The first approach is to return the other register's value. The second approach is to return all zeroes.

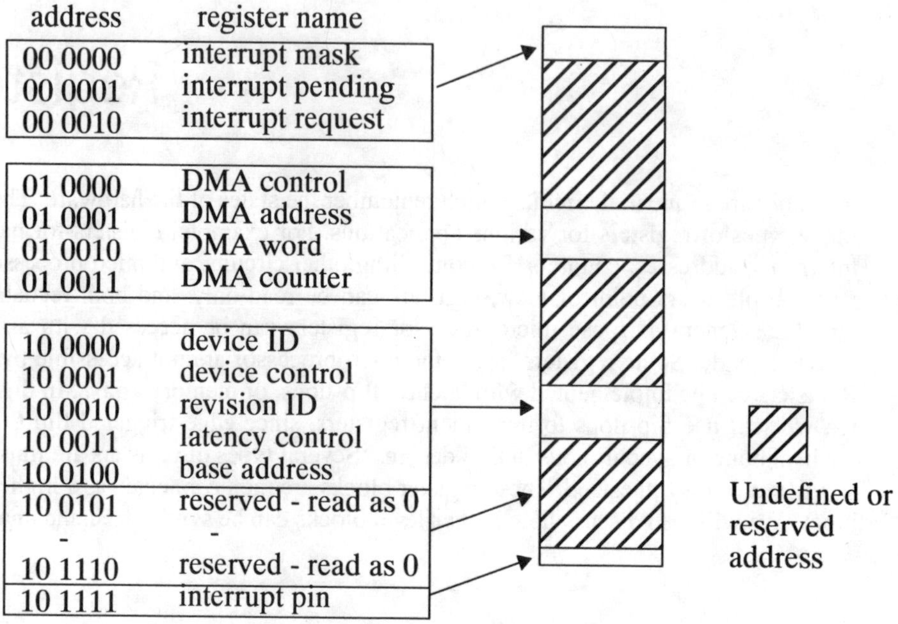

Figure 7-1 Registers memory map.

A general register block organization can be illustrated in Figure 7-2. In the center, there are registers implemented with flip-flops. The D-input of the flip-flops are controlled by the "write control" circuit. The "write control" circuit selects which register to be written according to the write address input. The update of the registers is controlled by a write enable input. There are cases when some registers are not implemented inside the register block. They are implemented in other blocks. These register flip-flop outputs are brought in to the register block. They go directly to the "register selection" block. They may be combined with other registers to form a full 32-bit word register. The register selection block uses the read address to select a particular register value or a fixed value (e.g., all zeroes, when the read address does not specify a register) to be used by other blocks or to be read outside of the design. Note that some registers may go directly out of the register block to be used by other blocks.

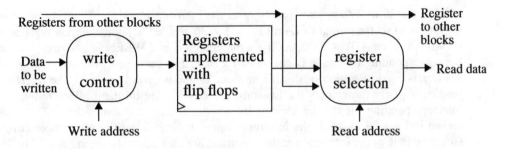

Figure 7-2 General register block organization.

7.2 INTERRUPT REGISTERS

*Interrupt register*s are registers that store the interrupt sources, and control how interrupts can be generated. There are three types of registers: interrupt mask register, interrupt pending register, and interrupt request register that can be defined. The following VHDL code shows how these three registers can be used to control the interrupt. Lines 5 and 6 declare reset and clock inputs RSTn0 and CLK0. Line 7 declares the 6-bit address input ADDR. Line 8 declares a low-active 4-bit input signal Command Byte Enable (CBEn). Each bit of CBEn controls a byte lane. Data to be written to the register comes from input ADi as specified in line 9. Lines 11 to 14 are interrupt sources to indicate situations when FIFO is overflow, DMA is done, timers are time out, or external interrupts. Output INTAn is a low-active output to indicate an interrupt is generated. Lines 16 to 18 are three interrupt registers. Each has 10 bits. These outputs are routed to the multiplexer circuit to provide values when registers are read.

```
1    library IEEE;
2    use IEEE.std_logic_1164.all;
3    entity ITREG is
4      port (
5        RSTn0       : in  std_logic;
6        CLK0        : in  std_logic;
7        ADDR        : in  std_logic_vector( 5 downto 0);
8        CBEn        : in  std_logic_vector( 3 downto 0);
9        ADi         : in  std_logic_vector(31 downto 0);
10       WREN        : in  std_logic;
11       FIFO_OVF    : in  std_logic;
12       DMA_DONE    : in  std_logic;
13       TIMER_INT   : in  std_logic_vector( 3 downto 0);
14       INTn        : in  std_logic_vector( 3 downto 0);
15       INTAn       : out std_logic;
16       INT_REQ     : out std_logic_vector( 9 downto 0);
```

```
17          INT_MASK    : out std_logic_vector( 9 downto 0);
18          INT_PEND    : out std_logic_vector( 9 downto 0));
19      end ITREG;
```

Lines 26 to 29 asynchronously reset the interrupt mask and interrupt pending registers and set the output INTAn. Lines 31 to 36 enable the interrupt mask register to be written when ADDR = "000000", write enable signal WREN is asserted '1', and the corresponding byte lane is enabled by CBEn. The purpose of the *interrupt mask register* is used to enable or disable any one of the interrupt sources. The *interrupt pending register* remembers the interrupt sources that are in queue. Each individual interrupt pending bit can be cleared by writing the corresponding bit with '1' as shown in lines 39 to 44. If the interrupt request register is written, the whole corresponding byte of the interrupt pending registers are cleared as shown in lines 37 to 38. If neither the interrupt request register nor the interrupt pending register is written, the interrupt pending register remembers its states and records the incoming interrupts with lines 46 and 47. Lines 49 to 60 repeat the same procedure for bits 8 and 9.

```
20      architecture RTL of ITREG is
21        signal REQ,MASKFF,PENDFF : std_logic_vector(9 downto 0);
22      begin
23        fgen : process (RSTn0, CLK0)
24          variable INT_OR : std_logic;
25        begin
26          if (RSTn0 = '0') then
27            MASKFF <= (MASKFF'range => '0');
28            PENDFF <= (PENDFF'range => '0');
29            INTAn  <= '1';
30          elsif (CLK0'event and CLK0 = '1') then
31            if (WREN='1' and ADDR="000000" and CBEn(0)='0') then
32              MASKFF(7 downto 0) <= ADi(7 downto 0);
33            end if;
34            if (WREN='1' and ADDR="000000" and CBEn(1)='0') then
35              MASKFF(9 downto 8) <= ADi(9 downto 8);
36            end if;
37            if (WREN='1' and ADDR="000010" and CBEn(0)='0') then
38              PENDFF(7 downto 0) <= "00000000";
39            elsif (WREN='1' and ADDR="000001" and CBEn(0)='0') then
40              for j in 0 to 7 loop
41                if (ADi(j) = '1') then
42                  PENDFF(j) <= '0';
43                end if;
44              end loop;
45            else
46              PENDFF(7 downto 0) <= (TIMER_INT & (not INTn)) or
47                                    PENDFF(7 downto 0);
48            end if;
49            if (WREN='1' and ADDR="000010" and CBEn(1)='0') then
50              PENDFF(9 downto 8) <= "00";
51            elsif (WREN='1' and ADDR="000001" and CBEn(1)='0') then
```

```
52                for j in 8 to 9 loop
53                   if (ADi(j) = '1') then
54                      PENDFF(j) <= '0';
55                   end if;
56                end loop;
57             else
58                PENDFF(9 downto 8) <= (FIFO_OVF & DMA_DONE) or
59                                        PENDFF(9 downto 8);
60             end if;
61             INT_OR := REQ(0);
62             for i in 1 to REQ'length-1 loop
63                INT_OR := INT_OR or REQ(i);
64             end loop;
65             INTAn <= not INT_OR;
66          end if;
67       end process;
68       INT_MASK    <= MASKFF;
69       INT_PEND    <= PENDFF;
70       REQ         <= PENDFF and MASKFF;
71       INT_REQ     <= REQ;
72    end RTL;
```

The *interrupt request register* is the product of a logical AND of the interrupt mask register and the interrupt pending register as shown in line 70. Note that the interrupt request register is not implemented with flip-flops. Lines 61 to 65 generate interrupt output INTAn. INTAn is asserted whenever there is a '1' in the interrupt request register. Lines 68, 69, and 71 connect the flip-flops to outputs. The following synthesis script can be used to synthesize the VHDL code. Note that lines 8 and 9 set the input delay for late arrival signals ADi, CBEn, DMA_DONE, and FIFO_OVF. The verification and synthesis results will be analyzed later. Figure 7-3 shows the synthesized schematic.

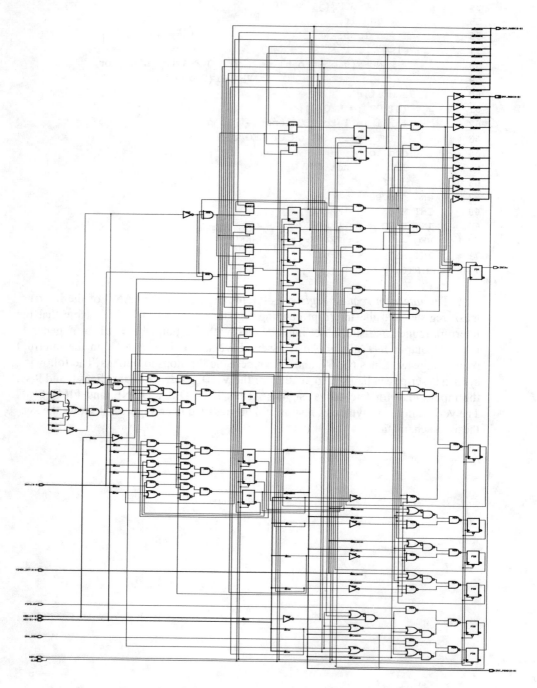

Figure 7-3 Synthesized schematic for ITREG VHDL code.

```
1    design = itreg
2    analyze -f vhdl VHDLSRC + "ch7/" + design + ".vhd"
3    elaborate ITREG -architecture RTL
4    include compile.common
5    create_clock "CLK0" -name clk -period CLKPERIOD -waveform {0.0
HALFCLK}
6    set_clock_skew -uncertainty CLKSKEW clk
7    set_input_delay  INPDELAY -add_delay -clock clk all_inputs()
8    set_input_delay  CLKPER8  -add_delay -clock clk {ADi CBEn}
9    set_input_delay  CLKPER9  -add_delay -clock clk {DMA_DONE
FIFO_OVF}
10   set_output_delay CLKPER6  -add_delay -clock clk all_outputs()
11   set_dont_touch_network {CLK0 RSTn0}
12   compile -map_effort medium
13   include compile.report
```

7.3 DMA AND CONTROL REGISTERS

A direct memory access (DMA) is a design that can transfer data without the CPU's control. To set up the DMA to work, a starting address and a word count are usually given. The DMA controller will start to transfer the data to the starting address. The address is automatically incremented to transfer the next data. This process is repeated until all the number of words are transferred. The following VHDL code shows examples of DMA and control registers. Lines 5 and 6 declare reset and clock signals. Lines 7 to 11 are the same as before in ITREG VHDL code. DMA_EN and FIFO_EN outputs control whether the DMA or the FIFO are enabled. Line 14 declares a low active output SOFT_RSTn so that the design can be *software reset*. Line 15 declares output DMAAD, which is a register for the DMA to use as the starting address for the DMA transfer. Line 16 specifies another output DMAWORD to tell the DMA controller the number of words to transfer. Outputs DMA_EN, DMAAD, and DMAWORD go to the DMA block that is not inside the register block.

```
1    library IEEE;
2    use IEEE.std_logic_1164.all;
3    entity SPREG is
4      port (
5        RSTn0, RSTn1 : in  std_logic;
6        CLK0, CLK1   : in  std_logic;
7        ADDR         : in  std_logic_vector( 5 downto 0);
8        CBEn         : in  std_logic_vector( 3 downto 0);
9        ADi          : in  std_logic_vector(31 downto 0);
10       WREN         : in  std_logic;
11       DMA_DONE     : in  std_logic;
12       DMA_EN       : out std_logic;
13       FIFO_EN      : out std_logic;
14       SOFT_RSTn    : out std_logic;
15       DMAAD        : out std_logic_vector(31 downto 2);
16       DMAWORD      : out std_logic_vector(31 downto 0));
17   end SPREG;
```

The first DMA and control register has bit 0 as the software reset bit. When '1' tries to write to that bit, Output SOFT_RSTn is asserted. This signal can then be combined in the reset control circuit that is not inside the register block. SOFT_RSTn is asynchronously set in line 23. The default for every clock is done in line 27. When it tries to write '1', SOFT_RSTn is asserted one clock cycle long. This bit should always read as '0' when the register is read. We will show how this can be done in the next section. DMA and control register bit 1 is the first in, first out (FIFO) enable bit. DMA_EN is asynchronously reset in line 24. It is set when the corresponding bit is written as shown in line 29. DMA_EN is automatically cleared when DMA is done, which is signaled by the situation when DMA_DONE is '1' as shown in lines 32 and 33. DMA_DONE is coming from the DMA controller circuit. The DMA control register has another 6-bit read-only register in bits (13 downto 8) as the FIFO word count to indicate the number of words inside the FIFO. The FIFO word count comes from the FIFO controller that already has the counter implemented as flip-flops. There is no need to have another set of flip-flops here. This FIFO word count read-only register will be shown in the next section.

```
18    architecture RTL of SPREG is
19    begin
20      fgen : process (RSTn0, CLK0)
21      begin
22        if (RSTn0 = '0') then
23          SOFT_RSTn <= '1';
24          DMA_EN    <= '0';
25          FIFO_EN   <= '0';
26        elsif (CLK0'event and CLK0 = '1') then
27          SOFT_RSTn <= '1';
28          if (WREN='1' and ADDR="010000" and CBEn(0)='0') then
29            DMA_EN    <= ADi(2);
30            FIFO_EN   <= ADi(1);
31            SOFT_RSTn <= not ADi(0);
32          elsif (DMA_DONE = '1') then
33            DMA_EN <= '0';
34          end if;
35        end if;
36      end process;
```

The following process implements the 30-bit DMA starting word address register. The two least significant bits are not used. They will be read as "00". This will be shown in the next section. Note that a **for loop** statement is used in lines 43 to 51. Note that lines 48 to 50 are independent of the loop identifier j. They should have been taken out of the loop statement. However, does that affect the synthesis results? This is left as Exercise 7.1.

```
37    dmasgen : process (RSTn0, CLK0)
38    begin
39      if (RSTn0 = '0') then
```

```
40              DMAAD <= (DMAAD'range => '0');
41          elsif (CLK0'event and CLK0 = '1') then
42            if (WREN = '1') and (ADDR = "010001") then
43              for j in 1 to 3 loop
44                if (CBEn(j) = '0') then
45                  DMAAD((j+1)*8-1 downto j*8) <=
46                    ADi((j+1)*8-1 downto j*8);
47                end if;
48                if (CBEn(0) = '0') then
49                  DMAAD(7 downto 2) <= Adi(7 downto 2);
50                end if;
51              end loop;
52            end if;
53          end if;
54        end process;
55      dmacntgen : process (RSTn1, CLK1)
56      begin
57          if (RSTn1 = '0') then
58            DMAWORD <= (DMAWORD'range => '0');
59          elsif (CLK1'event and CLK1 = '1') then
60            if (WREN = '1') and (ADDR = "010010") then
61              for j in 0 to 3 loop
62                if (CBEn(j) = '0') then
63                  DMAWORD((j+1)*8-1 downto j*8) <=
64                    ADi((j+1)*8-1 downto j*8);
65                end if;
66              end loop;
67            end if;
68          end if;
69        end process;
70      end RTL;
```

Lines 55 to 69 implement the DMA WORD register for DMA to know how many words to transfer before DMA_DONE is asserted. Note that line 60 checks the address "010010", which is not the same address as in lines 28 and 42. It is easy to cut and paste the VHDL code and forget to change the address. The DMA counter register is a read-only register that is not shown here. We assume that the DMA counter is implemented in the DMA controller which has the flip-flops for the counter. The output of the counter will be connected to the register selection multiplexer circuit to be discussed in the next section. The following synthesis script can be used to synthesized the VHDL code. Figure 7-4 shows part of the synthesized schematic. The bottom of the schematic shows three flip-flops for DMA_EN, FIFO_EN, and SOFT_RSTn, which are directly connected to outputs.

```
1     design = spreg
2     analyze -f vhdl VHDLSRC + "ch7/" + design + ".vhd"
3     elaborate SPREG -architecture RTL
4     include compile.common
5     create_clock "CLK*" -name clk -period CLKPERIOD -waveform {0.0
```

HALFCLK}

```
6    set_clock_skew -uncertainty CLKSKEW clk
7    set_input_delay  INPDELAY -add_delay -clock clk all_inputs()
8    set_input_delay  CLKPER8  -add_delay -clock clk {ADi CBEn}
9    set_input_delay  CLKPER9  -add_delay -clock clk {DMA_DONE}
10   set_output_delay CLKPER6  -add_delay -clock clk all_outputs()
11   set_dont_touch_network {CLK* RSTn*}
12   compile -map_effort medium
13   include compile.report
```

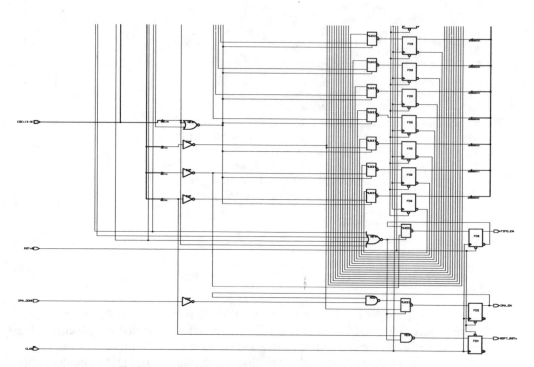

Figure 7-4 Synthesized schematic for SPREG VHDL code.

7.4 CONFIGURATION REGISTERS

This part of *configuration registers* definitions are taken from the Peripheral Component Interconnect (PCI) specification. For more detail on each register, please refer to the PCI specification. Here, we will use these registers as examples to show how various types of registers can be implemented. The configuration registers are shown in the following VHDL code. Note that this VHDL code implements configuration registers and also the multiplexer circuit. The read-only registers such as FIFOCNT (line 19), and DMACNTR (line 22) are brought in here as inputs. The multiplexer circuit will be discussed in the next section. Lines 5 to 9 are the same as in design SPREG. Line 10 declares a special signal

CFG_WRn, which is not the same as WREN, since configuration registers can only be updated in the configuration cycle. Line 11 declares an input signal M_ABORT that is set by the PCI protocol controller to indicate that the PCI master aborts the transaction. Line 12 declares an input signal T_ABORT, which is also coming from the PCI protocol controller to indicate when the PCI target aborts the current transaction. All other registers implemented in ITREG and SPREG are brought in here as shown in lines 14 to 23.

```
1      library IEEE;
2      use IEEE.std_logic_1164.all;
3      entity REGMUX is
4        port (
5          CLK0              : in  std_logic;
6          RSTn0             : in  std_logic;
7          ADDR              : in  std_logic_vector( 3 downto 0);
8          ADi               : in  std_logic_vector(31 downto 0);
9          CBEn              : in  std_logic_vector( 3 downto 0);
10         CFG_WRn           : in  std_logic;
11         M_ABORT           : in  std_logic;
12         T_ABORT           : in  std_logic;
13         BIST_CODE         : in  std_logic_vector( 4 downto 0);
14         INT_REQ           : in  std_logic_vector( 9 downto 0);
15         INT_MASK          : in  std_logic_vector( 9 downto 0);
16         INT_PEND          : in  std_logic_vector( 9 downto 0);
17         DMA_EN            : in  std_logic;
18         FIFO_EN           : in  std_logic;
19         FIFOCNT           : in  std_logic_vector( 5 downto 0);
20         DMAAD             : in  std_logic_vector(31 downto 2);
21         DMAWORD           : in  std_logic_vector(31 downto 0);
22         DMACNTR           : in  std_logic_vector(31 downto 0);
23         RDADDR            : in  std_logic_vector( 5 downto 0);
24         BIST_START        : out std_logic;
25         TARGET_B2B        : out std_logic;
26         MASTER_EN         : out std_logic;
27         RESPOND_EN        : out std_logic;
28         BASE_ADDR         : out std_logic_vector(11 downto 0);
29         LATENCY_PCI       : out std_logic_vector( 4 downto 0);
30         DOUT              : out std_logic_vector(31 downto 0));
31       end REGMUX;
```

Line 24 declares output signal BIST_START to start the built-in-self-test circuit. Lines 25 to 29 specify output signals to the PCI controller. TARGET_B2B indicates whether the PCI agent can generate fast back-to-back transactions. MASTER_EN enables the design to be a PCI bus master. RESPOND_EN tells the PCI protocol controller whether the design should respond or not to the PCI bus transactions. BASE_ADDR is the base address register as the design is seen from the PCI bus address map. The base address value is used for the address decoding circuit in the PCI protocol controller to determine whether the PCI bus address matched the base address so that the PCI protocol controller can respond accordingly. LATENCY_PCI tells the latency timer, in the PCI protocol controller, how long the design should continue for the transaction. If the latency timer times out, the design should try to get out of the bus as soon as possible. DOUT is the output of the multiplexer output that is selected from all registers using the read address RDADDR in line 23. Lines 33 to 44 define constants for the registers. These registers are said to be hardwired. There are no flip-flops for these registers. Therefore, their values cannot be changed.

```
32    architecture RTL of REGMUX is
33       constant ASIC_ID      : std_logic_vector(15 downto 0)
34                                  := "0101010100100100";
35       constant COMP_ID       : std_logic_vector(15 downto 0)
36                                  := "0001010101011100";
37       constant CLASS_CODE    : std_logic_vector(23 downto 0)
38                                  := "111111110000000000000000";
39       constant REVISION      : std_logic_vector(7 downto 0)
40                                  := "00000010";
41       constant BASEREG19_0   : std_logic_vector(19 downto 0)
42                                  := "00000000000000000000";
43       constant INT_PIN       : std_logic_vector(7 downto 0)
44                                  := "00000001";
45       signal DEVID_REG     : std_logic_vector(31 downto 0);
46       signal DEVCTL_REG    : std_logic_vector(31 downto 0);
47       signal REVID_REG     : std_logic_vector(31 downto 0);
48       signal LATENCY_REG   : std_logic_vector(31 downto 0);
49       signal BASE_REG      : std_logic_vector(31 downto 0);
50       signal BASEREG31_20: std_logic_vector(11 downto 0);
51       signal INT_PIN_REG   : std_logic_vector(31 downto 0);
52       signal MABORT_FF       : std_logic;
53       signal TABORT_FF       : std_logic;
54       signal TARGET_B2B_FF   : std_logic;
55       signal MASTER_EN_FF    : std_logic;
56       signal RESPOND_EN_FF   : std_logic;
57       signal LATENCY_FF      : std_logic_vector(4 downto 0);
58       signal INT_LINE_FF     : std_logic_vector(7 downto 0);
59       signal BIST_CODE_FF    : std_logic_vector(3 downto 0);
60       signal BIST_START_FF, BIST_START_DL  : std_logic;
61       signal AD3_0           : std_logic_vector(3 downto 0);
62       signal AD1_0, AD5_4    : std_logic_vector(1 downto 0);
63       signal ITOUT, MASK_REG, PEND_REG, REQ_REG, DMAOUT,
64       DMAAD_REG,CFGOUT,DMACTL_REG:std_logic_vector(31 downto 0);
```

Lines 45 to 64 declare signals. Note that lines 45 to 49, 51, 63 and 64 are full 32-bit-wide signals. They are used to form the full 32-bit-wide registers as seen from the user when they are read. These signals can be traced in the VHDL simulation waveform. This design approach has the advantage of fewer signals in the trace waveform window, and that unused bits can be seen together. Lines 69 to 70 implements the device ID and the revision ID registers for the design. They are constants. No flip-flops are needed. The device control register is implemented in lines 73 to 104. Note that lines 66 to 68 connect the flip-flops to outputs. The entire device control register is formed in lines 105 to 107. Note that the bit position for each individual register is defined in the PCI specification. To implement this type of register, caution should be taken so that the bit positions are exactly as shown in lines 84, 87, 88, 92, and 98, and lines 105 to 107. Common mistakes may be caused by writing with the correct bit positions, but reading the wrong bit positions, or vice versa. Lines 91 to 102 implement the master and target abort registers. They are cleared when '1' tries to be written, otherwise, the register clocks in the M_ABORT or T_ABORT from the inputs when its value is '0' (lines 94 and 100).

```
65    begin
66        TARGET_B2B   <= TARGET_B2B_FF;
67        MASTER_EN    <= MASTER_EN_FF;
68        RESPOND_EN   <= RESPOND_EN_FF;
69        DEVID_REG    <= ASIC_ID & COMP_ID;
70        REVID_REG    <= CLASS_CODE & REVISION;
71        BASE_ADDR    <= BASEREG31_20;
72        LATENCY_PCI  <= LATENCY_FF;
73        devctl_gen : process (RSTn0, CLK0)
74        begin
75          if (RSTn0 = '0') then
76            MABORT_FF        <= '0';
77            TABORT_FF        <= '0';
78            TARGET_B2B_FF  <= '0'; -- default no back to back
79            MASTER_EN_FF   <= '0'; -- default master not enabled
80            RESPOND_EN_FF  <= '0'; -- default no respond
81          elsif (CLK0'event and CLK0 = '1') then
82            if (CFG_WRn = '0') and (ADDR = "0001") then
83              if (CBEn(1) = '0') then
84                TARGET_B2B_FF  <= ADi(9);
85              end if;
86              if (CBEn(0) = '0') then
87                MASTER_EN_FF   <= ADi(2);
88                RESPOND_EN_FF  <= ADi(1);
89              end if;
90            end if;
91            if (CFG_WRn = '0') and (ADDR = "0001") and
92              (CBEn(3) = '0') and (ADi(29) = '1') then
93              MABORT_FF <= '0';
94            elsif (MABORT_FF = '0') then
95              MABORT_FF <= M_ABORT;
96            end if;
97            if (CFG_WRn = '0') and (ADDR = "0001") and
98              (CBEn(3) = '0') and (ADi(28) = '1') then
99              TABORT_FF <= '0'; -- clear by writing 1
100           elsif (TABORT_FF = '0') then -- when 0 clock in
101             TABORT_FF <= T_ABORT;
102           end if;
103         end if;
104       end process;
```

```
105     DEVCTL_REG <= "00" & MABORT_FF & TABORT_FF &
106              "001010000000" & "000000" & TARGET_B2B_FF &
107              "000000" & MASTER_EN_FF & RESPOND_EN_FF & '0';
```

Lines 109 to 142 implement the base address register, latency register, and the interrupt pin register. The base register is implemented with 12 flip-flops. The default is set in line 112 and written in lines 119 to 126. The latency register is implemented in lines 113 and 127 to 129. The bit 30 of the latency register is the BIST start register. When it is written as '1', it will stay as '1' until the BIST circuit completes the self-test. This is done with the scheme such that when the BIST start register is written '1', output signal BIST_START is asserted '1' in one clock cycle. When the BIST circuit receives the BIST_START signal as '1', it will start the self-check. The BIST circuit also provides 5 bits of the self-test status BIST_CODE. Bit 4 of the BIST_CODE is default to be '1'. When the BIST self-test is done, bit 4 is cleared to '0' by the BIST self-test circuit. Bits *3 downto 0* of the BIST_CODE are the self-test status to be remembered so that the user can read the status and take appropriate action. The BIST_START output can be implemented as the signal chopper as discussed before. The BIST_START_FF is delayed by one clock cycle in line 136 and chopped in line 108. Lines 138 to 140 implement the interrupt line register.

```
108     BIST_START     <= BIST_START_FF and (not BIST_START_DL);
109     basereg_gen : process (RSTn0, CLK0)
110     begin
111       if (RSTn0 = '0') then
112         BASEREG31_20   <= "111111000000";
113         LATENCY_FF     <= "00000";
114         INT_LINE_FF    <= "00000000";
115         BIST_START_FF <= '0';
116         BIST_START_DL <= '0';
117         BIST_CODE_FF  <= "0000";
118       elsif (CLK0'event and CLK0 = '1') then
119         if (CFG_WRn = '0') and ADDR = "0100" then
120           if (CBEn(3) = '0') then
121             BASEREG31_20(11 downto 4) <= ADi(31 downto 24);
122           end if;
123           if (CBEn(2) = '0') then
124             BASEREG31_20(3 downto 0) <= ADi(23 downto 20);
125           end if;
126         end if;
127         if (CFG_WRn='0' and ADDR="0011" and CBEn(1)='0') then
128           LATENCY_FF    <= ADi(15 downto 11);
129         end if;
130         if (CFG_WRn = '0') and (ADDR = "0011") and
131           (CBEn(3) = '0') and (ADi(30) = '1') then
132           BIST_START_FF   <= '1'; -- when write '1' start BIST
133         elsif (BIST_CODE(4) = '0') then
134           BIST_START_FF   <= '0'; -- clear when BIST done.
135         end if;
```

```
136           BIST_START_DL <= BIST_START_FF;
137           BIST_CODE_FF <= BIST_CODE(3 downto 0);
138           if (CFG_WRn='0' and ADDR="1111" and CBEn(0)='0') then
139             INT_LINE_FF  <= ADi(7 downto 0);
140           end if;
141         end if;
142      end process;
143      BASE_REG     <= BASEREG31_20 & BASEREG19_0;
144      LATENCY_REG <= '1' & BIST_START_FF & "00" & BIST_CODE_FF &
145                        "00000000" & LATENCY_FF & "00000000000";
146      INT_PIN_REG <= "0000000000000000" & INT_PIN & INT_LINE_FF;
```

Line 143 forms the base address register. The least significant 20 bits are set to '0'. Lines 144 to 145 form the full 32-bit latency register. Line 146 forms the interrupt pin register. Now, we have all the registers we wanted. The next section discusses the multiplexer circuit to select the register to be read.

7.5 READING REGISTERS

Lines 148 to 149 form the address to select various registers. Lines 150 to 162 use the least significant 4 bits of the address to select the configuration registers. Note that other reserved registers should be read as zeroes as shown in line 160. Lines 163 to 165 form the full 32-bit interrupt mask, interrupt pending, and interrupt request registers. The least significant 2 bits of the address are used to select the three registers as shown in lines 166 to 169. Note that the interrupt request register will be selected when AD1_0 is either "10" or "11".

```
147    ----------------------------------------------------------------
148    AD3_0  <= RDADDR(3 downto 0); AD1_0 <= RDADDR(1 downto 0);
149    AD5_4 <= RDADDR(5 downto 4);
150    cfggen : process (AD3_0, DEVID_REG, DEVCTL_REG, REVID_REG,
151                        LATENCY_REG, BASE_REG, INT_PIN_REG)
152    begin
153      case AD3_0 is
154        when "0000" => CFGOUT <= DEVID_REG;
155        when "0001" => CFGOUT <= DEVCTL_REG;
156        when "0010" => CFGOUT <= REVID_REG;
157        when "0011" => CFGOUT <= LATENCY_REG;
158        when "0100" => CFGOUT <= BASE_REG;
159        when "1111" => CFGOUT <= INT_PIN_REG;
160        when others => CFGOUT <= (others => '0');
161      end case;
162    end process;
163    MASK_REG <= "000000000000000000000" & INT_MASK;
164    PEND_REG <= "000000000000000000000" & INT_PEND;
165    REQ_REG  <= "000000000000000000000" & INT_REQ;
166    itmux : with AD1_0 select
```

```
167        ITOUT <= MASK_REG when "00",
168                 PEND_REG when "01",
169                 REQ_REG  when others;
170     DMAAD_REG  <= DMAAD & "00";
171     DMACTL_REG <= "000000000000000000" & FIFOCNT &
172                   "00000" & FIFO_EN & DMA_EN & '0';
173     dmamux : with AD1_0 select
174       DMAOUT <= DMACTL_REG when "00",
175                 DMAAD_REG  when "01",
176                 DMAWORD    when "10",
177                 DMACNTR    when others;
178     fmux : with AD5_4 select
179       DOUT <= ITOUT    when "00",
180               DMAOUT   when "01",
181               CFGOUT   when others;
182   end RTL;
```

Line 170 forms the DMA address register. Lines 171 to 172 form the DMA and control register. Note that the read-only register FIFOCNT is attached here. Lines 174 to 177 select DMA registers. The DMA counter DMACNTR is directly coming from the input. Lines 178 to 181 use bits 5 and 4 of the address to select the output of the register group. The following synthesis script can be used to synthesize the REGMUX VHDL code. Note that wildcard * can be used in line 5, 9, and 11. The synthesized schematic is not shown here due to its size.

```
1     design = regmux
2     analyze -f vhdl VHDLSRC + "ch7/" + design + ".vhd"
3     elaborate REGMUX -architecture RTL
4     include compile.common
5     create_clock "CLK*" -name clk -period CLKPERIOD -waveform {0.0
HALFCLK}
6     set_clock_skew -uncertainty CLKSKEW clk
7     set_input_delay   INPDELAY -add_delay -clock clk all_inputs()
8     set_input_delay   CLKPER8  -add_delay -clock clk {ADi CBEn}
9     set_input_delay   CLKPER9  -add_delay -clock clk {*_ABORT}
10    set_output_delay CLKPER6  -add_delay -clock clk all_outputs()
11    set_dont_touch_network {CLK* RSTn*}
12    compile -map_effort medium
13    include compile.report
```

7.6 REGISTER BLOCK PARTITIONING AND SYNTHESIS

The register blocks ITREG, SPREG, and REGMUX can be connected together with the following VHDL code. Writing VHDL for a higher level block using existing blocks can be time consuming. However, the entity declarations of the existing blocks can be easily modified to become the component declarations. These component declarations can also be used as templates for the component instantiation statements. Note that the component instantiation statements use the name mapping so that the code can be modified easier when ports are added, deleted, or changed. Internal signals are declared for interblock connections. Internal signals are also declared for the interconnects that go to outputs and between blocks so that "buffer" port type is not used. Note also that almost each port occupies a single line, which makes the code longer. However, it is easier to read and maintain the code.

```
1       library IEEE;
2       use IEEE.std_logic_1164.all;
3       entity REG is
4         port (
5           RSTn             : in  std_logic_vector( 3 downto 0);
6           CLK              : in  std_logic_vector( 3 downto 0);
7           WRADDR           : in  std_logic_vector( 5 downto 0);
8           CBEn             : in  std_logic_vector( 3 downto 0);
9           ADi              : in  std_logic_vector(31 downto 0);
10          WREN             : in  std_logic;
11          FIFO_OVF         : in  std_logic;
12          DMA_DONE         : in  std_logic;
13          TIMER_INT        : in  std_logic_vector( 3 downto 0);
14          INTn             : in  std_logic_vector( 3 downto 0);
15          INTAn            : out std_logic;
16          DMA_EN           : out std_logic;
17          FIFO_EN          : out std_logic;
18          SOFT_RSTn        : out std_logic;
19          DMAAD            : out std_logic_vector(31 downto 2);
20          DMAWORD          : out std_logic_vector(31 downto 0);
21          CFG_WRn          : in  std_logic;
22          M_ABORT          : in  std_logic;
23          T_ABORT          : in  std_logic;
24          BIST_CODE        : in  std_logic_vector( 4 downto 0);
25          FIFOCNT          : in  std_logic_vector( 5 downto 0);
26          DMACNTR          : in  std_logic_vector(31 downto 0);
27          RDADDR           : in  std_logic_vector( 5 downto 0);
28          BIST_START       : out std_logic;
29          TARGET_B2B       : out std_logic;
30          MASTER_EN        : out std_logic;
31          RESPOND_EN       : out std_logic;
32          BASE_ADDR        : out std_logic_vector(11 downto 0);
33          LATENCY_PCI      : out std_logic_vector( 4 downto 0);
34          DOUT             : out std_logic_vector(31 downto 0));
35        end REG;
36        architecture RTL of REG is
37          component ITREG
38          port (
39            RSTn0          : in  std_logic;
40            CLK0           : in  std_logic;
41            ADDR           : in  std_logic_vector( 5 downto 0);
```

```
42        CBEn          : in  std_logic_vector( 3 downto 0);
43        ADi           : in  std_logic_vector(31 downto 0);
44        WREN          : in  std_logic;
45        FIFO_OVF      : in  std_logic;
46        DMA_DONE      : in  std_logic;
47        TIMER_INT     : in  std_logic_vector( 3 downto 0);
48        INTn          : in  std_logic_vector( 3 downto 0);
49        INTAn         : out std_logic;
50        INT_REQ       : out std_logic_vector( 9 downto 0);
51        INT_MASK      : out std_logic_vector( 9 downto 0);
52        INT_PEND      : out std_logic_vector( 9 downto 0));
53    end component;
54    component SPREG
55    port (
56      RSTn0, RSTn1 : in  std_logic;
57      CLK0, CLK1   : in  std_logic;
58      ADDR          : in  std_logic_vector( 5 downto 0);
59      CBEn          : in  std_logic_vector( 3 downto 0);
60      ADi           : in  std_logic_vector(31 downto 0);
61      WREN          : in  std_logic;
62      DMA_DONE      : in  std_logic;
63      DMA_EN        : out std_logic;
64      FIFO_EN       : out std_logic;
65      SOFT_RSTn     : out std_logic;
66      DMAAD         : out std_logic_vector(31 downto 2);
67      DMAWORD       : out std_logic_vector(31 downto 0));
68    end component;
69    component REGMUX
70    port (
71      CLK0          : in  std_logic;
72      RSTn0         : in  std_logic;
73      ADDR          : in  std_logic_vector( 3 downto 0);
74      ADi           : in  std_logic_vector(31 downto 0);
75      CBEn          : in  std_logic_vector( 3 downto 0);
76      CFG_WRn       : in  std_logic;
77      M_ABORT       : in  std_logic;
78      T_ABORT       : in  std_logic;
79      BIST_CODE     : in  std_logic_vector( 4 downto 0);
80      INT_REQ       : in  std_logic_vector( 9 downto 0);
81      INT_MASK      : in  std_logic_vector( 9 downto 0);
82      INT_PEND      : in  std_logic_vector( 9 downto 0);
83      DMA_EN        : in  std_logic;
84      FIFO_EN       : in  std_logic;
85      FIFOCNT       : in  std_logic_vector( 5 downto 0);
86      DMAAD         : in  std_logic_vector(31 downto 2);
87      DMAWORD       : in  std_logic_vector(31 downto 0);
88      DMACNTR       : in  std_logic_vector(31 downto 0);
89      RDADDR        : in  std_logic_vector( 5 downto 0);
90      BIST_START    : out std_logic;
91      TARGET_B2B    : out std_logic;
```

```
92          MASTER_EN        : out std_logic;
93          RESPOND_EN       : out std_logic;
94          BASE_ADDR        : out std_logic_vector(11 downto 0);
95          LATENCY_PCI      : out std_logic_vector( 4 downto 0);
96          DOUT             : out std_logic_vector(31 downto 0));
97       end component;
98       signal INT_REQ          : std_logic_vector( 9 downto 0);
99       signal INT_MASK         : std_logic_vector( 9 downto 0);
100      signal INT_PEND         : std_logic_vector( 9 downto 0);
101      signal DMA_EN_S         : std_logic;
102      signal FIFO_EN_S        : std_logic;
103      signal DMAAD_S          : std_logic_vector(31 downto 2);
104      signal DMAWORD_S        : std_logic_vector(31 downto 0);
105      signal TARGET_B2B_S     : std_logic;
106      signal MASTER_EN_S      : std_logic;
107      signal RESPOND_EN_S     : std_logic;
108      signal BASE_ADDR_S      : std_logic_vector(11 downto 0);
109      signal LATENCY_PCI_S : std_logic_vector( 4 downto 0);
110   begin
111      DMA_EN           <= DMA_EN_S;
112      FIFO_EN          <= FIFO_EN_S;
113      DMAAD            <= DMAAD_S;
114      DMAWORD          <= DMAWORD_S;
115      TARGET_B2B       <= TARGET_B2B_S;
116      MASTER_EN        <= MASTER_EN_S;
117      RESPOND_EN       <= RESPOND_EN_S;
118      BASE_ADDR        <= BASE_ADDR_S;
119      ------------------------------------------------------
120      itreg0 : ITREG
121      port map (
122      RSTn0        => RSTn(0),
123      CLK0         => CLK(0),
124      ADDR         => WRADDR,
125      CBEn         => CBEn,
126      ADi          => ADi,
127      WREN         => WREN,
128      FIFO_OVF     => FIFO_OVF,
129      DMA_DONE     => DMA_DONE,
130      TIMER_INT    => TIMER_INT,
131      INTn         => INTn,
132      INTAn        => INTAn,
133      INT_REQ      => INT_REQ,
134      INT_MASK     => INT_MASK,
135      INT_PEND     => INT_PEND);
136      spreg0 : SPREG
137      port map (
138      RSTn0 => RSTn(1), RSTn1 => RSTn(2),
139      CLK0  => CLK(1),  CLK1  => CLK(2),
140      ADDR            => WRADDR,
141      CBEn            => CBEn,
```

```
142        ADi              => ADi,
143        WREN             => WREN,
144        DMA_DONE         => DMA_DONE,
145        DMA_EN           => DMA_EN_S,
146        FIFO_EN          => FIFO_EN_S,
147        SOFT_RSTn        => SOFT_RSTn,
148        DMAAD            => DMAAD_S,
149        DMAWORD          => DMAWORD_S);
150     regmux0 : REGMUX
151     port map (
152        CLK0             => CLK(3),
153        RSTn0            => RSTn(3),
154        ADDR             => WRADDR(3 downto 0),
155        ADi              => ADi,
156        CBEn             => CBEn,
157        CFG_WRn          => CFG_WRn,
158        M_ABORT          => M_ABORT,
159        T_ABORT          => T_ABORT,
160        BIST_CODE        => BIST_CODE,
161        INT_REQ          => INT_REQ,
162        INT_MASK         => INT_MASK,
163        INT_PEND         => INT_PEND,
164        DMA_EN           => DMA_EN_S,
165        FIFO_EN          => FIFO_EN_S,
166        FIFOCNT          => FIFOCNT,
167        DMAAD            => DMAAD_S,
168        DMAWORD          => DMAWORD_S,
169        DMACNTR          => DMACNTR,
170        RDADDR           => RDADDR,
171        BIST_START       => BIST_START,
172        TARGET_B2B       => TARGET_B2B_S,
173        MASTER_EN        => MASTER_EN_S,
174        RESPOND_EN       => RESPOND_EN_S,
175        BASE_ADDR        => BASE_ADDR_S,
176        LATENCY_PCI      => LATENCY_PCI,
177        DOUT             => DOUT);
178     end RTL;
```

The following synthesis script can be used to generate the schematic. Note that there is no compile command. The REG VHDL code contains signal and components connections. There is no extra circuit required. Figure 7-5 shows the schematic.

```
1     design = reg
2     read ../db/itreg.db
3     read ../db/spreg.db
4     read ../db/regmux.db
5     analyze -f vhdl VHDLSRC + "ch7/" + design + ".vhd"
6     elaborate REG -architecture RTL
7     include compile.common
8     create_clock "CLK*" -name clk -period CLKPERIOD -waveform {0.0
```

```
HALFCLK}
9       set_clock_skew -uncertainty CLKSKEW clk
10      set_input_delay  INPDELAY -add_delay -clock clk all_inputs()
11      set_input_delay  CLKPER8  -add_delay -clock clk {ADi CBEn}
12      set_input_delay  CLKPER9  -add_delay -clock clk {*_ABORT}
13      set_input_delay  CLKPER9  -add_delay -clock clk {DMA_DONE
FIFO_OVF}
14      set_output_delay CLKPER6  -add_delay -clock clk all_outputs()
15      set_dont_touch_network {CLK* RSTn*}
```

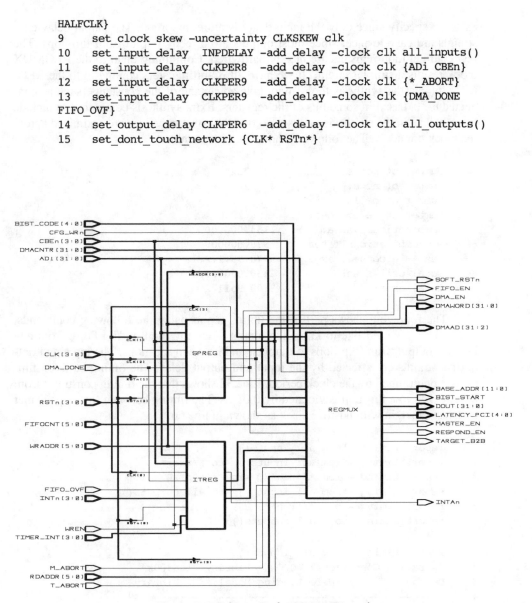

Figure 7-5 Schematic for REG VHDL code.

The schematic shows the register block partition. There are many other ways to design this register block. In this case, ITREG and SPREG are designed as separate blocks. Their outputs are almost direct output of flip-flops, except for the interrupt request register. Some of these registers do not use all 32 bits. The individual registers are implemented and grouped in the REGMUX block. This allows registers to go out of the register block and into the REGMUX block. The read-only register can come in directly to the REGMUX block. The synthesis constraints of ITREG and SPREG are easier to specify since only the input delay is more important. The output delay constraints are easy to meet since most outputs are directly output from flip-flops. The few configuration registers are mostly hardwired. Combining them to the REGMUX block results a simpler circuit. If we have all 32-bit registers coming in to the REG-MUX block, the synthesis tool cannot take advantage of the constants (such as connected to '0') used in some bits of the registers. Extra synthesis commands to include other blocks in the synthesis are used so that the constant values are optimized across the block boundary. The following shows the area summary.

```
1    Number of ports:              232
2    Number of nets:               262
3    Number of cells:                3
4    Number of references:           3
5    Combinational area:       1154.000000
6    Noncombinational area:     976.000000
7    Net Interconnect area:    5978.951172
8    Total cell area:          2130.000000
9    Total area:               8108.951172
```

The following shows part of the timing report using the following commands. Basically, we would like to know the timing from any input to flip-flops, from any input to output, from flip-flops to outputs, and between flip-flops. The input and output constraints are affected by the input and output delay. The timing between flip-flops is determined by the clock period. Line 32 shows that the timing constraint from input CBEn to a flip-flop is violated by 0.01 ns. The timing between flip-flops is met by a large margin with positive slack as shown in line 56.

```
1    report_timing -from all_inputs()
2    report_timing -from all_inputs() -to */*reg*/D
3    report_timing -from */*reg*/Q -to */*reg*/D
4    report_timing -from all_inputs() -to all_outputs()
5    report_timing -to    */*reg*/D
6    report_timing -to    all_outputs()

1    report_timing -from all_inputs()
2    Operating Conditions: WCCOM   Library: lca300k
3    Design              Wire Loading Model        Library
4    -------------------------------------------------
5    REG                     B9X9               lca300k
6    ITREG                   B9X9               lca300k
7    SPREG                   B9X9               lca300k
8    REGMUX                  B9X9               lca300k
9    -------------------------------------------------
10     clock clk (rise edge)           0.00        0.00
11     clock network delay (ideal)     0.00        0.00
12     input external delay           24.00       24.00 f
13     CBEn[0] (in)                    0.00       24.00 f
14     itreg0/CBEn[0] (ITREG)          1.17       25.17 f
```

```
15      itreg0/U254/Z (IVP)                    0.58        25.75 r
16      itreg0/U223/Z (AN2)                    1.53        27.28 r
17      itreg0/U283/Z (ND3)                    0.88        28.16 f
18      itreg0/U284/Z (ND2)                    0.33        28.49 r
19      itreg0/PENDFF_reg[4]/D (FD2)           0.00        28.49 r
20      data arrival time                                  28.49
21
22      clock clk (rise edge)                 30.00        30.00
23      clock network delay (ideal)            0.00        30.00
24      clock uncertainty                     -1.00        29.00
25      itreg0/PENDFF_reg[4]/CP (FD2)          0.00        29.00 r
26      library setup time                    -0.52        28.48
27      data required time                                 28.48
28      ------------------------------------------------------------
29      data required time                                 28.48
30      data arrival time                                 -28.49
31      ------------------------------------------------------------
32      slack (VIOLATED)                                   -0.01
33
34   report_timing -from */*reg*/Q -to */*reg*/D
35      ------------------------------------------------------------
36      clock clk (rise edge)                  0.00         0.00
37      clock network delay (ideal)            0.00         0.00
38      itreg0/PENDFF_reg[4]/CP (FD2)          0.00         0.00 r
39      itreg0/PENDFF_reg[4]/Q (FD2)           2.10         2.10 f
40      itreg0/U242/Z (ND2)                    0.52         2.62 r
41      itreg0/U306/Z (AN5)                    0.78         3.41 r
42      itreg0/U307/Z (AN4P)                   0.85         4.26 r
43      itreg0/INTAn_reg/D (FD4)               0.00         4.26 r
44      data arrival time                                   4.26
45
46      clock clk (rise edge)                 30.00        30.00
47      clock network delay (ideal)            0.00        30.00
48      clock uncertainty                     -1.00        29.00
49      itreg0/INTAn_reg/CP (FD4)              0.00        29.00 r
50      library setup time                    -0.51        28.49
51      data required time                                 28.49
52      ------------------------------------------------------------
53      data required time                                 28.49
54      data arrival time                                  -4.26
55      ------------------------------------------------------------
56      slack (MET)                                        24.23
```

A useful command is the "report_net", which produces a summary of interconnect information. Suppose we assume that DMA_EN output has a higher load. We use the "set_load 50 DMA_EN" command. The following shows the timing violation and part of the net report. Line 29 shows that net DMA_EN_S has a loading of more than 50 units, which is much larger than the rest. This causes the slow timing of the flip-flop as shown in line 2, which takes 74.65 ns to drive this huge load.

```
1        spreg0/DMA_EN_reg/CP (FD2)        0.00          0.00 r
2        spreg0/DMA_EN_reg/Q (FD2)        74.65         74.65 r
3        spreg0/DMA_EN (SPREG)             0.00         74.65 r
4        regmux0/DMA_EN (REGMUX)           0.00         74.65 r
5        regmux0/U661/Z (IVP)              7.98         82.63 f
6        regmux0/U662/Z (AO7)              1.27         83.90 r
7        regmux0/U670/Z (AO1)              0.92         84.82 f
8        regmux0/U671/Z (ND4)              0.83         85.65 r
9        regmux0/DOUT[1] (REGMUX)          0.00         85.65 r
10       DOUT[1] (out)                     0.00         85.65 r
11       data arrival time                             85.65
12
13       clock clk (rise edge)            30.00         30.00
14       clock network delay (ideal)      0.00         30.00
15       clock uncertainty                -1.00         29.00
16       output external delay           -18.00         11.00
17       data required time                            11.00
18       ------------------------------------------------------
19       data required time                            11.00
20       data arrival time                            -85.65
21       ------------------------------------------------------
22       slack (VIOLATED)                             -74.65
23
24   report_net
25   Net              Fanout Fanin    Load    Resistance Pins
26   DMAWORD_S[30]      2      1       0.48      0.02       3
27   DMAWORD_S[31]      2      1       0.53      0.02       3
28   DMA_DONE           2      1       0.24      0.02       3
29   DMA_EN_S           3      1      50.29      0.02       4
30   DOUT[0]            1      1       0.38      0.01       2
```

The schematic shown in Figure 7-5 can be synthesized with the following two commands. Line 1 tells the synthesis tool that sub-blocks ITREG, SPREG, and REG-MUX are not to be touched. They need not be synthesized. Figure 7-6 shows the synthesized schematic. Note that buffers are inserted. The timing has largely improved. Part of the timing report follows.

```
1    set_dont_touch {ITREG SPREG REGMUX}
2    compile
```

```
1        spreg0/DMA_EN_reg/CP (FD2)        0.00          0.00 r
2        spreg0/DMA_EN_reg/Q (FD2)         2.04          2.04 f
3        spreg0/DMA_EN (SPREG)             0.00          2.04 f
4        U2/Z (BUF8A)                     13.21         15.25 f
5        DMA_EN (out)                      0.42         15.67 f
6        data arrival time                             15.67
7        clock clk (rise edge)            30.00         30.00
8        clock network delay (ideal)      0.00         30.00
```

9	clock uncertainty	-1.00	29.00
10	output external delay	-18.00	11.00
11	data required time		11.00
12	---		
13	data required time		11.00
14	data arrival time		-15.67
15	slack (VIOLATED)		-4.67

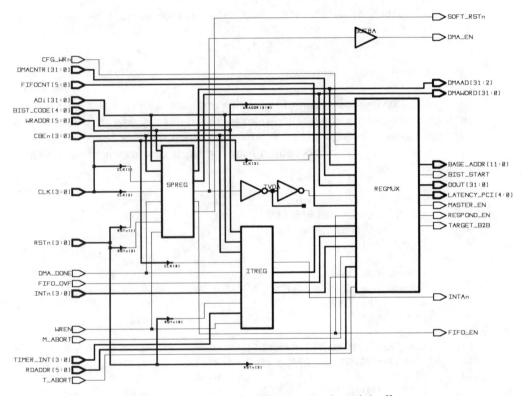

Figure 7-6 Schematic for REG VHDL code with buffers.

The REG block is synthesized from bottom up. Sub-blocks are synthesized and then designated "dont_touch". The following synthesis script shows another way to synthesize the REG block. All VHDL code is read in lines 2 to 5. Line 16 ungroups the sub-blocks so that the hierarchy is removed. The whole REG block is synthesized together.

```
1    design = reg
2    read -f vhdl VHDLSRC + "ch7/" + "itreg.vhd"
3    read -f vhdl VHDLSRC + "ch7/" + "spreg.vhd"
```

```
4      read -f vhdl VHDLSRC + "ch7/" + "regmux.vhd"
5      analyze -f vhdl VHDLSRC + "ch7/" + design + ".vhd"
6      elaborate REG -architecture RTL
7      include compile.common
8      create_clock "CLK*" -name clk -period CLKPERIOD -waveform {0.0
HALFCLK}
9      set_clock_skew -uncertainty CLKSKEW clk
10     set_input_delay  INPDELAY -add_delay -clock clk all_inputs()
11     set_input_delay  CLKPER8  -add_delay -clock clk {ADi CBEn}
12     set_input_delay  CLKPER9  -add_delay -clock clk {*_ABORT}
13     set_input_delay  CLKPER9  -add_delay -clock clk {DMA_DONE
FIFO_OVF}
14     set_output_delay CLKPER6  -add_delay -clock clk all_outputs()
15     set_dont_touch_network {CLK* RSTn*}
16     ungroup -all
17     compile
18     include compile.report
```

The following shows part of the timing and area summary. Note that all input timing constraints are met with smaller areas. This is due to the fact that generally, the synthesis tools can better optimize the circuit if there is no block boundary. However, synthesis tools may not be efficient in handling larger designs where all sub-blocks have been flattened.

```
1      clock clk (rise edge)           0.00        0.00
2      clock network delay (ideal)     0.00        0.00
3      input external delay            27.00       27.00 f
4      DMA_DONE (in)                   0.00        27.00 f
5      U201/Z (OR2)                    0.98        27.98 f
6      U202/Z (MUX21L)                 0.26        28.25 r
7      spreg0/DMA_EN_reg/D (FD2)       0.00        28.25 r
8      data arrival time                           28.25
9      slack (MET)                                  0.24
10     ----------------
11     itreg0/PENDFF_reg[0]/CP (FD2)   0.00        0.00 r
12     itreg0/PENDFF_reg[0]/QN (FD2)   2.57        2.57 r
13     U260/Z (ND2)                    0.55        3.12 f
14     U262/Z (ND2)                    0.29        3.41 r
15     U264/Z (ND2)                    0.40        3.81 f
16     itreg0/PENDFF_reg[0]/D (FD2)    0.00        3.82 f
17     data arrival time                           3.82
18     slack (MET)                                 24.89
19     --------------
20     RDADDR[5] (in)                  0.00        2.50 f
21     U28/Z (OR2P)                    2.57        5.07 f
22     U35/Z (NR2)                     1.01        6.08 r
23     U195/Z (B4I)                    1.53        7.60 f
24     U425/Z (OR2P)                   1.53        9.13 f
25     U427/Z (ND2)                    0.27        9.41 r
26     U431/Z (NR2)                    0.30        9.71 f
```

```
27      U432/Z (ND4)                         0.78      10.49 r
28      DOUT[24] (out)                       0.00      10.49 r
29      data arrival time                              10.49
30      slack (MET)                                     0.51
31      --------------------------------------------------------
32   Number of ports:                232
33   Number of nets:                 847
34   Number of cells:                642
35   Number of references:            32
36   Combinational area:          1051.000000
37   Noncombinational area:        976.000000
38   Net Interconnect area:       5593.105957
39   Total cell area:             2027.000000
40   Total area:                  7620.105957
```

7.7 TESTING REGISTERS

Testing registers can be complicated. There are different types of registers with different addresses and widths. Some are not easy to test without other blocks that accept or generate signals from and to the register block. However, let's try the following simple test bench. The idea of the simple test bench is to somewhat randomly exercise all input signals to test as much as possible. Complete tests are usually done for the whole design so that the interactions among blocks can be tested. The random input pattern is achieved by shifting a linear feedback value as shown in line 81. The only other feature is to control the input signal delays. For example, RDADDR and WRADDR are set with half of the clock input delay as shown in lines 71 to 72. They are implemented as a counter so that the corresponding registers can be followed much more easily. Read-only registers such as DMACNTR and FIFOCNT are set to have input delay of 1 ns so that the value will line up with the rising edge of the clock better in the simulation waveform as shown in Figure 7-7.

```
1      library IEEE;
2      use IEEE.std_logic_1164.all;
3      use IEEE.std_logic_unsigned.all;
4      entity TBREG is
5      end TBREG;
6      architecture BEH of TBREG is
7        component REG
8        port(
9          RSTn            : in  std_logic_vector( 3 downto 0);
10         CLK             : in  std_logic_vector( 3 downto 0);
11         WRADDR          : in  std_logic_vector( 5 downto 0);
12         CBEn            : in  std_logic_vector( 3 downto 0);
13         ADi             : in  std_logic_vector(31 downto 0);
14         WREN            : in  std_logic;
15         FIFO_OVF        : in  std_logic;
16         DMA_DONE        : in  std_logic;
17         TIMER_INT       : in  std_logic_vector( 3 downto 0);
```

```
18        INTn              : in  std_logic_vector( 3 downto 0);
19        INTAn             : out std_logic;
20        DMA_EN            : out std_logic;
21        FIFO_EN           : out std_logic;
22        SOFT_RSTn         : out std_logic;
23        DMAAD             : out std_logic_vector(31 downto 2);
24        DMAWORD           : out std_logic_vector(31 downto 0);
25        CFG_WRn           : in  std_logic;
26        M_ABORT           : in  std_logic;
27        T_ABORT           : in  std_logic;
28        BIST_CODE         : in  std_logic_vector( 4 downto 0);
29        FIFOCNT           : in  std_logic_vector( 5 downto 0);
30        DMACNTR           : in  std_logic_vector(31 downto 0);
31        RDADDR            : in  std_logic_vector( 5 downto 0);
32        BIST_START        : out std_logic;
33        TARGET_B2B        : out std_logic;
34        MASTER_EN         : out std_logic;
35        RESPOND_EN        : out std_logic;
36        BASE_ADDR         : out std_logic_vector(11 downto 0);
37        LATENCY_PCI       : out std_logic_vector( 4 downto 0);
38        DOUT              : out std_logic_vector(31 downto 0));
39      end component;
40      constant PERIOD2 : time := 50 ns;
41      constant DELAY   : time := 80 ns;
42      signal RSTn, CLK : std_logic;
43       signal WREN, FIFO_OVF, DMA_DONE, INTAn : std_logic;
44      signal DMA_EN, FIFO_EN, SOFT_RSTn        : std_logic;
45      signal CFG_WRn, M_ABORT, T_ABORT         : std_logic;
46      signal BIST_START, TARGET_B2B            : std_logic;
47      signal MASTER_EN, RESPOND_EN             : std_logic;
48      signal RSTnS, CLKS, CBEn, TIMER_INT, INTn :
49        std_logic_vector( 3 downto 0);
50      signal FIFOCNT, WRADDR, RDADDR :
51        std_logic_vector( 5 downto 0);
52      signal ADi, DMAWORD, DMACNTR, DOUT :
53        std_logic_vector(31 downto 0);
54      signal DMAAD        : std_logic_vector(31 downto 2);
55      signal BIST_CODE    : std_logic_vector( 4 downto 0);
56      signal LATENCY_PCI  : std_logic_vector( 4 downto 0);
57      signal BASE_ADDR    : std_logic_vector(11 downto 0);
58    begin
59      RSTn <= '0', '1' after 10 * PERIOD2;
60      clkgen : process
61      begin
62        CLK <= '0'; wait for PERIOD2;
63        CLK <= '1'; wait for PERIOD2;
64      end process;
65      CLKS  <= CLK & CLK & CLK & CLK;
66      RSTnS <= RSTn & RSTn & RSTn & RSTn;
67      adgen : process
68        variable D6 : std_logic_vector(5 downto 0) := "000000";
69      begin
```

```
70        D6 := D6 + 1;
71        RDADDR <= D6 after PERIOD2;
72        WRADDR <= D6 after PERIOD2;
73        wait until (CLK'event) and (CLK = '1');
74      end process;
75      dgen : process
76        variable D32 :std_logic_vector(31 downto 0) :=
77          (others => '0');
78      begin
79        BIST_CODE <= D32(31 downto 27) after DELAY;
80        for i in 0 to 31 loop
81          D32 := D32(30 downto 0) & (not D32(31) xor D32(2));
82          DMACNTR   <= D32(20 downto 1) & D32(11 downto 0)
83                      AFTER 1 ns;
84          ADi       <= D32 AFTER DELAY;
85          CBEn      <= D32(27 downto 24) after DELAY;
86          FIFOCNT   <= D32(17 downto 12) after 1 ns;
87          INTn      <= D32(23 downto 20) after DELAY;
88          TIMER_INT <= D32(19 downto 16) after DELAY;
89          CFG_WRn   <= D32(12) after DELAY;
90          WREN      <= D32(13) or D32(12) after DELAY;
91          M_ABORT   <= D32(11) after DELAY;
92          T_ABORT   <= D32(12) after DELAY;
93          DMA_DONE  <= D32(9)  after DELAY;
94          FIFO_OVF  <= D32(10) after DELAY;
95          wait until (CLK'event) and (CLK = '1');
96        end loop;
97      end process;
98      reg0 : REG port map (
99        RSTn        => RSTnS,      CLK         => CLKS,
100       WRADDR      => WRADDR,     CBEn        => CBEn,
101       ADi         => ADi,       WREN        => WREN,
102       FIFO_OVF    => FIFO_OVF,   DMA_DONE    => DMA_DONE,
103       TIMER_INT   => TIMER_INT,  INTn        => INTn,
104       INTAn       => INTAn,      DMA_EN      => DMA_EN,
105       FIFO_EN     => FIFO_EN,    SOFT_RSTn   => SOFT_RSTn,
106       DMAAD       => DMAAD,      DMAWORD     => DMAWORD,
107       CFG_WRn     => CFG_WRn,    M_ABORT     => M_ABORT,
108       T_ABORT     => T_ABORT,    BIST_CODE   => BIST_CODE,
109       FIFOCNT     => FIFOCNT,    DMACNTR     => DMACNTR,
110       RDADDR      => RDADDR,     BIST_START  => BIST_START,
111       TARGET_B2B  => TARGET_B2B, MASTER_EN   => MASTER_EN,
112       RESPOND_EN  => RESPOND_EN, BASE_ADDR   => BASE_ADDR,
113       LATENCY_PCI => LATENCY_PCI, DOUT       => DOUT);
114     end BEH;
115     configuration CFG_TBREG of TBREG is
116       for BEH
117       end for;
118     end CFG_TBREG;
```

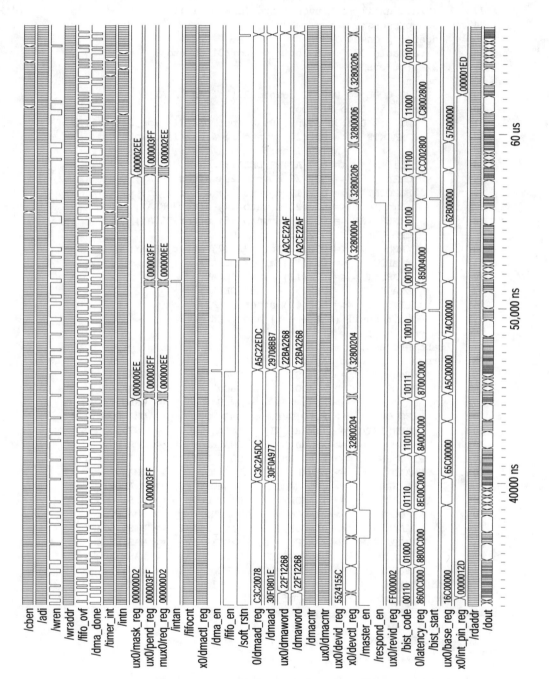

Figure 7-7 Simulation waveform for TBREG.

The above simulation waveform is not meant to illustrate the detailed values of each signal. The purpose is to show that all signals are toggled. This is important to observe as it indicates that the randomness of inputs exercises all signals and all signals are connected even if they are not correct. After this is achieved, zoom in to examine various functions. Figure 7-8 shows a zoomed-in waveform. At the center of the rising edge clock, ADi is 68B3A62B hexadecimal. CBEn is "1000". WRADDR is "010000". WREN is '1'. Therefore, the DMA and control register is written. Note the ADi(0) is '1' and CBEn(0) is '0'. Software reset SOFT_RSTn is asserted one clock cycle. The DMA control register has the value of 00003404. Its bit 0 is '0' even '1' is tried to be written. This is correct as the bit for SOFT_RSTn should always be read as '0'. The FIFOCNT has value "110100" which is hexadecimal 34. This is correct. However, ADi(3 downto 0) is "1011" or hexadecimal B. According to line 29 and 30 of SPREG VHDL code, FIFO_EN register is written with ADi(1), DMA_EN register is written with ADi(2). DMA_EN is not asserted while FIFO_EN is asserted. This is correct. However, the DMACTL_REG signal shows the bits 3 *downto* 0 as hexadecimal 4 (or "0100"). Oops! This is caused by line 172 in the REGMUX VHDL code. The DMA control register is implemented wrong with the orders of FIFO_EN and DMA_EN reversed when they are concatenated.

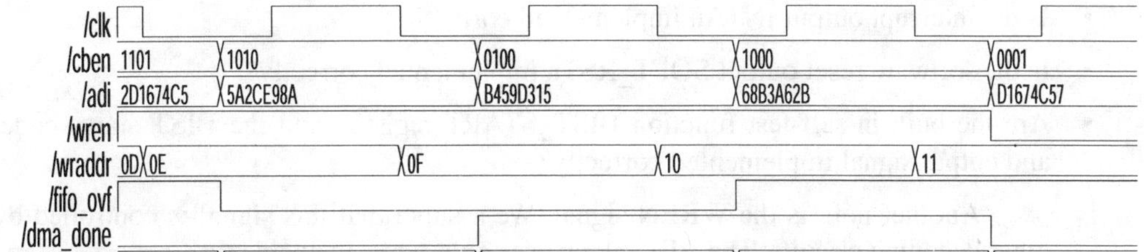

Signal					
/clk					
/cben	1101	1010	0100	1000	0001
/adi	2D1674C5	5A2CE98A	B459D315	68B3A62B	D1674C57
/wren					
/wraddr	0D	0E	0F	10	11
/fifo_ovf					
/dma_done					
/timer_int	0110	1100	1001	0011	0111
/intn	0001	0010	0101	1011	0110
x0/mask_reg	000000EE				
x0/pend_reg	000003FF				
ux0/req_reg	000000EE				
/intan	0				
/fifocnt	001110	011101	111010	110100	101001
0/dmactl_reg	00000E00	00001D00	00003A00	00003404	00002904
/dma_en					
/fifo_en					
/soft_rstn					
0/dmaad_reg	A5C22EDC				
/dmaad	29708BB7				
x0/dmaword	22BA2268				
/dmaword	22BA2268				
/dmacntr	674C598A	CE98A315	9D31562B	3A62BC57	74C578AF
ux0/dmacntr	674C598A	CE98A315	9D31562B	3A62BC57	74C578AF

52500 ns   52,600 ns   52700 ns   52800 ns

Figure 7-8 Simulation waveform for TBREG, zoomed in.

The bottom of the simulation waveform shows that input signal DMACNTR is equal to the DMACNTR signal inside the REGMUX block. This verifies the connection from REG block input DMACNTR to the REGMUX block. There are many functions that the REG block design should verify. Here are some:

- Is the register being written identified by the WRADDR?

- Does the DOUT value correspond to the register identified by the RDADDR?

- Is the DOUT value correct when the RDADDR address is in the memory address gap or in the reserved region?

- Does each register have the right positions if the register is formed by more than one register?

- Does the control register DMA_EN and FIFO_EN perform as expected?

- Does the CBEn perform as intended?

- Is it possible that more than one registers is written at one time?

- Are the interrupt pending, mask, and request registers implemented correctly?

- Is the interrupt output INTAn implemented correctly?

- Is the software reset output SOFT_RSTn implemented correctly?

- Are the built-in-self-test function BIST_START register, and the BIST status code, and output signal implemented correctly?

Another note is the WREN signal. We assume that this signal is controlled by other decoding circuits. The REG block uses only 6 bits to enable writing to registers. If this signal is not controlled correctly, registers may be overwritten by mistake. For example, the address usually contains more than 6 bits. The same 32-bit address intended for another region of the memory map may be used in the REG block also. The decoding circuit may need to decode the address bits 31 downto 7 to enable the WREN signal. To facilitate tracing signals, most VHDL simulators allow signals of interest to be explicitly enumerated in a command file. For example, interested signals to be traced can be put into a file. The file can then be read using the command "include tracefile" for Synopsys VSS or "do tracefile" for QuickVHDL. The following shows the trace file which traces the signals as shown in the above simulation waveforms. Note that related signals are placed together in the file that also defines the display order. The clock signal is traced in line 1 and line 17 since the clock signal is usually a reference signal. Tracing the clock more than once reduces the number of times to scroll back and forth to reference the clock signal. In practice, several trace files can be created for different simulation and signal review purposes.

```
1    wave /RSTn /CLK /CBEn /ADi /WREN /WRADDR
2    wave /FIFO_OVF /DMA_DONE /TIMER_INT /INTn
3    wave /reg0/regmux0/MASK_REG   /reg0/regmux0/PEND_REG
4    wave /reg0/regmux0/REQ_REG /INTAn /FIFOCNT
```

```
5    wave /reg0/regmux0/DMACTL_REG /DMA_EN /FIFO_EN /SOFT_RSTn
6    wave /reg0/regmux0/DMAAD_REG /DMAAD
7    wave /reg0/regmux0/DMAWORD    /DMAWORD
8    wave /DMACNTR /reg0/regmux0/DMACNTR
9    wave /reg0/regmux0/DEVID_REG
10   wave /M_ABORT /T_ABORT /reg0/regmux0/DEVCTL_REG
11   wave /TARGET_B2B /MASTER_EN /RESPOND_EN
12   wave /reg0/regmux0/REVID_REG
13   wave /BIST_CODE /reg0/regmux0/LATENCY_REG
14   wave /BIST_START /LATENCY_PCI
15   wave /reg0/regmux0/BASE_REG /BASE_ADDR
16   wave /reg0/regmux0/INT_PIN_REG
17   wave /CLK /RDADDR /DOUT
```

7.8 MICROPROCESSOR REGISTERS

Microprocessor registers are usually used to store computational results from the arithmetic logic unit (ALU). All microprocessor registers are usually the same width. The ALU may require more than two operands. The following VHDL code shows a register file with sixteen 32-bit registers with one write port and two read ports (to support two operands for the ALU). REGH VHDL code implements one 32-bit register with write enable signal EN.

```
1    library IEEE;
2    use IEEE.std_logic_1164.all;
3    entity REGH is
4      port (
5        RSTn    : in  std_logic;
6        CLK     : in  std_logic;
7        EN      : in  std_logic;
8        DIN     : in  std_logic_vector(31 downto 0);
9        DOUT    : out std_logic_vector(31 downto 0));
10   end REGH;
11   architecture RTL of REGH is
12   begin
13     rgen : process (RSTn, CLK)
14     begin
15       if (RSTn = '0') then
16         DOUT <= (DOUT'range => '0');
17       elsif (CLK'event and CLK = '1') then
18         if (EN = '1') then
19           DOUT <= DIN;
20         end if;
21       end if;
22     end process;
23   end RTL;
```

The sixteen 32-bit registers are modeled in the following block of code. Lines 29 and 30 declare reset and clock signal. Each has 16 bits. Each 32-bit register has individual RSTn and CLK bit. WREN is the write enable signal. In line 32, WAD specifies which register to be written. RAD1 and RAD2 are the input addresses used to specify any one of the 16 registers which map to DOUT1 and DOUT2, respectively.

```
24    library IEEE;
25    use IEEE.std_logic_1164.all;
26    use IEEE.std_logic_unsigned.all;
27    entity REGS is
28      port (
29        RSTn    : in  std_logic_vector(15 downto 0);
30        CLK     : in  std_logic_vector(15 downto 0);
31        WREN    : in  std_logic;
32        WAD     : in  std_logic_vector( 3 downto 0);
33        DIN     : in  std_logic_vector(31 downto 0);
34        RAD1    : in  std_logic_vector( 3 downto 0);
35        RAD2    : in  std_logic_vector( 3 downto 0);
36        DOUT1   : out std_logic_vector(31 downto 0);
37        DOUT2   : out std_logic_vector(31 downto 0));
38    end REGS;
```

Component REGH is declared in line 44 to 51. Lines 53 to 57 generate the EN signal of 16 bits. Based on WAD and WREN, at most one of the EN bits can be asserted '1'. Lines 58 to 62 instantiate component REGH 16 times. Note that individual bits of RSTn, CLK, and EN are mapped. REGH DIN input is mapped with the same input. REGH DOUT is mapped to sixteen 32-bit signal arrays as declared in lines 40, 41, and 43.

```
39    architecture RTL of REGS is
40      type regary is array(0 to 15) of
41        std_logic_vector(31 downto 0);
42      signal EN : std_logic_vector(15 downto 0);
43      signal FF : regary;
44      component REGH
45      port (
46        RSTn    : in  std_logic;
47        CLK     : in  std_logic;
48        EN      : in  std_logic;
49        DIN     : in  std_logic_vector(31 downto 0);
50        DOUT    : out std_logic_vector(31 downto 0));
51      end component;
52    begin
53      p0 : process (WAD, WREN)
54      begin
55        EN <= (EN'range => '0');
56        EN(conv_integer(WAD)) <= WREN;
57      end process;
```

```
58        g0 : for j in 0 to 15 generate
59          regh0 : REGH
60            port map (RSTn => RSTn(j), CLK => CLK(j), EN => EN(j),
61              DIN => DIN, DOUT => FF(j));
62          end generate;
63          DOUT1 <= FF(conv_integer(RAD1));
64          DOUT2 <= FF(conv_integer(RAD2));
65        end RTL;
```

The selection of DOUT1 and DOUT2 are done in lines 63 to 64. The ALU usually takes two operands and generates one result. The above register file has two read ports. Each can be connected to the two operand inputs of the ALU. The result of the ALU can be connected to the DIN input of the REGS block. The WREN signal is usually controlled by the instruction decoder and controller. Some instructions require the storage of the result such as an ADD instruction while others cannot update the register result, such as a branch instruction. Verification and synthesis of this code are left as Exercises 7.4 and 7.5.

7.9 EXERCISES

7.1   Refer to the SPREG VHDL code lines 48 to 50. These three lines should have been taken out of the **loop statement**. Modify the VHDL code to do that. Synthesize the VHDL code. Does the new VHDL code produce a better (smaller) schematic? Does it affect the VHDL simulation whether those three lines are inside or outside of the **loop statement**?

7.2   Refer to REGMUX VHDL code. We have identified that line 172 is not correct. Update the VHDL code and rerun the test bench. Can you verify that the register block performs correctly on the questions of possible errors listed in Section 7.7?

7.3   Refer to the TBREG VHDL code. What types of errors cannot be verified with the current test bench? Modify the test bench to test those errors and verify the updated code.

7.4   Refer to REGS VHDL code. Develop a test bench to verify the REGS VHDL code.

7.5   Refer to REGS VHDL code. Develop synthesis scripts to synthesize the VHDL code. Report synthesis results in the class.

7.6   Refer to REGS VHDL code. Only one register can be updated at any given clock cycle. Two registers can be read at the same time. This is often referred as one write port and two read ports. Modify the VHDL code so that the registers can have two write ports and three read ports. Verify and synthesize the VHDL code.

Chapter 8

Clock and Reset Circuits

Controlling the clock and reset circuit is fundamental to the design. Without proper clock and reset circuits, the circuit will not function as expected. In this chapter, we will study basic clock and reset circuits and discuss how to implement them with VHDL and synthesis.

8.1 CLOCK BUFFER AND CLOCK TREE

Each flip-flop requires a clock input. For synchronous designs, a clock signal may connect to many flip-flops so that they can change states synchronously. A design with 1,000 flip-flops will be too slow for the clock signal to drive all the flip-flops directly. Therefore, special clock buffers and clock buffer trees are usually used. Depending on the process technology, a clock buffer may drive up to a certain number of flip-flops. The length of the clock buffer delay is usually calculated by means of an equation. The simplest form of a delay equation may be $T = G + C*L$, where T is the total delay, G is the gate delay, C is a constant factor, and L is the amount of load. The load L is a function of the interconnect and number of gates that the clock buffer needs to drive. A clock may be able to drive more loads compared to regular gates. However, it has its own limit. Therefore, a clock signal is usually connected to a clock buffer tree as shown in Figure 8-1, which illustrates a clock buffer tree of two levels. The clock signal source drives six clock buffers. Each of the first-level clock buffers in turn drives six clock buffers. Each of the second-level clock buffers are used to drive flip-flop clock inputs. The number of levels and the span of each level depend on the process technology and the design requirements. However, the clock buffer tree has two primary goals:

- To reduce the total *clock delay* from the design clock input signal going through the clock buffers tree to the clock input of all flip-flops.

- To minimize the *clock skew* among the flip-flops that are driven by the same clock signal source. The *clock skew* is defined as the timing difference among the flip-flop clock signals.

The impact on the clock skew will be discussed in a later section. The clock delay is reduced by the fact that a clock buffer drives only a certain number of flip-flops. The clock skew is minimized by the greater flexibility in placing the clock buffers in the right place, and by optimizing the number of flip-flops for each clock buffer to drive.1

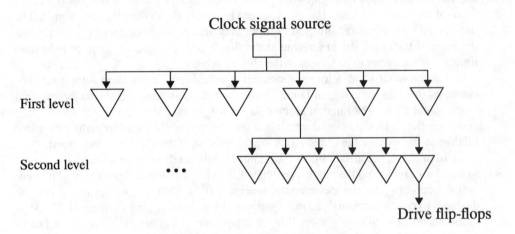

Figure 8-1 Clock buffer tree.

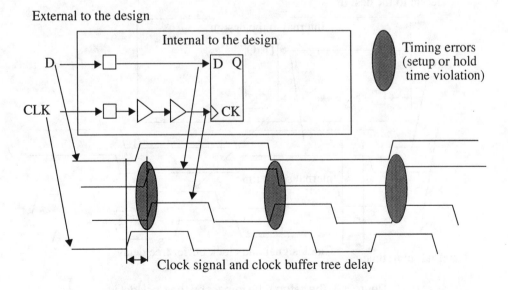

Figure 8-2 The hold time impact by the clock delay.

The design impact for the clock delay can be described in Figure 8-2. Data signal D is synchronized to the clock signal external to the design. Suppose that signal D satisfies the minimum hold time of the external clock signal CLK. Signal D is connected to a pad in the design. It does not go through other gates before it reaches the D-input of a flip-flop. The clock signal is connected to a pad, too. Since the clock signal needs to drive many flip-flops, a clock buffer tree is used. The figure shows that the clock signal goes through two-stage of clock buffers before it reaches the clock input of the same flip-flop. The clock signal has more delay than the data signal. The data signal arriving at the flip-flop D-input may arrive earlier or very close to the rising edge of the clock signal arriving at the flip-flop clock input. A setup or hold time timing error may occur as shown in the shaded region.

On the other hand, a longer clock delay would help in the short setup time. For example, the data input signal D may have a small setup time relative to the external clock signal CLK. The internal clock delay is longer than the data signal delay to the same flip-flop. The data signal arriving at the D-input of the flip-flop gains more time relative to the rising edge of the clock signal arriving at the flip-flop clock input.

In practice, the arrival time of all input signals needs to be analyzed with respect to the clock signal and the clock tree delay. Likewise, all output signals should be generated according to the design requirements. It is always good practice for the designer to control and analyze input/outputs (I/O) signal timing thoroughly. For synchronous designs, allowing a small input setup time and generating outputs as fast as possible would make the interface to other designs much easier.

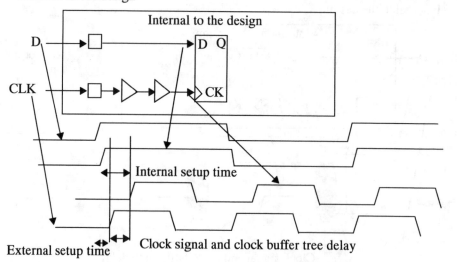

Figure 8-3 The setup time impact by the clock delay.

8.2 CLOCK TREE GENERATION

There are basically two ways that a clock tree can be generated.

- *The clock tree is generated by the place and route tool.* The target process technology may be standard cells, gate arrays, or FPGA. The particular place and route tool has better control of the actual interconnect delay and the placing and connecting of clock buffers. The clock buffers are inserted to reduce the clock delay and clock skew by the place and route tool. After the place and route process, a netlist file and a timing file are generated so that designers can run the postlayout timing simulation. The postlayout netlist, which is not the same as the original netlist generated by the synthesis tool, has clock buffers. The timing file is usually a Standard Delay Format (SDF) file. In general, this is a better approach since the place and route tool have better control of timing than the synthesis tools. This also reduces the burden on the designers to design the clock buffer trees and to connect a certain number of flip-flops to each clock buffer.

- *The clock tree is generated before the place and route process.* This is usually done when the place and route tool does not generate clock buffer trees to reduce the clock delay and clock skew. In general, this results in worse clock delays and clock skews since the exact locations of gate are not known in the VHDL design and synthesis process. However, in certain rare cases the designer may want to purposely control the clock network. For example, a certain set of flip-flops may require that special clock timing control be a little bit earlier or later than is the case for other flip-flops.

In Chapter 10, a design case study is done to show that the clock tree is generated before the place and route process. In Chapter 11, a design case study shows the design process with the clock buffer trees generated by the place and route tool. Here we will show how the clock tree can be generated with the VHDL and synthesis process. The following VHDL codes show how a clock buffer trees with two levels of inverted clock buffers can be generated with VHDL and synthesis. The first VHDL code is a package that contains a component declaration. The component name should match exactly the clock buffer to be used in the synthesis library. In this example, an inverted clock buffer with a buffer size of 60 is used.

```
1    library IEEE;
2    use IEEE.std_logic_1164.all;
3    package GATECOMP is
4      component STDINV_60x
5      port (
6        IN0 : in  std_logic;
7        Y   : out std_logic);
8      end component;
9    end GATECOMP;
```

The following VHDL shows how the inverted clock buffers can be instantiated with generate statements. The entity also has generics WIDE1 and BUF as shown in lines 6 and 7. They are used to specify how many clock buffers in the first and the second level of the clock buffer trees, respectively. Note in line 22 that the second level of clock buffers are more or less evenly distributed to connect to the first level of clock buffers.

```
1     library IEEE;
2     use IEEE.std_logic_1164.all;
3     use work.GATECOMP.all;
4     entity TREE is
5       generic(
6         WIDE1   : integer range 1 to 20  := 4;
7         BUF     : integer range 1 to 400 := 20);
8       port(
9         CLKIN : in  std_logic; -- input clock
10        CLK   : out std_logic_vector(BUF-1 downto 0));
11    end TREE;
12    architecture RTL of TREE is
13      signal CLKtemp1 : std_logic_vector(WIDE1-1 downto 0);
14      signal CLK1     : std_logic_vector(BUF-1 downto 0);
15    begin
16      genclk1 : for j in 0 to WIDE1-1 generate
17        u0 : stdinv_60x
18          port map (IN0 => CLKIN, Y => CLKtemp1(j));
19      end generate;
20      genclk4 : for j in 0 to BUF-1 generate
21        u2 : stdinv_60x
22          port map (IN0=>CLKtemp1(j mod (WIDE1)), Y=>CLK1(j));
23      end generate;
24      CLK <= CLK1;
25    end RTL;
```

Figure 8-4 shows the schematic results after the following synthesis commands. Note that the VHDL codes are read in without the "compile" command.

```
1     read -f vhdl ../vhdl/gatecomp.vhd
2     read -f vhdl ../vhdl/tree.vhd
```

Figure 8-5 shows the result schematic when the compile command is used to optimize the Figure 8-4 schematic. The clock buffers are removed since all the outputs are functioning the same as the input signal. The result schematic is just like a bus ripper.

In practice, the clock buffers tree has its own entity and design block. It can be instantiated with generics mapped to the desired number of clock buffers in the first and the second level of the clock buffers tree. It is easy to change the VHDL code to have three or four levels of clock buffers. This is left as Exercise 8.1.

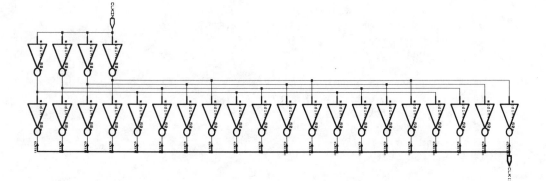

Figure 8-4 A clock tree example.

Figure 8-5 A clock tree after the "compile" command.

The following VHDL code shows the simulation model for the clock buffers. This code is used for simulation purposes, not for synthesis. Another file for simulation is the configuration file for the TREE entity shown below.

```
1    library IEEE;
2    use IEEE.std_logic_1164.all;
3    entity STDINV_60x is
4      port (
5        IN0 : in  std_logic;
6        Y   : out std_logic);
7    end STDINV_60x;
8    architecture RTL of STDINV_60x is
9    begin
10     Y <= not IN0;
11   end RTL;
```

```
1      configuration CFG_TREE of TREE is
2        for RTL
3          for genclk1
4            for all : STDINV_60x
5              use entity WORK.STDINV_60x(RTL);
6            end for;
7          end for;
8          for genclk4
9            for all : STDINV_60x
10             use entity WORK.STDINV_60x(RTL);
11           end for;
12         end for;
13       end for;
14     end CFG_TREE;
```

Each VHDL code block has its own file. This allows the design to be easily ported across process technologies with minimal changes since different technology libraries have their own clock buffer naming schemes. It is also a good practice to separate the clock circuitry from the majority of the design that is to be synthesized as shown in Figure 8-6, an example of a hierarchical partition. The design case study in Chapters 10, 11, and 12 will illustrate the hierarchical partition of the VHDL code for ease of synthesis and verification.

8.3 RESET CIRCUITRY

A design usually has a reset input. The reset can be caused when the design is starting up from the power up sequence or caused by a pushbutton reset switch. These two kinds of resets are hardware reset. It is controlled with an input signal such as RESETn for many integrated circuits (ICs). Another type of reset is the software reset. The reset is caused by the software writing to a particular register (called the software reset register) as shown in Chapter 7. The hardware needs to handle both the hardware and the software reset. The hardware reset input may be asynchronous since there is no control governing when the reset button is pushed and released. The reset input signal is synchronized and then combined (logical AND) with the software reset. The result is then connected to a reset buffers tree which is the same as a clock buffer trees. The reset circuitry VHDL code is shown below. RESETn is the reset input signal as declared in line 6. In line 7, SOFT_RSTn is coming from the register block when the software reset register is written. Output signal RESETn2TREE is generated to connect to the reset buffers tree. The RESETn input is synchronized through two stages of flip-flops as shown in lines 16 and 17. Line 19 generates the RESETn2TREE output signal. Figure 8-7 shows the synthesized schematic.

```
1      library IEEE;
2      use IEEE.std_logic_1164.all;
3      entity RESETSYNC is
4        port(
5          CLOCK       : in  std_logic;
```

```
6          RESETn      : in  std_logic;
7          SOFT_RSTn   : in  std_logic;
8          RESETn2TREE : out std_logic);
9     end RESETSYNC;
10    architecture RTL of RESETSYNC is
11      signal FF1, FF2 : std_logic;
12    begin
13      sync : process
14      begin
15        wait until (CLOCK'event and CLOCK = '1');
16        FF1   <= RESETn;
17        FF2   <= FF1;
18      end process;
19      RESETn2TREE <= SOFT_RSTn and FF2;
20      end RTL;
```

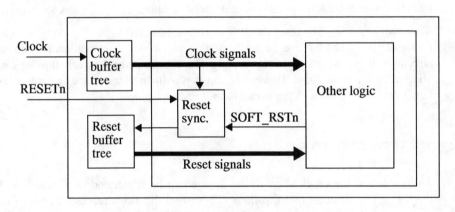

Figure 8-6 Hierarchical partition example.

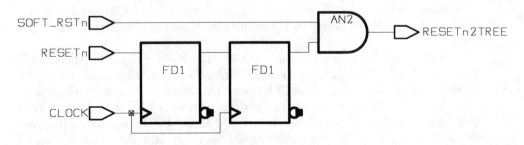

Figure 8-7 Synthesized schematic for RESETSYNC.

It is important to note that the clock input to the RESETSYNC block is coming from the output of the clock buffers tree, not directly from the design input. This is because signal SOFT_RSTn is generated, based on the clock coming from the output of the clock buffers tree. Also, we want to ensure that the reset signals generated from the reset buffers tree does not have setup or hold time violation relative to the clock signals coming out of the clock buffers tree.

The reset signals after the reset buffers tree can be used to reset or set the flip-flops synchronously or asynchronously. For a complete treatment of the differences between synchronous or asynchronous reset or set in finite state machines see Chapter 9 of K.C. Chang, *Digital Design and Modeling with VHDL and Synthesis*, IEEE Computer Society Press, 1997. In general, the asynchronous reset (set) signals are connected directly to the asynchronous reset (set) input of the flip-flops. The synchronous reset (set) signals usually go through combinational gates to the D-input of the flip-flops. Therefore, their differences can be summarized as follows:

- *Synchronous reset.* The reset signal is combined with the combinational circuits and the synthesis to generate D-input for the flip-flop unless the synthesis library has synchronous reset (set) flip-flops. More combinational circuits and longer delay paths are expected.

- *Asynchronous reset.* The reset signal is directly connected to the flip-flops. In synthesis, the reset signal should also be set "set_dont_touch" so that additional combinational circuits will not be added in the design. The timing paths between flip-flops are not affected by the reset signal. However, a flip-flop with an asynchronous reset (set) input is larger than a flip- flop with the asynchronous reset (set) input.

8.4 CLOCK SKEW AND FIXES

The impact of the clock skew to the design can be illustrated in Figure 8-8. The D input is going through two flip-flops. These two flip-flops represent general circuits such as synchronization circuits, shift registers, or delays of a signal for a full clock cycle. The clock signal CLK is going through a clock buffers tree to reach F1/CK and F2/CK flip-flops clock inputs as represented by CLK1 and CLK2 signals. Both CLK1 and CLK2 have clock delays relative to the input clock signal CLK. Due to different interconnect paths and clock buffers, CLK1 and CLK2 will arrive at flip-flops F1 and F2 at different times. Figure 8-8 shows that the clock skew between CLK1 and CLK2 is measured as the time difference between the rising edge of CLK1 and CLK2. The clock skew may also be caused by the falling edge if the negative edge of the clock is used.

Input signal D changes states at about the middle of the clock. Flip-flop F1 will catch the change which generates a signal F1/Q. The rising edge of signal F1/Q is behind the rising edge of the clock CLK1, which is often referred to as the clock-to-Q-delay for the flip-flop. F1/Q takes a short time to arrive at F2/D. In a typical circuit, there may be combinational circuits between F1/Q and F2/D. This is the same issue when the timing delay from the combinational circuits is short. Due to clock skew, F2/D may arrive at F2 about the same time as the rising edge of CLK2. A setup timing error occurs when F2/D arrives at F2 before the rising edge of CLK2 and the timing difference is less than the required flip-flop setup time. A hold timing error occurs when F2/D arrives at F2 after the rising edge of CLK2 and the timing difference is less than the required flip-flop hold time. When setup or hold time violations occur, the F2/Q value is unknown and 'X' is generated in the simulation. In actual circuit, the value can be '0', '1' and the function of the circuit is unpredictable.

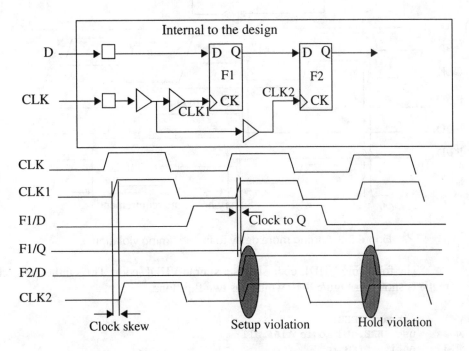

Figure 8-8 Timing violations caused by the clock skew.

The timing violation shown in Figure 8-8 is caused by the timing delay from F1/Q to F2/D being too short. This is often referred to as the short paths problem. To correct this type of timing problem, more delay is added between F1/Q to F2/D as shown in Figure 8-9. By adding more delay, F2/D signal changes follow the rising edge of CLK2 longer than the required hold time for the flip-flop. The timing violation is thus eliminated. This short path problem often causes the hold time violation. Adding delay to correct this problem is often referred to as the "fix_hold" process.

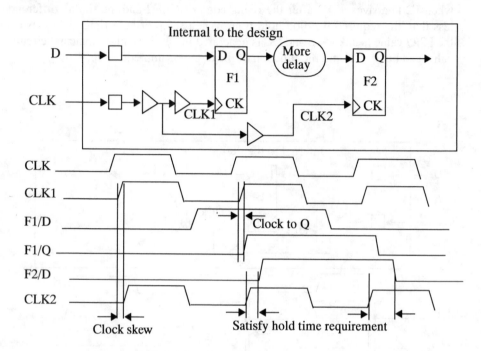

Figure 8-9 Adding more delay to fix the timing violation.

The following VHDL code shows a simple VHDL code. The synthesized schematic is shown in Figure 8-10 which has two flip-flops.

```
1     library IEEE;
2     use IEEE.std_logic_1164.all;
3     entity FIXB is
4       port (
5         CLK        : in  std_logic;
6         RSTn       : in  std_logic;
7         DIN        : in  std_logic;
8         DOUT       : out std_logic);
9     end FIXB;
10    architecture RTL of FIXB is
11      signal F1 : std_logic;
12    begin
13      ffgen1 : process (RSTn, CLK)
14      begin
15        if (RSTn = '0') then
16          DOUT  <= '0';
17          F1    <= '0';
18        elsif (CLK'event and CLK = '1') then
```

```
19            F1   <= DIN;
20            DOUT <= F1;
21         end if;
22      end process;
23   end RTL;
```

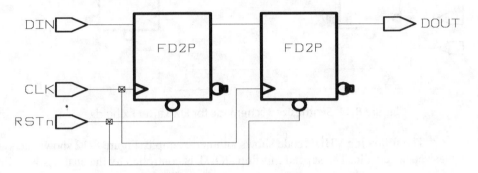

Figure 8-10 Synthesized schematic for FIXB.

To fix the short path problem, the following synthesis script can be used. Line 1 set the operating condition to be the best commercial condition. Lines 2, 5, and 6 tell the synthesis tool not to touch the clock, reset, input, and output signals. Line 4 declares the clock. Line 3 declared the clock skew as 0.8 ns. Line 8 updates the timing. Line 9 specifies that the clock clk is to be considered for fixing the short path problem. Line 10 synthesizes the circuit to add more delay. Figure 8-11 shows the schematic after the synthesis script is run. Note that the gate IVDA is added between two flip-flops.

```
1     set_operating_conditions BCCOM
2     set_dont_touch_network {CLK* RSTn*}
3     CLKSKEW   = 0.8
4     create_clock "CLK*" –name clk –period CLKPERIOD –waveform {0.0
HALFCLK}
5     set_dont_touch_network all_inputs()
6     set_dont_touch_network all_outputs()
7     set_clock_skew –ideal –uncertainty CLKSKEW clk
8     update_timing
9     set_fix_hold clk
10    compile –prioritize_min_paths –only_design_rule >
fix_hold.trans
```

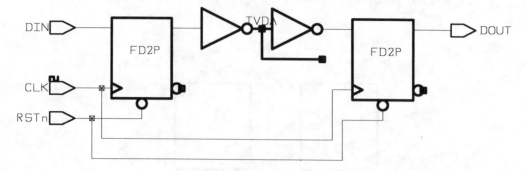

Figure 8-11 Synthesized schematic for FIXB after fix hold.

The following VHDL code shows another example. Figure 8-12 shows the synthesized schematic. The second flip-flop DOUT is controlled by the enable signal EN.

```
1    library IEEE;
2    use IEEE.std_logic_1164.all;
3    entity FIXA is
4      port (
5        CLK        : in  std_logic;
6        RSTn       : in  std_logic;
7        EN         : in  std_logic;
8        DIN        : in  std_logic;
9        DOUT       : out std_logic);
10   end FIXA;
11   architecture RTL of FIXA is
12     signal F1 : std_logic;
13   begin
14     ffgen1 : process (RSTn, CLK)
15     begin
16       if (RSTn = '0') then
17         DOUT  <= '0';
18         F1    <= '0';
19       elsif (CLK'event and CLK = '1') then
20         F1    <= DIN;
21         if (EN = '1') then
22           DOUT <= F1;
23         end if;
24       end if;
25     end process;
26   end RTL;
```

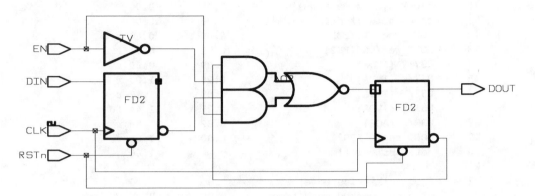

Figure 8-12 Synthesized schematic for FIXA.

The following synthesis script can also be used to correct the short-path problem. The script is approximately the same as above. In line 3, the clock skew is increased to be 1.5 ns. Line 5 specifies input delays for inputs EN and DIN. Figure 8-13 shows the schematic after buffers are added to correct the short-path problems.

```
1     set_operating_conditions BCCOM
2     set_dont_touch_network {CLK* RSTn*}
3     CLKSKEW    = 1.5
4     create_clock "CLK*" -name clk -period CLKPERIOD -waveform {0.0
HALFCLK}
5     set_input_delay CLKPER5 -add_delay -clock clk {EN DIN}
6     set_dont_touch_network all_outputs()
7     set_clock_skew -ideal -uncertainty CLKSKEW clk
8     update_timing
9     set_fix_hold clk
10    compile -prioritize_min_paths -only_design_rule >
fix_hold.trans
```

The following shows the timing report with the command in line 1. Line 2 shows the library and operating condition. Lines 12 and 13 show the clock to Q delay for the flip-flop. Lines 19 and 20 show that the data arrival time is 1.66 ns. It is compared with the clock skew (1.5 ns) and the flip-flop hold time (0.05 ns) as shown in lines 24 to 32.

```
1     report_timing -delay min -from */QN -to */D
2
3     Operating Conditions: BCCOM    Library: lca300k
4
5     Design           Wire Loading Model       Library
6     FIXA                  B9X9                 lca300k
```

	Point	Incr	Path	
7				
8	Point	Incr	Path	
9	---			
10	clock clk (rise edge)	0.00	0.00	
11	clock network delay (ideal)	0.00	0.00	
12	DOUT_reg/CP (FD2)	0.00	0.00	r
13	DOUT_reg/QN (FD2)	0.71	0.71	f
14	U37/Y (IVDA)	0.09	0.80	r
15	U37/Z (IVDA)	0.28	1.08	f
16	U35/Y (IVDA)	0.09	1.17	r
17	U35/Z (IVDA)	0.27	1.44	f
18	U32/Z (AO2)	0.22	1.66	r
19	DOUT_reg/D (FD2)	0.00	1.66	r
20	data arrival time		1.66	
21				
22	clock clk (rise edge)	0.00	0.00	
23	clock network delay (ideal)	0.00	0.00	
24	clock uncertainty	1.50	1.50	
25	DOUT_reg/CP (FD2)	1.50	1.50	r
26	library hold time	0.05	1.55	
27	data required time		1.55	
28	---			
29	data required time		1.55	
30	data arrival time		-1.66	
31	---			
32	slack (MET)		0.11	

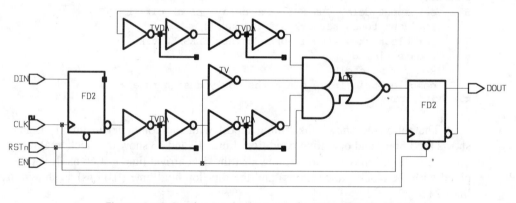

Figure 8-13 Synthesized schematic for FIXA after fix hold.

The following VHDL code shows another example where the first flip-flop is clocked with the falling edge of the clock. Figure 8-14 shows the synthesized schematic.

```
1    library IEEE;
2    use IEEE.std_logic_1164.all;
3    entity FIX is
4      port (
5        CLK          : in  std_logic;
6        RSTn         : in  std_logic;
7        EN           : in  std_logic;
8        DIN          : in  std_logic;
9        DOUT         : out std_logic);
10   end FIX;
11   architecture RTL of FIX is
12     signal F1 : std_logic;
13   begin
14     ffgen0 : process (RSTn, CLK)
15     begin
16       if (RSTn = '0') then
17         F1  <= '0';
18       elsif (CLK'event and CLK = '0') then
19         F1 <= DIN;
20       end if;
21     end process;
22     ffgen1 : process (RSTn, CLK)
23     begin
24       if (RSTn = '0') then
25         DOUT  <= '0';
26       elsif (CLK'event and CLK = '1') then
27         if (EN = '1') then
28           DOUT <= F1;
29         end if;
30       end if;
31     end process;
32   end RTL;
```

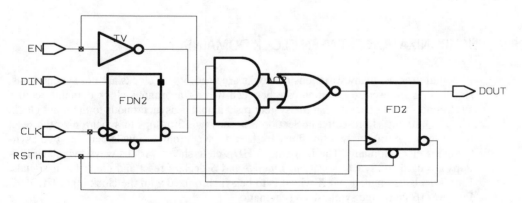

Figure 8-14 Synthesized schematic for FIX.

Using the same fix hold synthesis script as shown for the FIXA, Figure 8-15 shows the resulting schematic. Note that no extra delay is added between the two flip-flops. This is because the first flip-flop can only change value at the falling edge of the clock. It has a half clock cycle to arrive at the second flip-flop. Extra delay is added on the feedback loop for the second flip-flop. Is this necessary? This is left as Exercise 8.15.

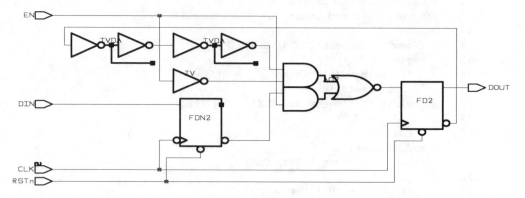

Figure 8-15 Synthesized schematic for FIX after fix hold.

There remains an issue — how large a number for the clock skew? As we have shown above, increasing the clock skew value from 0.8 ns to 1.5 ns causes more delay buffers to be added. The clock skew value is determined by how well the place and route tools can control the clock skew, the design size, and the number of flip-flops. Increasing the clock skew in the fix hold process will cause more buffers to be added, which makes the design bigger. Reducing the clock skew may make it harder for the place and route process to satisfy the timing constraints and may increase the number of place and route iterations, which extends the schedule.

8.5 SYNCHRONIZATION BETWEEN CLOCK DOMAINS

In digital designs, many variations of clocks are used. When clocks are not synchronous, signals that are used to communicate between two asynchronous clock domains require synchronization. It may be that the system reset signal is asynchronous to the input clock. The reset circuit as presented in Section 8.3 uses two flip-flops to synchronize the reset input signal. Similarly, two flip-flops are used to synchronize signals between two asynchronous clock domains. The following VHDL code shows various ways two-stage flip-flops are used for synchronization. Lines 5 and 6 declare reset and clock input signals. Line 7 declares input signal X which is to be synchronized with the clock input CLOCK. Figure 8-16 shows the synthesized schematic.

```
1    library IEEE;
2    use IEEE.std_logic_1164.all;
3    entity SYNC2 is
4      port(
5        RSTn     : in  std_logic;
6        CLOCK    : in  std_logic;
7        X        : in  std_logic;
8        Y        : out std_logic;
9        YS       : out std_logic;
10       YA       : out std_logic;
11       Y_I      : out std_logic;
12       YS_I     : out std_logic;
13       YA_I     : out std_logic);
14   end SYNC2;
15   architecture RTL of SYNC2 is
16     signal Y_FF, YS_FF, YA_FF       : std_logic;
17     signal Y_I_FF, YS_I_FF, YA_I_FF : std_logic;
18   begin
19     clk0 : process (CLOCK)
20     begin
21       if (CLOCK'event and CLOCK = '1') then
22         Y_FF     <= X;
23         Y        <= Y_FF;
24       end if;
25     end process;
26     clks : process
27     begin
28       wait until (CLOCK'event and CLOCK = '1');
29       if (RSTn = '0') then
30         YS_FF   <= '0';
31         YS      <= '0';
32         YS_I    <= '0';
33       else
34         YS_FF   <= X;
35         YS      <= YS_FF;
36         YS_I    <= YS_I_FF;
37       end if;
38       Y_I   <= Y_I_FF;
39     end process;
40     clkr : process (RSTn, CLOCK)
41     begin
42       if (RSTn = '0') then
43         YA_FF   <= '0';
44         YA      <= '0';
45         YA_I    <= '0';
46       elsif (CLOCK'event and CLOCK = '1') then
47         YA_FF   <= X;
48         YA      <= YA_FF;
49         YA_I    <= YA_I_FF;
```

```
50        end if;
51      end process;
52      clkfa : process (RSTn, CLOCK)
53      begin
54        if (RSTn = '0') then
55          YA_I_FF <= '0';
56        elsif (CLOCK'event and CLOCK = '0') then
57          YA_I_FF <= X;
58        end if;
59      end process;
60      clkfs : process
61      begin
62        wait until (CLOCK'event and CLOCK = '0');
63        if (RSTn = '0') then
64          YS_I_FF <= '0';
65        else
66          YS_I_FF <= X;
67        end if;
68        Y_I_FF  <= X;
69      end process;
70    end RTL;
```

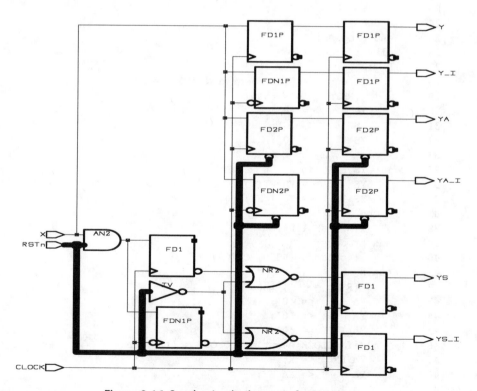

Figure 8-16 Synthesized schematic for SYNC2.

Refer to Figure 8-16. Output signal Y is obtained by two flip-flops that are both clocked with the rising edge. Output signal Y_I is similar to Y except that the first flip-flop is clocked by the falling edge of the clock. Signals YA and YA_I are similar to signals Y and Y_I, respectively, except that signals YA and YA_I are obtained with asynchronous reset flip-flops. Signals YS and YS_I are similar to signals YA and YA_I, respectively, except that synchronous reset is used for signals YS and YS_I.

Figure 8-17 shows a simulation waveform. Signal X is generated (somewhat randomly) based on the SRCCLK clock signal. The SRCCLK and CLOCK signals are asynchronous to each other. The first pulse of signal X is not caught by the rising edge of CLOCK. Signals Y, YS, and YA do not show any pulse. However, the first 3 X pulses are all caught by the falling edge of CLOCK. Signals Y_I, YS_I, and YA_I have a signal pulse of three clock cycles wide. The fourth X pulse is wide enough to cover both the rising and falling edge of the CLOCK. A pulse is generated for all six output signals. The fifth X pulse covers only the falling edge of CLOCK signal. Signals Y, YS, and YA have one clock cycle wide pulse where signals Y_I, YS_I, and YA_I have a pulse with two clock cycles wide. Figure 8-18 shows another portion of the SYNC2 simulation waveform. This is left for the reader as an Exercise 8.7 to verify the differences.

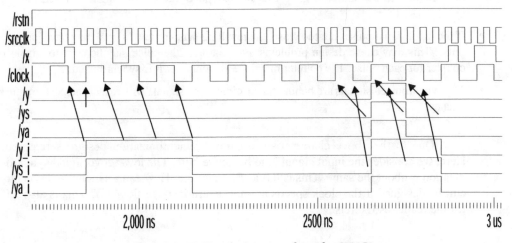

Figure 8-17 Simulation waveform for SYNC2.

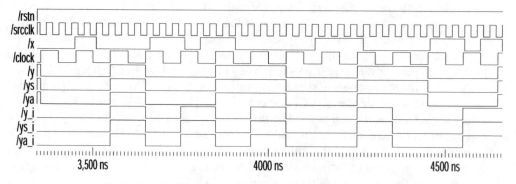

Figure 8-18 Simulation waveform for SYNC2 — continued.

From Figure 8-17 and Figure 8-18, the resulting synchronized signals can have the following characteristics:

• If the input signal pulse is too narrow, the synchronizing flip-flops may not catch it. Therefore, no pulse is generated. This implies that the synchronizing clock should be at least as fast as the clock that is used to generate the signal to be synchronized. If the relative clock periods between the source clock and the synchronizing clock are known, the length of the source signal can be extended before the signal is sent to be synchronized.

• If the input signal pulses are too close, the resulting synchronized signal may have just one pulse. This may cause design problems. For example, the pulse may be used to enable a counter for each pulse. The counter may lose a count after it is synchronized.

• The synchronized signals are behind the original input signal. This causes the performance latency.

Due to the above characteristics, careful synchronization design is required. Here, we consider the input signal X to be active high. The low-active signal synchronization is about the same. This is left as Exercise 8.8. How can a synchronization circuit be designed if the clock speeds are not determined (or flexible)? This is also left for the reader as Exercise 8.9.

8.6 CLOCK DIVIDER

In digital designs, a clock is often divided to get a slower speed clock. For example, an input clock may be divided by 2 to get a clock whose clock period is twice the original clock. The following VHDL code shows an example of a clock divider. Line 5 declares the input clock to be divided. The output signal D2CLK2TREE is the clock signal divided by 2. This D2CLK2TREE is then connected to the clock buffers tree for the flip-flops that use the divided-by-2 clock. One of the divided-by-2 clocks after the clock buffers tree is coming in as an input signal D2CLK that is used to synchronize the input RESETn signal to generate the output signal RSTn2TREE. The RSTn2TREE signal is then connected to the reset buffers tree to generate many reset signals.

Lines 15 to 20 perform the synchronization for the reset signal. Lines 21 to 25 synchronize the reset input signal with the falling edge of the original input clock to get the signal RSTn_sync0. Signal RSTn_sync0 is then going through another two stages of flip-flops as shown in lines 26 to 31. Line 32 generates an active high signal RSTn_FIRST to indicate the first cycle of the asserted reset signal. The RSTn_FIRST signal is used to control the state of the divided clock after the reset signal is first asserted. Line 33 shows the inverse of the divided clock to be output. In practice, both the divided clock and its inverse are generated and used, for example, as in TI TMS320C30 and TMS320C40 microprocessors.

```
1    library IEEE;
2    use IEEE.std_logic_1164.all;
3    entity CLKDIV is
4      port (
5        CLKIN        : in  std_logic;
6        D2CLK        : in  std_logic; -- after buffer tree
7        RESETn       : in  std_logic;
8        RSTn2TREE    : out std_logic;
9        D2CLK2TREE   : out std_logic);
10   end CLKDIV;
11   architecture RTL of CLKDIV is
12     signal RSTn_D2sync, FF, RSTn_DLY        : std_logic;
13     signal RSTn_sync0, RSTn_sync1, RSTn_FIRST : std_logic;
14   begin
15     rsyncd2 : process
16     begin
17       wait until (D2CLK'event and D2CLK = '1') ;
18       RSTn_D2sync    <= RESETn;
19       RSTn2TREE      <= RSTn_D2sync;
20     end process;
21     rsync0 : process
22     begin
23       wait until (CLKIN'event and CLKIN = '0') ;
24       RSTn_sync0 <= RESETn;
25     end process;
26     rsync1 : process
27     begin
28       wait until (CLKIN'event and CLKIN = '1') ;
29       RSTn_sync1 <= RSTn_sync0;
30       RSTn_DLY   <= RSTn_sync1;
31     end process;
32     RSTn_FIRST <= (not RSTn_DLY) nor (RSTn_sync1);
33     D2CLK2TREE <= not FF;
34     gen : process
35     begin
36       wait until (CLKIN'event and CLKIN = '1') ;
```

```
37          if (RSTn_FIRST = '1') then
38            FF <= '1';
39          else
40            FF <= not FF;
41          end if;
42       end process;
43     end RTL;
```

The following test bench is used to show how the clock divider works. Line 27
is used to model the clock buffers tree delay.

```
1     library IEEE;
2     use IEEE.std_logic_1164.all;
3     entity TBCLKDIV is
4     end TBCLKDIV;
5     architecture BEH of TBCLKDIV is
6       component CLKDIV
7       port (
8         CLKIN        : in  std_logic;
9         D2CLK        : in  std_logic; -- after buffer tree
10        RESETn       : in  std_logic;
11        RSTn2TREE    : out std_logic;
12        D2CLK2TREE   : out std_logic);
13      end component;
14      signal CLKIN, D2CLK, RESETn  : std_logic;
15      signal RSTn2TREE, D2CLK2TREE : std_logic;
16      constant HALFCLK : time := 50 ns;
17      constant DELAY   : time := HALFCLK / 2;
18      constant CLK_DLY : time := HALFCLK / 10;
19    begin
20      RESETn <= '1', '0' after HALFCLK*13 + DELAY,
21                     '1' after HALFCLK*27 + DELAY;
22      clkp : process
23      begin
24        CLKIN <= '0'; wait for HALFCLK;
25        CLKIN <= '1'; wait for HALFCLK;
26      end process;
27      D2CLK <= D2CLK2TREE after CLK_DLY;
28      clkdiv0 : CLKDIV port map (
29        CLKIN          => CLKIN,      D2CLK          => D2CLK,
30        RESETn         => RESETn,     RSTn2TREE      => RSTn2TREE,
31        D2CLK2TREE     => D2CLK2TREE);
32    end BEH;
33    configuration CFG_TBCLKDIV of TBCLKDIV is
34      for BEH
35      end for;
36    end CFG_TBCLKDIV;
```

Figure 8-19 shows the simulation waveform. The reset RESETn signal is asserted before the falling edge of the clock CLKIN signal. It is synchronized with the CLKIN signal and the RSTn_FIRST pulse is generated. The divided-by-2 clock is generated as shown in D2CLK2TREE. It changes from the unknown state 'U' to low right at the rising edge of CLKIN signal. D2CLK is behind D2CLK2TREE signal to model the clock buffers delay. The D2CLK clock is coming in to be used to synchronize the reset signal to generate the RSTn2TREE signal. Note that RSTn2TREE signal changes states at the same time as the rising edge of D2CLK, not CLKIN.

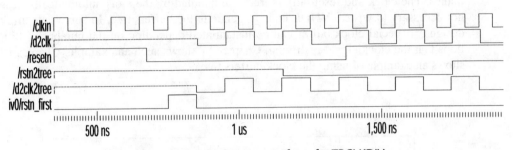

Figure 8-19 Simulation waveform for TBCLKDIV.

Figure 8-20 shows the synthesized schematic. Note that the schematic has not been fixed for the short path problem.

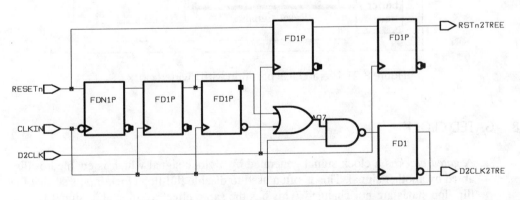

Figure 8-20 Synthesized schematic for CLKDIV.

Figure 8-21 shows a hierarchical partition circuit with the clock divider and the reset circuit in a block. This partition allows the clock and reset circuit to be synthesized. The block is kept distinct so that the design can be verified more easily. The clock buffers tree and the reset buffers tree are usually instantiated with synthesis library- specific components. The other synthesized circuits are done in one block that may be further partitioned horizontally and hierarchically. Putting the rest of the synthesized circuit in one block makes the synthesis and verification process easier. In FPGA designs, the clock and reset signals are usually routed with dedicated interconnects. A dedicated global set and reset (GSR) signal is usually required to take advantage of the dedicated interconnet resource for the reset signal. A special GSR component from the library can be instantiated to serve this purpose. This GSR signal may be the result of the synchronized power-up reset and the software reset. The FPGA synthesis and place and route tools require that the clock and reset signals be specified either by the VHDL attribute in the VHDL code or by the synthesis commands. The clock and reset buffers trees are handled by the tools automatically. It is not necessary to write a VHDL code to generate the buffers tree. In some gate array designs, a special clock buffer can be instantiated depending on the number of flip-flops that the clock signal is driving. Chapter 11 shows such an example. Chapter 10 shows an example of using the clock buffers tree.

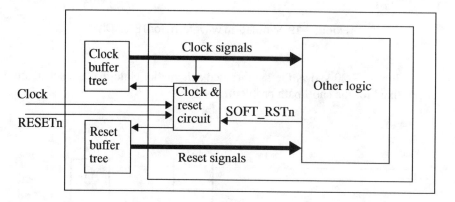

Figure 8-21 Hierarchical partition with divided clock.

8.7 GATED CLOCK

A *gated clock* is a clock signal generated by a clock signal which is enabled or disabled by another signal. This is often used to disable flip-flop clock inputs so that the flip-flop states are not changed. This has the same effect as an enable signal for the flip-flop. When the enable signal is deasserted, the flip-flop retains its state since the flip-flop is clocking in the old state value. However, by disabling the clock for the flip-flop, the power consumption for the flip-flop is reduced. This technique has been widely used in low-power designs.

The following VHDL code shows an example of a gated clock. Lines 6 and 7 declare the clock and reset signals. Line 8 declares an input signal PSAVE to tell the circuit to be in a power-save mode when this signal is asserted. Line 9 declares an output signal CNTR as an 8-bit counter. When PSAVE is asserted, the clock input of the 8-bit flip-flops for the counter is disabled and the 8-bit counter remains its value. The power control is usually controlled by a finite-state machine. Lines 17 to 33 model a state machine combinational circuit. Lines 34 to 41 model the finite state machine flip- flops.

```vhdl
1    library IEEE;
2    use IEEE.std_logic_1164.all;
3    use IEEE.std_logic_unsigned.all;
4    entity GATECLK is
5      port (
6         CLOCK        : in  std_logic;
7         RSTn         : in  std_logic;
8         PSAVE        : in  std_logic;
9         CNTR         : out std_logic_vector(7 downto 0));
10   end GATECLK;
11   architecture RTL of GATECLK is
12      type state_type is (NORMAL, SLEEP);
13      signal STATE, NX_STATE   : state_type;
14      signal GATE_CLK, CNTR_EN : std_logic;
15      signal CNTR_FF  : std_logic_vector(7 downto 0);
16   begin
17      comb : process (STATE, PSAVE)
18      begin
19        case STATE is
20          when NORMAL =>
21            if (PSAVE = '1') then
22              NX_STATE <= SLEEP;
23            else
24              NX_STATE <= NORMAL;
25            end if;
26          when SLEEP  =>
27            if (PSAVE = '1') then
28              NX_STATE <= SLEEP;
29            else
30              NX_STATE <= NORMAL;
31            end if;
32        end case;
33      end process;
34      seq : process (RSTn, CLOCK)
35      begin
36        if (RSTn = '0') then
37          STATE <= NORMAL;
38        elsif (CLOCK'event and CLOCK = '1') then
39          STATE <= NX_STATE;
40        end if;
41      end process;
42      seq0 : process
```

```
43       begin
44         wait until (CLOCK'event and CLOCK = '0') ;
45         if (STATE = SLEEP) then
46           CNTR_EN <= '0';
47         else
48           CNTR_EN <= '1';
49         end if;
50       end process;
51       GATE_CLK <= CNTR_EN and CLOCK;
52       gclk : process
53       begin
54         wait until (GATE_CLK'event and GATE_CLK = '1') ;
55         if (RSTn = '0') then
56           CNTR_FF <= "00000000";
57         else
58           CNTR_FF <= CNTR_FF + 1;
59         end if;
60       end process;
61       CNTR <= CNTR_FF;
62     end RTL;
63
```

Lines 42 to 52 use a negative-edge-triggered flip-flop to generate the CNTR_EN signal. The CNTR_EN signal is ANDed with the clock signal CLOCK in line 51. Lines 52 to 60 model the 8-bit counter. Figure 8-22 shows the simulation waveform. The CNTR_EN signal is used to mask out the rising edge of the clock CLOCK.

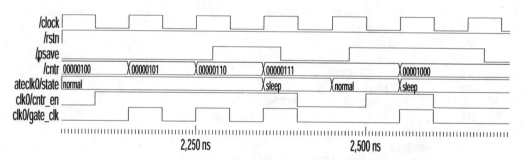

Figure 8-22 Simulation waveform for GATECLK.

Figure 8-23 shows the synthesized schematic. Note that CNTR_EN signal can be generated simply by delaying the inverted PSAVE signal by one half clock signal. The finite-state machine is not necessary in this case. Figure 8-23 shows the general framework of a finite state machine controller that can be used to control various gated clock signals in practice.

It is important to ensure that the gated clock signal contains no glitches. For example, the original clock pulse may last longer to overlap the CNTR_EN signal. A glitch will be generated at the time close to the falling edge of the clock signal. The counter will thus not operate correctly. The inverted clock signal to the flip-flop that generates the CNTR_EN signal may need to be delayed to ensure the glitch will not happen. The gated clock signal may be used to drive many flip-flops. It is recommended that the circuit generating the gated clock be treated in the same way as the clock circuit with its own block. Unless really necessary, a gated clock should be avoided. Thorough timing verification is required to ensure proper operation of the gated clock.

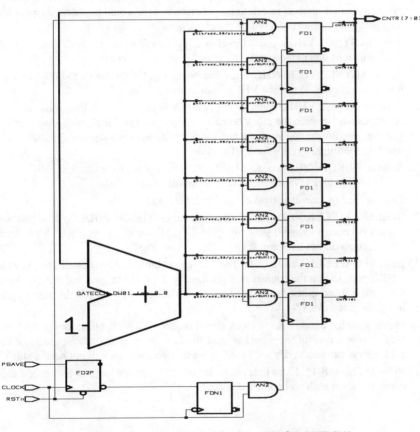

Figure 8-23 Synthesized schematic for GATECLK.

8.8 EXERCISES

8.1   Refer to TREE VHDL code. Modify the VHDL code to generate a clock buffers tree with three levels of clock buffers. Use a synthesis library you have access to. Develop synthesis script to generate the schematic.

8.2   What is the impact of the clock delay?

8.3   What problems can clock skew cause? How can these problems be resolved?

8.4   What are the advantages of having reset and clock circuitry in its own block separated from the rest of the circuits?

8.5   Compare the schematics in Figure 8-13 and Figure 8-15. The first stage flip-flop is clocked with the falling edge in the schematic. No extra buffers are inserted after the first flip-flop. What is the reason for this?

8.6   Figure 8-16 shows six ways of synchronizing circuits with two flip-flops. Analyze each synchronization method to determine whether buffers would be inserted or not to resolve the short path problems caused by the clock skew.

8.7   Refer to SYNC2 VHDL code. Develop a test bench to verify the SYNC2 VHDL code. Find situations where the synchronization behaviors are not desirable.

8.8   Refer to SYNC2 VHDL code. What can be changed if the signal to be synchronized is a low-active signals. Verify the VHDL code with a test bench.

8.9   Refer to SYNC2 VHDL code. Using two flip-flops, there is a limitation on the relative clock periods between the clock domains. Develop a synchronization circuit in VHDL so that the synchronization works at any clock period between the clock domains. Use a test bench to thoroughly verify the VHDL code.

8.10 Refer to CLKDIV VHDL code. Modify the VHDL code to generate the divided-by-2 and divided-by-4 clocks. Verify the VHDL code with simulations.

8.11 Can you design a divided-by-3 clock divider circuit?

8.12 Refer to GATECLK VHDL code. Redesign the GATECLK VHDL code so that the design margin is enough to avoid generating a glitch. How can you verify with confidence that your design will not have a glitch?

8.13 Refer to GATECLK VHDL code. The counter is controlled by the gated clock. Modify the VHDL code so that the counter is controlled by the enable signal and that the flip-flops do not have a gated clock input. Verify your VHDL code. What are the advantages of not using the gated clock?

8.14 In many design cases, a gated clock signal is used as a write pulse for the RAM or FIFO. How can you control the signal so that the high pulse is about 25 percent of the clock cycle? How can you verify with confidence that your design will not have a glitch?

8.15 Refer to Figure 8-15. Extra delay is added on the feedback loop for the second flip-flop to corret the short path problem. Is this necessary?

Chapter 9

Dual-Port RAM, FIFO, and DRAM Modeling

Basic ROM and RAM modeling has been illustrated in K.C. Chang, *Digital Design and Modeling with VHDL and Synthesis* (IEEE Computer Society Press, 1997). In this chapter, we will concentrate on modeling dual-port RAM, synchronous and asynchronous first in, first out (FIFO), and dynamic RAM (DRAM). DPR, FIFO, and DRAM, along with ROM and static RAM (SRAM) are typical custom blocks. They are usually provided by the foundry vendor, and they should never be synthesized with random gates. They should be instantiated in the design. These models can also represent actual integrated circuits (ICs). They are outside and to be interfaced or controlled by your design. They may be part of the test bench for the design. Learning to model these types of blocks allows designers to develop more robust test benches.

9.1 DUAL-PORT RAM

A dual-port RAM (DPR) is a RAM that has two access ports A and B. Figure 9-1 shows a general DPR block diagram. It consists of a memory array to store data. The memory array can be accessed (read or write) by two separate sets of control signals, address, and data buses. Port A address bus is A, data bus coming in is ADIN, data bus output from the memory is ADOUT, and the write enable pulse is AWR. Port B has the corresponding signals B, BDIN, BDOUT, and BWR. The data is stored in a single memory array. Therefore, an arbitration circuit is required to determine which port has the right to write to the memory when both ports are trying to update the data in the same address at the same time. The arbitration circuit will allocate control to one port, and display a busy signal to the other (ABUSYn and BBUSYn). The port which gains the control can proceed to complete the write operation. The write operation for the other port is ignored and discarded. Both ports can read data from the same address at the same time.

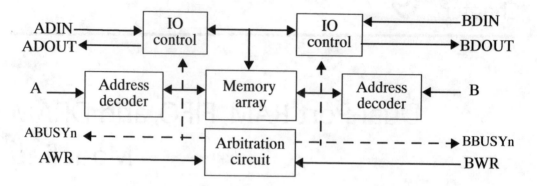

Figure 9-1 A Dual-Port RAM block diagram.

A DPR block is usually a custom design. There is no need to write synthesizable VHDL code to synthesize it into schematics of logic gates. However, to verify the correct operations with other circuits, a behavioral VHDL model is necessary. This behavioral VHDL model can be developed or provided by a vendor. The behavioral VHDL model can range from simple to very complex depending on timing and checking requirements. The following shows a simple DPR VHDL code to be used to verify the DPR functionality and its interface circuits. The port signals from lines 13 to 20 are described above. Generic N is the width of each data in the memory. Generic WORDS is the total number of memory words. Generic M is the width of the address bus (2**M = WORDS). Generics BPC, FLOORPLAN, and BUFFER_SIZE are used for the layout tool that does not affect the DPR functions.

```
1    library IEEE;
2    use IEEE.std_logic_1164.all;
3    use IEEE.std_logic_arith.all;
4    entity DUALRAM is
5      generic (
6        N           : integer := 4;
7        WORDS       : integer := 128;
8        M           : integer := 7;
9        BPC         : integer := 2;
10       FLOORPLAN   : integer := 0;
11       BUFFER_SIZE : string := "DEFAULT");
12     port (
13       A    : in  std_logic_vector(M-1 downto 0);
14       ADIN : in  std_logic_vector(N-1 downto 0);
15       AWR  : in  std_logic; -- low pulse to write
16       B    : in  std_logic_vector(M-1 downto 0);
17       BDIN : in  std_logic_vector(N-1 downto 0);
18       BWR  : in  std_logic; -- low pulse to write
```

```
19            ADOUT : out std_logic_vector(N-1 downto 0);
20            BDOUT : out std_logic_vector(N-1 downto 0));
21     end DUALRAM;
```

Line 23 declares a time constant DLY. Lines 24 and 25 define array subtype and types to be used in line 26 for the memory array. Line 27 declares signal ADDR_MATCH to indicate that the addresses in both ports are identical. Line 28 and declares signals ABUSYnS and BBUSYnS to indicate the busy situation for port A or port B, respectively. Line 29 declares signals writeAS and writeBS to indicate that the write operation for port A or port B, respectively.

```
22     architecture BEH of DUALRAM is
23       constant DLY : time := 3 ns;
24       subtype bitsn is std_logic_vector(N-1 downto 0);
25       type arytype is array(natural range <>) of bitsn;
26       signal MEM : arytype (0 to WORDS-1);
27       signal ADDR_MATCH : std_logic;
28       signal ABUSYnS, BBUSYnS : std_logic := '1';
29       signal writeAS, writeBS : std_logic := '1';
30     begin
31     ----------------------------------------------------------------
```

Line 32 assigns signal ADDR_MATCH. Lines 33 to 79 use a process statement with A, AWR, B, BWR, ADDR_MATCH, and MEM as the sensitivity list. Lines 34 to 35 declare local variables to indicate the write and busy status on both ports. Lines 37 to 46 check the status when AWR is asserted low. If port A is busy (as checked in line 38), the AWR write pulse is ignored. Line 39 checks whether the addresses are identical. It sets the ABUSYn, BBUSYn, and writeA variables. Line 40 indicates that ABUSYn is asserted when port B has not completed its write operation. In line 41, if B does not have a pending write operation, BBUSYn is asserted low and variable writeA is set when port B does not have a pending write operation. Line 43 always sets writeA variable when addresses do not match. This allows both ports to perform write operation, independently since their addresses are not the same.

```
32     ADDR_MATCH <= '1' when (A = B) and (not IS_X(A)) else '0';
33     p0 : process (A, AWR, B, BWR, ADDR_MATCH, MEM)
34       variable writeA, writeB : std_logic := '0';
35       variable ABUSYn, BBUSYn : std_logic := '1';
36     begin
37       if (AWR'event and AWR = '0') then
38         if (ABUSYn = '1') then
39           if (ADDR_MATCH = '1') then
40             ABUSYn := not writeB;
41             BBUSYn := writeB ; writeA := not writeB;
42           else
43             writeA := '1';
44           end if;
45         end if;
```

```
46        end if;
47        if (BWR'event and BWR = '0') then
48          if (BBUSYn = '1') then
49            if (ADDR_MATCH = '1') then
50              BBUSYn := not writeA; ABUSYn := writeA;
51              writeB := not writeA;
52            else
53              writeB := '1';
54            end if;
55          end if;
56        end if;
```

Similiarly, lines 47 to 56 handle the function when BWR is asserted low. Lines 57 to 63 check the situation when AWR is deasserted to '1'. Input data ADIN is written to the memory array addressed by A when port A is not busy and the writeA flag is set. In line 60, flag writeA is cleared to indicate that the write operation is completed. Lines 64 to 70 are similar to lines 57 to 63 that model the situation when BWR is deasserted.

```
57        if (AWR'event and AWR = '1') then
58          if (ABUSYn = '1') and (writeA = '1') then
59            MEM(conv_integer(unsigned(A))) <= ADIN;
60            writeA := '0'; BBUSYn := '1';
61          end if;
62          ABUSYn := '1';
63        end if;
64        if (BWR'event and BWR = '1') then
65          if (BBUSYn = '1') and (writeB = '1') then
66            MEM(conv_integer(unsigned(B))) <= BDIN;
67            writeB := '0'; ABUSYn := '1';
68          end if;
69          BBUSYn := '1';
70        end if;
71        if (ABUSYn = '1') then
72          ADOUT <= MEM(conv_integer(unsigned(A))) after DLY;
73        end if;
74        if (BBUSYn = '1') then
75          BDOUT <= MEM(conv_integer(unsigned(B))) after DLY;
76        end if;
77        ABUSYnS <= ABUSYn; BBUSYnS <= BBUSYn;
78        writeAS <= writeA; writeBS <= writeB;
79      end process;
80    end BEH;
```

Lines 71 to 76 model the read operation to select data from the memory to ADOUT and BDOUT. Lines 77 and 78 assign variables to signals.

To test the DPR model, the following test bench can be used. Lines 1 to 3 reference IEEE library and packages. Lines 4 to 5 declare entity TBDUALRAM without any port signals. Lines 7 to 24 declare the DUALRAM component. Lines 25 to 38 declare constants and signals to be used.

```
1     library IEEE;
2     use IEEE.std_logic_1164.all;
3     use IEEE.std_logic_unsigned.all;
4     entity TBDUALRAM is
5     end TBDUALRAM;
6     architecture BEH of TBDUALRAM is
7       component DUALRAM
8       generic (
9         N            : integer := 4;
10        WORDS        : integer := 128;
11        M            : integer := 7;
12        BPC          : integer := 2;
13        FLOORPLAN    : integer := 0;
14        BUFFER_SIZE  : string := "DEFAULT");
15      port (
16        A     : in  std_logic_vector(M-1 downto 0);
17        ADIN  : in  std_logic_vector(N-1 downto 0);
18        AWR   : in  std_logic; -- low pulse to write
19        B     : in  std_logic_vector(M-1 downto 0);
20        BDIN  : in  std_logic_vector(N-1 downto 0);
21        BWR   : in  std_logic; -- low pulse to write
22        ADOUT : out std_logic_vector(N-1 downto 0);
23        BDOUT : out std_logic_vector(N-1 downto 0));
24      end component;
25      constant  DLY     : time := 10 ns;
26      constant  PERIOD2 : time := 50 ns;
27      constant  N       : integer := 8;
28      constant  WORDS   : integer := 16;
29      constant  M       : integer := 4;
30      signal  A     : std_logic_vector(M-1 downto 0);
31      signal  ADIN  : std_logic_vector(N-1 downto 0);
32      signal  AWR   : std_logic; -- low pulse to write
33      signal  B     : std_logic_vector(M-1 downto 0);
34      signal  BDIN  : std_logic_vector(N-1 downto 0);
35      signal  BWR   : std_logic; -- low pulse to write
36      signal  ADOUT : std_logic_vector(N-1 downto 0);
37      signal  BDOUT : std_logic_vector(N-1 downto 0);
38      signal  CLK   : std_logic;
```

Lines 44 to 50 perform initialization. Lines 52 to 60 generate signals for write from port A with port B idle. The purpose of lines 60 to 62 is to produce a time gap in the simulation waveform so that each test interval is easily identified. Lines 64 to 73 repeat a similar process for port B.

```
39    begin
40    ----------------------------------------------------------------
41      p0 : process
42        variable D : std_logic_vector(N-1 downto 0);
43      begin
44        A    <= (A'range => '0');
45        ADIN <= (ADIN'range => '0');
46        AWR  <= '1';
47        B    <= (B'range => '0');
48        BDIN <= (BDIN'range => '0');
49        BWR  <= '1';
50        D    := (D'range => '0');
51        ---------------------------- write from A
52        wait until CLK'event and CLK = '1';
53        for j in 1 to WORDS loop
54          A    <= A + 1;
55          ADIN <= D;
56          AWR  <= '0' after DLY, '1' after PERIOD2;
57          wait until CLK'event and CLK = '1';
58          D    := D(N-2 downto 0) & not (D(N-1) xor D(0));
59        end loop;
60        for j in 0 to 4 loop
61          wait until CLK'event and CLK = '1';
62        end loop;
63        ---------------------------- write from B
64        for j in 1 to WORDS loop
65          B    <= B + 1;
66          BDIN <= D;
67          BWR  <= '0' after DLY, '1' after PERIOD2;
68          wait until CLK'event and CLK = '1';
69          D    := D(N-2 downto 0) & not (D(N-1) xor D(0));
70        end loop;
71        for j in 0 to 4 loop
72          wait until CLK'event and CLK = '1';
73        end loop;
```

Lines 75 to 115 is a loop statement that loops forever. The purpose of this section is to write to the memory with two ports at the same time. Lines 76 to 79 set up values for A, ADIN, B, and BDIN signals. Based on value D(0), B is set to the same value as A so that the arbitration can be tested. Lines 83 to 112 is a case statement to generate various combinations of AWR and BWR waveforms. For example, both AWR and BWR can be asserted at the same time or separately. They can also be deasserted at the same time or separately. They may be overlapping such that AWR is asserted (asserted) earlier than BWR is asserted (deasserted). AWR write pulse may be completely contained in the BWR pulse. An AWR pulse may overlap two BWR pulses.

```
74              -------------------------- write from both sides
75          loop
76            A    <= D(A'range);
77            ADIN <= not (D(N-1) xor D(0)) & D(N-1 downto 1);
78            B    <= B + 1;
79            BDIN <= D;
80            if (D(0) = '0') then -- force the same address
81              B <= D(B'range);
82            end if;
83            case D(2 downto 0) is
84              when "000" =>                      -- AWR first
85                AWR <= '0' after DLY,   '1' after PERIOD2;
86                BWR <= '0' after 2*DLY, '1' after PERIOD2;
87              when "001" =>                      -- BWR first
88                AWR <= '0' after 2*DLY, '1' after PERIOD2;
89                BWR <= '0' after DLY,   '1' after PERIOD2;
90              when "010" =>                       -- same time
91                AWR <= '0' after DLY, '1' after PERIOD2;
92                BWR <= '0' after DLY, '1' after PERIOD2;
93              when "011" =>     -- AWR covers BWR
94                AWR <= '0' after DLY,   '1' after PERIOD2 + DLY;
95                BWR <= '0' after 2*DLY, '1' after PERIOD2;
96              when "100" =>     -- BWR covers AWR
97                BWR <= '0' after DLY,   '1' after PERIOD2 + DLY;
98                AWR <= '0' after 2*DLY, '1' after PERIOD2;
99              when "101" =>     -- BWR overlap AWR
100               BWR <= '0' after DLY,   '1' after PERIOD2 - DLY;
101               AWR <= '0' after 2*DLY, '1' after PERIOD2;
102             when "110" =>     -- AWR overlap 2 BWR
103               AWR <= '0' after DLY,   '1' after 2*PERIOD2 - DLY;
104               BWR <= '0' after 2*DLY, '1' after PERIOD2,
105                            '0' after PERIOD2+DLY,
106                            '1' after 2*PERIOD2-2*DLY;
107             when others =>    -- BWR overlap 2 AWR
108               BWR <= '0' after DLY,   '1' after 2*PERIOD2 - DLY;
109               AWR <= '0' after 2*DLY, '1' after PERIOD2,
110                            '0' after PERIOD2+DLY,
111                            '1' after 2*PERIOD2-2*DLY;
112           end case;
113           wait until CLK'event and CLK = '1';
114           D    := D + 1;
115         end loop;
116       end process;
117       clkgen : process
118       begin
119         CLK <= '0'; wait for PERIOD2;
120         CLK <= '1'; wait for PERIOD2;
121       end process;
122       dualram0 : DUALRAM
123       generic map (
```

```
124      N              => N, WORDS      => WORDS,
125      M              => M, BPC        => 2,
126      FLOORPLAN   => 0, BUFFER_SIZE => "2")
127    port map (
128      A    => A,      ADIN  => ADIN,    AWR   => AWR,
129      B    => B,      BDIN  => BDIN,    BWR   => BWR,
130      ADOUT => ADOUT, BDOUT => BDOUT);
131    end BEH;
132    configuration CFG_TBDUALRAM of TBDUALRAM is
133      for BEH
134      end for;
135    end CFG_TBDUALRAM;
```

Line 113 proceeds one clock period. Line 114 increases D value by 1. Lines 117 to 121 generate the clock signal CLK. The clock signal CLK is not used in the design. It is used merely as a timing reference. Lines 122 to 130 instantiate the DUALRAMM component. Lines 132 to 135 set the configuration for the test bench.

Figure 9-2 shows the simulation waveform where the DPR is written with port A and port B is idle. Note that values XX appear before the data is written, since no data has been written to the memory array.

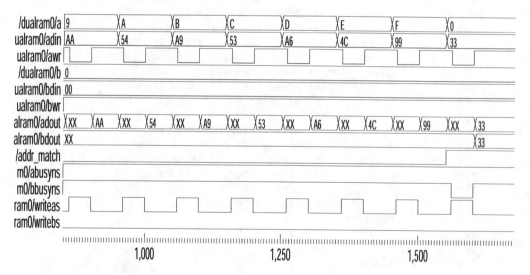

Figure 9-2 TBDUALRAM simulation waveform — 1.

Figure 9-3 is similar to Figure 9-2. Here, port B is used to write to the memory array. Figure 9-4 shows a few sequences where both ports are used to write. Note that AWR and BWR have various timing relationships. You are encouraged to verify the flag signals writeAS, writeBS, ABUSYnS, and BBUSYnS and the ADOUT and

BDOUT values. Figure 9-5 shows two sequences where a write pulse is wider than the other two write pulses.

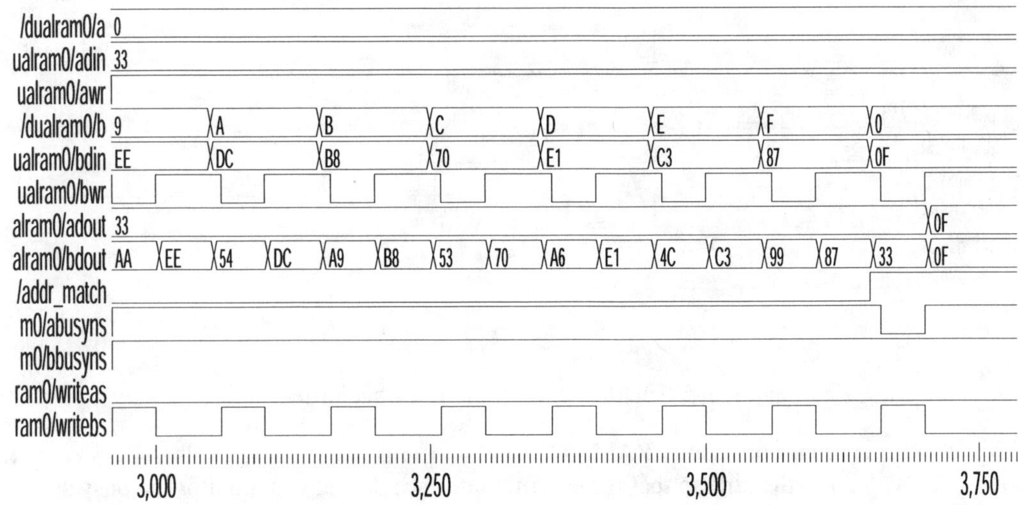

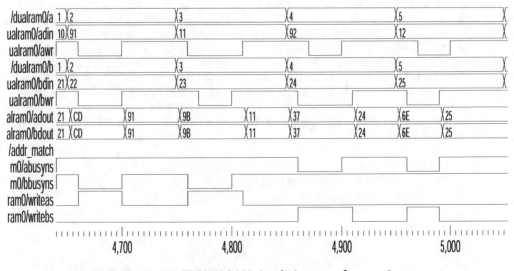

Figure 9-3 TBDUALRAM simulation waveform — 2.

Figure 9-4 TBDUALRAM simulation waveform — 3.

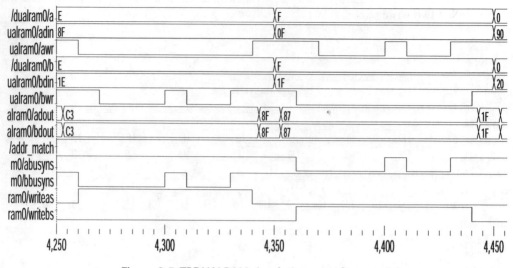

Figure 9-5 TBDUALRAM simulation waveform — 4.

A dual-port static RAM IC usually brings the busy signals ABUSYn and BBUSYn to the pins so that the external interface design can monitor the busy status. This is left as Exercise 9.4.

9.2 SYNCHRONOUS FIFO

A synchronous first in, first out (FIFO) is a data storage to model a queue so that the data that comes in first will be read out first. The synchronous behavior is ensured by writing (storing) and reading data synchronized with a clock. A general synchronous FIFO can be described in Figure 9-6. The data is pushed (written) through one side and read through the other side. The data is to be stored in data registers. A read pointer is used to select the data that is to be read when the read pulse signal Rn is asserted. A write pointer is used to select the data register to store the input data when the write pulse Wn is asserted. There is also a circuit to generate output signals FULLn and EMPTYn to tell the write side and the read side that the FIFO is full or empty. The following shows the VHDL entity of the synchronous FIFO.

```
1    library IEEE;
2    use IEEE.std_logic_1164.all;
3    use IEEE.std_logic_unsigned.all;
4    entity FIFOSYN is
5      port (
6        RSTn, CLK, Rn, Wn : in  std_logic;
7        EMPTYn, FULLn     : out std_logic;
8        D                 : in  std_logic_vector(7 downto 0);
9        Q                 : out std_logic_vector(7 downto 0));
10   end FIFOSYN;
```

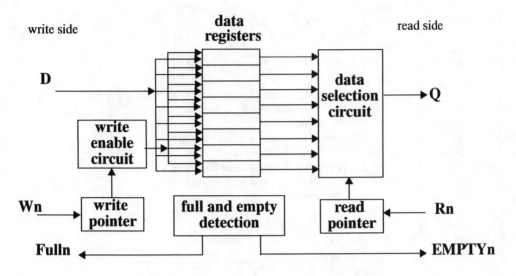

Figure 9-6 Synchronous FIFO block diagram.

The following shows the VHDL architecture of the FIFO. Lines 12 and 13 declare constant, M and N to represent the width of the data and the number of words in the FIFO, respectively. Line 14 declares signals RCNTR and WCNTR to be used as the read and write counters. Since there are eight words in the FIFO, 3 bits are required for the pointer. Lines 15 and 16 define subtype and type for the data registers. Line 17 declares a signal REG for all the data registers with the regtype as defined in line 16. Line 18 declares a signal RW to be used as the concatenation of Rn and Wn input signals as shown in line 21. Line 19 declares signals FULLnF and EMPTYnF for the full and empty flags, respectively. Lines 24 to 31 asynchronously reset the data registers, read pointer, write pointer, and empty and full flags.

```
11    architecture RTL of FIFOSYN is
12       constant M : integer := 8;
13       constant N : integer := 8;
14       signal RCNTR, WCNTR : std_logic_vector(2 downto 0);
15       subtype wrdtype is std_logic_vector(N-1 downto 0);
16       type regtype is array(0 to M-1) of wrdtype;
17       signal REG : regtype;
18       signal RW : std_logic_vector(1 downto 0);
19       signal FULLnF, EMPTYnF : std_logic;
20    begin
21       RW <= Rn & Wn;
22       seq : process (RSTn, CLK)
23       begin
24          if (RSTn = '0') then
```

```
25              RCNTR <= (RCNTR'range => '0');
26              WCNTR <= (RCNTR'range => '0');
27              EMPTYnF <= '0';
28              FULLnF  <= '1';
29              for j in 0 to M-1 loop
30                REG(j) <= (REG(j)'range => '0');
31              end loop;
32            elsif (CLK'event and CLK = '1') then
33              case RW is
34                when "00" => -- read and write at the same time
35                  RCNTR <= RCNTR + 1;
36                  WCNTR <= WCNTR + 1;
37                  REG(conv_integer(WCNTR)) <= D;
38                when "01" => -- only read
39                  if (EMPTYnF = '1') then -- not empty
40                    if (RCNTR+1) = WCNTR then
41                      EMPTYnF <= '0';
42                    end if;
43                    RCNTR <= RCNTR + 1;
44                  end if;
45                  FULLnF <= '1';
46                when "10" => -- only write
47                  EMPTYnF <= '1';
48                  if (FULLnF = '1') then -- not full
49                    REG(conv_integer(WCNTR)) <= D;
50                    if (WCNTR+1) = RCNTR then
51                      FULLnF <= '0';
52                    end if;
53                    WCNTR   <= WCNTR + 1;
54                  end if;
55                when others => null;
56              end case;
57          end if;
58        end process;
59        Q <= REG(conv_integer(RCNTR));
60        FULLn <= FULLnF;
61        EMPTYn <= EMPTYnF;
62      end RTL;
```

Line 32 checks the rising edge of the clock. Lines 33 to 56 use a case statement to describe the FIFO functions. Lines 34 to 37 implement the situation when read and write pulses are coming at the same clock cycle. Both read and write pointers are increased by 1 and the input data is stored in the data register pointer in the same clock cycle. Lines 38 to 45 implement the situation when only the read pulse is asserted. When the FIFO is not empty, empty flag EMPTYnF is set if the increase of the read pointer is the same as the write pointer. The read pointer is increased by one. The full flag FULLnF is deasserted to '1'. Lines 46 to 54 implement the situation when only the write pulse is asserted. The empty flag EMPTYnF is deasserted. If the FIFO is not full, the data is written into the data register. The FIFO is full when the write pointer is reaching the read pointer. Line 55 implements the situation when there is no read or write pulse asserted. The FIFO retains its status. Line 59 implements the data selection circuit to select a data from data registers to go to output A. Lines 60 to 61 connect empty and full flags to outputs FULLn and EMPTYn, respectively.

To verify the FIFO functionality, the following test bench can be used. Lines 6 to 12 declare component FIFOSYN. It is instantiated in lines 42 to 44. Line 20 generates the reset signal RSTn. Lines 21 to 25 generate the clock signal CLK. Lines 26 to 41 is a process statement that can continue to run. The process assigns values to D, Rn, and Wn signals. Line 28 declares a variable CNT to be used as a counter. Lines 31 to 32 write to the FIFO until it is full. This is achieved by deasserting Rn so that the data written to FIFO is not read out. Lines 33 to 34 read data out until the FIFO is empty. Otherwise, the Rn and Wn signals are randomly generated based on bit 2 and bit 6 of the variable V, respectively. Variable V is updated in line 39 using the linear feedback of its previous value. Lines 46 to 52 provides the VHDL configuration information.

```
1      library IEEE;
2      use IEEE.std_logic_1164.all;
3      entity TBFIFOSYN is
4      end TBFIFOSYN;
5      architecture BEH of TBFIFOSYN is
6        component FIFOSYN
7        port (
8          RSTn, CLK, Rn, Wn : in  std_logic;
9          EMPTYn, FULLn     : out std_logic;
10         D                 : in  std_logic_vector(7 downto 0);
11         Q                 : out std_logic_vector(7 downto 0));
12       end component;
13       constant  PERIOD2    : time := 25 ns;
14       constant  DLY        : time := 10 ns;
15       signal RSTn, CLK, Rn, Wn : std_logic;
16       signal EMPTYn, FULLn     : std_logic;
17       signal D, Q              : std_logic_vector(7 downto 0);
18     begin
19     ---------------------------------------------------------------
20       RSTn <= '0', '1' after 6*PERIOD2;
21       clkgen : process
22       begin
23         CLK <= '0'; wait for PERIOD2;
24         CLK <= '1'; wait for PERIOD2;
25       end process;
26       p0 : process
27         variable V : std_logic_vector(7 downto 0) := "00000000";
28         variable CNT : integer := 0;
29       begin
30         D <= V     after DLY;
31         if (CNT > 10) and (CNT < 20) then
32           Rn <= '1' after DLY; Wn <= '0' after DLY;
33         elsif (CNT > 20) and (CNT < 30) then
```

```
34              Rn <= '0' after DLY; Wn <= '1' after DLY;
35          else
36              Rn <= V(2) after DLY; Wn <= V(6) after DLY;
37          end if;
38          wait until CLK'event and CLK = '1';
39          V   := V(6 downto 0) & (V(7) xor (not V(2)));
40          CNT := CNT + 1;
41      end process;
42      fifosyn0 : FIFOSYN port map (
43          RSTn    => RSTn,   CLK   => CLK,    Rn => Rn, Wn => Wn,
44          EMPTYn => EMPTYn, FULLn => FULLn, D   => D, Q => Q);
45   end BEH;
46   configuration CFG_TBFIFOSYN of TBFIFOSYN is
47      for BEH
48          for all : FIFOSYN
49              use entity work.FIFOSYN(RTL);
50          end for;
51      end for;
52   end CFG_TBFIFOSYN;
```

Figure 9-7 shows part of the simulation waveform. After the RSTn is deasserted, the write pulse Wn is asserted. Four data values 07, 0E, 1C, and 38 are written into the data registers with the write pointers incrementing from 0 to 4. Note that the first value 07 has already shown up on the Q output. When the read pulse is asserted, the read pointer is increased by 1 and the next data is shown in output Q. This shows that the data on the output bus is available to be read. The read control circuit should get the data (before it is gone) and assert the read pulse in the same clock cycle. Three data are read. The read pointer is pointing to the unread data register indexed 3. Write pulse Wn is again asserted with seven data values 8D, 1B, 37, 6E, DC, B9, and 72 written into data registers. We have written four values, read three values, and written seven values. All the data in the data registers remains unread. The FIFO is full and the full flag FULLn is therefore asserted.

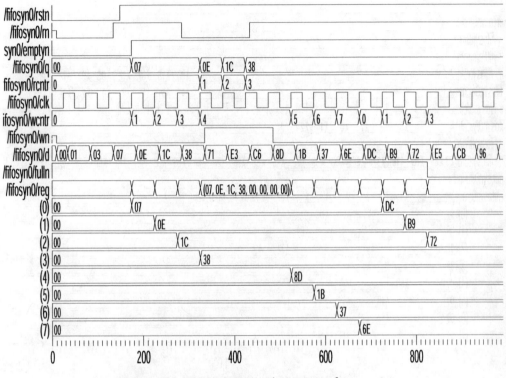

Figure 9-7 TBFIFOSYN simulation waveform — 1.

Figure 9-8 shows another part of the simulation waveform. It illustrates the conditions that cause the FIFO to be empty or full. The test bench also randomly generates Rn and Wn pulses to verify the FIFO operation. More simulation is left as Exercise 9.5. The synthesis of FIFOSYN VHDL code is also left as Exercise 9.6.

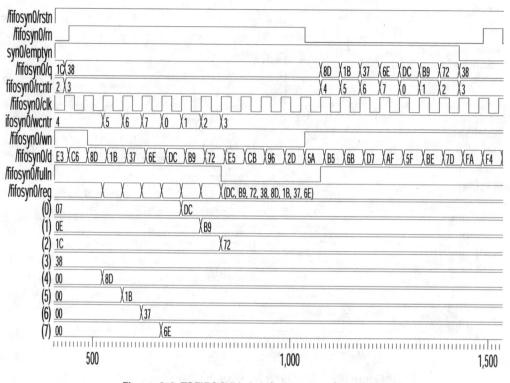

Figure 9-8 TBFIFOSYN simulation waveform — 2.

9.3 ASYNCHRONOUS FIFO

An asynchronous FIFO is similar to a synchronous FIFO except that the read and write pulses are not synchronous with the same clock. Figure 9-9 shows an asynchronous FIFO block diagram. A DPR is used to store the data. One port of the DPR is used to push the data. The data is read from the other port. There are also a read pointer and a write pointer to remember the read and write addresses for the DPR. The read and write addresses are compared to generate the full and empty flags. The input data D is connected to the DPR write port directly. The output data from the DPR read port is latched.

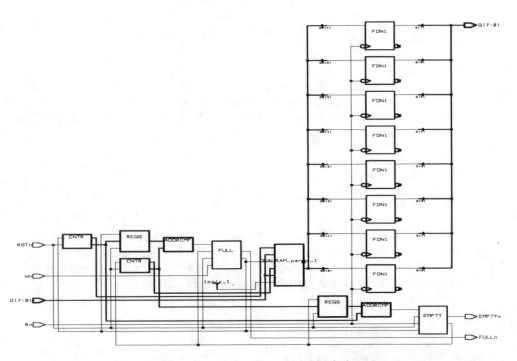

Figure 9-9 An Asynchronous FIFO block diagram.

We have discussed the DPR functionality and a VHDL model. Let's describe each component. The following VHDL code is used to model a counter CNTR to be used for the read or write pointer. A package CONST that defines constants M, N, and W to indicate the read (or write) pointer width, word width, and the number of words in the DPR is referenced in line 4. The counter is simple. The input clock CLK is delayed by 1 ns in line 14 to generate a signal CLKS. Signal CLKS is used to clock the counter. Both the counter value and its inverted value are output as shown in lines 23 to 24. The reason that the clock input is delayed by 1 ns is to allow the signals from the empty and full detection circuit to arrive. This is an asynchronous FIFO. Depending on the implementation process technology, the delay (of 1 ns) needs to be carefully designed to ensure the correct circuit operations.

```
1     library IEEE;
2     use IEEE.std_logic_1164.all;
3     use IEEE.std_logic_unsigned.all;
4     use work.CONST.all;
5     entity CNTR is
6       port (
7         CLK, RSTn : in  std_logic;
8         Q, Qn     : out std_logic_vector(M-1 downto 0));
```

```
9     end CNTR;
10    architecture RTL of CNTR is
11      signal FF : std_logic_vector(Q'range);
12      signal CLKS : std_logic;
13    begin
14      CLKS <= CLK after 1 ns;
15      ffs : process (RSTn, CLKS)
16      begin
17        if (RSTn = '0') then
18          FF  <= (FF'range => '0');
19        elsif (CLKS'event and CLKS = '1') then
20          FF <= FF + 1;
21        end if;
22      end process;
23      Q  <= FF;
24      Qn <= not FF;
25    end RTL;
```

The following VHDL code shows the register block REGS. Similarly, the clock input signal is delayed in line 13 by 2 ns. The flip-flop is inferred in lines 14 to 21. The REGS block is used to latch the read and write pointer value so that they are stable and can be compared in the address compare block ADDRCMP. The ADDRCMP VHDL is also shown below.

```
1     library IEEE;
2     use IEEE.std_logic_1164.all;
3     use work.CONST.all;
4     entity REGS is
5       port (
6         CLK, RSTn : in  std_logic;
7         RI        : in  std_logic_vector(M-1 downto 0);
8         RO        : out std_logic_vector(M-1 downto 0));
9     end REGS;
10    architecture RTL of REGS is
11      signal CLK_DLY : std_logic;
12    begin
13      CLK_DLY <= CLK after 2 ns;
14      ffs : process (RSTn, CLK_DLY)
15      begin
16        if (RSTn = '0') then
17          RO <= (RO'range => '0');
18        elsif (CLK_DLY'event and CLK_DLY = '1') then
19          RO <= RI;
20        end if;
21      end process;
22    end RTL;
```

```
1    library IEEE;
2    use IEEE.std_logic_1164.all;
3    use IEEE.std_logic_1164.all;
4    use work.CONST.all;
5    entity ADDRCMP is
6      port (
7        A, B : in  std_logic_vector(M-1 downto 0);
8        EQn  : out std_logic);
9    end ADDRCMP;
10   architecture RTL of ADDRCMP is
11   begin
12     EQn <= '0' when A = B else '1';
13   end RTL;
```

The full flag is generated by block FULL. The VHDL code controlling the full flag is shown below. The full flag generation circuit has input Wn (write pulse), read clock (RCLKn), and equal (EQn from the compare ADDRCMP block). It has outputs FULLn and write clock WCLKn. Lines 13 to 15 generate delayed signals RCLKn_DLY, FULLnF_SETn, and FULLnF_SETn_DLY. A latch enable signal LATCH_EN is used as the clock signal for a flip-flop, which is inferred in lines 17 to 24. A latch is also inferred in lines 25 to 32. Lines 33 to 35 generate signal WCLKnS and output signals WCLKn and FULLn.

```
1    library IEEE;
2    use IEEE.std_logic_1164.all;
3    entity FULL is
4      port (
5        RSTn, Wn, RCLKn, EQn : in  std_logic;
6        FULLn, WCLKn         : out std_logic);
7    end FULL;
8    architecture BEH of FULL is
9      signal FULLnF, FULLnL, WCLKnS : std_logic;
10     signal RCLKn_DLY, FULLnF_SETn : std_logic;
11     signal FULLnF_SETn_DLY, LATCH_EN : std_logic;
12   begin
13     RCLKn_DLY <= not RCLKn after 7 ns;
14     FULLnF_SETn <= RCLKn_DLY and RSTn;
15     FULLnF_SETn_DLY <= not FULLnF_SETn after 3 ns;
16     LATCH_EN <= FULLnF_SETn_DLY or WCLKnS;
17     ff : process (FULLnF_SETn, LATCH_EN)
18     begin
19       if (FULLnF_SETn = '0') then
20         FULLnF <= '1';
21       elsif (LATCH_EN'event and LATCH_EN = '0') then
22         FULLnF <= EQn;
23       end if;
24     end process;
25     latch : process (RSTn, FULLnF, LATCH_EN)
26     begin
```

```
27         if (RSTn = '0') then
28           FULLnL <= '1';
29         elsif (LATCH_EN = '1') then
30           FULLnL <= not FULLnF;
31         end if;
32       end process;
33     WCLKnS <= (not Wn) nand (not FULLnL);
34     WCLKn <= WCLKnS;
35     FULLn <= FULLnF;
36   end BEH;
```

The empty flag detection circuit VHDL code is shown below. It is very similar to the full flag detection circuit.

```
1     library IEEE;
2     use IEEE.std_logic_1164.all;
3     entity EMPTY is
4       port (
5         EQn, WCLKn, Rn, RSTn : in  std_logic;
6         RCLKn, EMPTYn        : out std_logic);
7     end EMPTY;
8     architecture BEH of EMPTY is
9       signal EMPTYnF, EMPTYnL : std_logic; -- flag ff and latch
10      signal WCLKn_DLY, EMPTYn_D, EMPTYn_D_DLY : std_logic;
11      signal LATCH_EN, EMPTYn_CP, RCLKnS       : std_logic;
12    begin
13      WCLKn_DLY    <= not WCLKn after 7 ns;
14      EMPTYn_D     <= WCLKn nand WCLKn_DLY;
15      EMPTYn_D_DLY <= not EMPTYn_D after 4 ns;
16      LATCH_EN     <= EMPTYn_D_DLY or RCLKnS;
17      EMPTYn_CP    <= not LATCH_EN after 3 ns;
18      emptyff : process (RSTn, EMPTYn_D, EMPTYn_CP)
19      begin
20        if (RSTn = '0') then
21          EMPTYnF <= '0';
22        elsif (EMPTYn_D = '0') then
23          EMPTYnF <= '1';
24        elsif (EMPTYn_CP'event and EMPTYn_CP = '0') then
25          EMPTYnF <= EQn;
26        end if;
27      end process;
28      emptylatch : process (RSTn, LATCH_EN, EMPTYnF)
29      begin
30        if (RSTn = '0') then
31          EMPTYnL <= '0';
32        elsif (LATCH_EN = '1') then
33          EMPTYnL <= EMPTYnF;
34        end if;
35      end process;
36      EMPTYn <= EMPTYnF;
```

```
37      RCLKnS <= not Rn nand EMPTYnL;
38      RCLKn  <= RCLKnS;
39    end BEH;
```

The following shows the complete asynchronous FIFO VHDL model. It instantiates components ADDRCMP, REGS, CNTR, FULL, EMPTY, and DUALRAM blocks as shown in the Figure 9-9 schematic. The output from the DUALRAM is registered 93 to 97. Note that the flip-flop clock signal is generated in line 92.

```
1     library IEEE;
2     use IEEE.std_logic_1164.all;
3     use work.CONST.all;
4     entity FIFOWXN is
5       port (
6         RSTn, Rn, Wn  : in  std_logic;
7         EMPTYn, FULLn : out std_logic;
8         D             : in  std_logic_vector(N-1 downto 0);
9         Q             : out std_logic_vector(N-1 downto 0));
10    end FIFOWXN;
11    architecture RTL of FIFOWXN is
12      component CNTR
13      port (
14        CLK, RSTn : in  std_logic;
15        Q, Qn     : out std_logic_vector(M-1 downto 0));
16      end component;
17      component REGS
18      port (
19        CLK, RSTn : in  std_logic;
20        RI        : in  std_logic_vector(M-1 downto 0);
21        RO        : out std_logic_vector(M-1 downto 0));
22      end component;
23      component ADDRCMP
24      port (
25        A, B : in  std_logic_vector(M-1 downto 0);
26        EQn  : out std_logic);
27      end component;
28      component EMPTY
29      port (
30        EQn, WCLKn, Rn, RSTn : in  std_logic;
31        RCLKn, EMPTYn        : out std_logic);
32      end component;
33      component FULL
34      port (
35        RSTn, Wn, RCLKn, EQn : in  std_logic;
36        FULLn, WCLKn         : out std_logic);
37      end component;
38      component DUALRAM
39      generic (
40        N         : integer := 4;
41        WORDS     : integer := 128;
```

```
42       M              : integer := 7;
43       BPC            : integer := 2;
44       FLOORPLAN      : integer := 0;
45       BUFFER_SIZE : string := "DEFAULT");
46    port (
47       A     : in  std_logic_vector(M-1 downto 0);
48       ADIN  : in  std_logic_vector(N-1 downto 0);
49       AWR   : in  std_logic; -- low pulse to write
50       B     : in  std_logic_vector(M-1 downto 0);
51       BDIN  : in  std_logic_vector(N-1 downto 0);
52       BWR   : in  std_logic; -- low pulse to write
53       ADOUT : out std_logic_vector(N-1 downto 0);
54       BDOUT : out std_logic_vector(N-1 downto 0));
55    end component;
56    signal QS                : std_logic_vector(N-1 downto 0);
57    signal RAD, WAD          : std_logic_vector(M-1 downto 0);
58    signal RADn, WADn        : std_logic_vector(M-1 downto 0);
59    signal RADn2CMP, WADn2CMP : std_logic_vector(M-1 downto 0);
60    signal EMPTYCMP, FULLCMP, WCLKn, RCLKn : std_logic;
61    signal TIEHIGH, RnS : std_logic;
62  begin
63    TIEHIGH <= '1';
64    rcntr : CNTR port map (
65      CLK => RCLKn, RSTn => RSTn, Q => RAD, Qn => RADn);
66    regsr : REGS port map (
67      CLK => RCLKn, RSTn => RSTn,
68      RI  => RADn,  RO   => RADn2CMP);
69    addrcmpr : ADDRCMP port map (
70      A => RADn2CMP, B => WADn, EQn => EMPTYCMP);
71    rcntw : CNTR port map (
72      CLK => WCLKn, RSTn => RSTn, Q => WAD, Qn => WADn);
73    regsw : REGS port map (
74      CLK => WCLKn, RSTn => RSTn,
75      RI  => WADn,  RO   => WADn2CMP);
76    addrcmpw : ADDRCMP port map (
77      A => WADn2CMP, B => RADn, EQn  => FULLCMP);
78    empty0 : EMPTY port map (
79      EQn => EMPTYCMP, WCLKn => WCLKn, Rn => Rn, RSTn => RSTn,
80      RCLKn => RCLKn, EMPTYn => EMPTYn);
81    full0 : FULL port map (
82      RSTn => RSTn, Wn => Wn, RCLKn => RCLKn, EQn => FULLCMP,
83      FULLn => FULLN, WCLKn => WCLKn);
84    dualram0 : DUALRAM generic map (
85      N              => N, WORDS        => W,
86      M              => M, BPC          => 2,
87      FLOORPLAN   => 0, BUFFER_SIZE => "DEFAULT")
88    port map (
89      A      => WAD,  ADIN => D, AWR   => WCLKn,
90      B      => RAD,  BDIN => D, BWR   => TIEHIGH,
91      ADOUT => open, BDOUT => QS);
```

```
92      RnS <= Rn after 1 ns;
93      rq : process
94      begin
95        wait until RnS'event and RnS = '0';
96        Q <= QS;
97      end process;
98    end RTL;
```

A VHDL test bench can be used to verify the asynchronous FIFO model. The test bench is not shown here and left as Exercise 9.7. Figures 9-10 and 9-11 show part of the simulation waveform. In Figure 9-10, the FIFO is first reset by asserting FIFO_RSTn. After the FIFO is reset, a valid value is written to the FIFO and then read right away so that the FIFO output Q will have a valid value. This is usually done by the FIFO controller, which is not described here. Note that we also use two clocks: one for the write side and one for the read side. They are asynchronous. The read and write pulses are somewhat randomly generated. The first two data written (after the first initialization Wn pulse) are 01 and 03 and 1C when Wn is asserted. They are read back when Rn is asserted. At the end of Figure 9-10, the empty flag EMPTYn is asserted since 6 data were written and 6 data were read (not counting the initialization write and read).

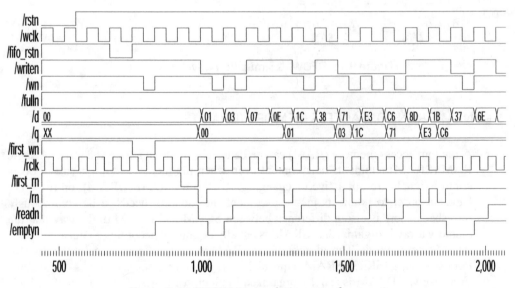

Figure 9-10 FIFOWXN simulation waveform — 1.

The test bench is changed by making the write clock faster than the read clock. The FIFO is eventually full, as the full flag FULLn is asserted at the end of Figure 9-11 simulation waveform.

Note also the timing relationship between the write clock Wn and the write data. Here the write data is stable and the falling edge of the Wn pulse is just about in the middle of the stable data. The read data comes out very soon after the read pulse Rn falling edge. These data are also taken into consideration for the FIFO controller design. More FIFO-related designs are left as Exercises 9.9 and 9.10.

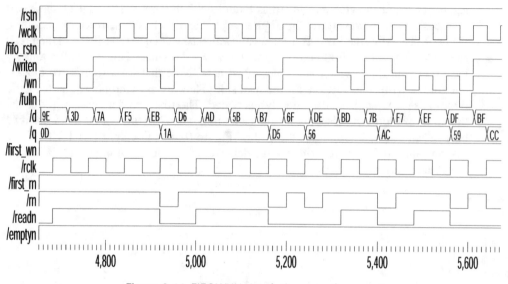

Figure 9-11 FIFOWXN simulation waveform — 2.

9.4 DYNAMIC RANDOM ACCESS MEMORY (DRAM)

A random access memory (RAM) can be categorized into dynamic and static. A static RAM (SRAM) retains its data content even when the memory is not accessed or its control signals are not active or changing. A dynamic RAM (DRAM), on the other hand, in order to retain its memory content the memory needs to be continuously refreshed. There is also a difference between SRAM and DRAM in the way that the memory array is organized. A SRAM is usually organized according to linear addressing scheme of words. The address of the SRAM is the full linear range of the SRAM. For example, a 16Kx32 SRAM requires 14 bits of address bus and 32 bits of data bus. Conversely, a DRAM is usually organized in blocks of pages. The address of an individual word is composed of the row address and the column address. These addresses are communicated on the same address bus. In practice, the size of the row address and the column address are usually the same in order to take full advantage of the address bus. The size of the column address depends on the page size, which is the number of individual addressable units (words, bytes, 36 bits). Figure 9-12 shows a DRAM block diagram. The memory is organized into 512 pages and each page has

512x16 bits. The address is 9 bits wide. The row address comes in first with control signal RASn asserted. The column address comes later with CASn asserted. The address goes through the address buffers before they are decoded. Write operation is enabled by the WEn pulse. The read operation is controlled by the OEn signal. Bidirectional data bus DQ gets the data output form the data output buffer. Input data from DQ goes to the data input buffer before writing into the memory through the sense amplifier.

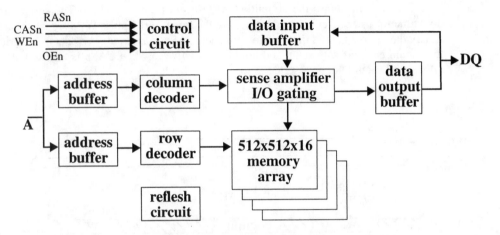

Figure 9-12 DRAM block diagram.

To further understand how a DRAM works, let's use the MT4C16256 256Kx16 DRAM manufactured by Micron as an example. Figure 9-13 shows the simplified timing diagram for the read cycle.

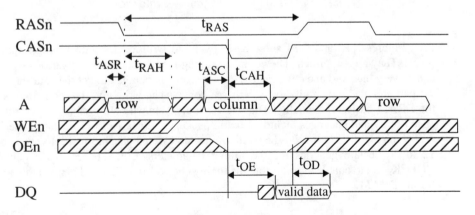

Figure 9-13 DRAM timing diagram — read cycle.

Address A is treated as a row address when RASn is asserted. The row address should satisfy the row address setup time t_{ASR} and hold time t_{RAH}. Similarly, address A is treated as a column address when CASn is asserted with RASn remaining asserted. The column address should satisfy the column address setup time t_{ASC} and hold time t_{CAH}. After the row and column addresses have been supplied, OEn signal can be asserted (with WEn asserted high) to read the data out. The valid data comes out of the data bus DQ after OEn is asserted with delay t_{OE}. The output data should hold stable t_{OD} after OEn is deasserted.

Figure 9-14 shows the simplified write cycle timing diagram. The setup and hold time for row and column addresses are the same. Write pulse WEn signal should satisfy the setup and hold time relative to when CASn is asserted. The data to be written should be valid with setup and hold time relative to when CASn is asserted. The OEn signal should not be asserted.

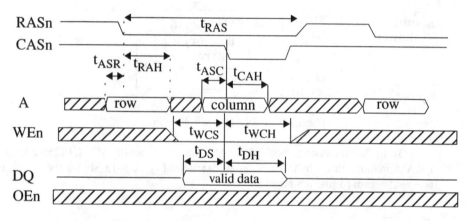

Figure 9-14 DRAM timing diagram — write cycle.

Figures 9-13 and 9-14 show basic read and write cycles with timing requirements. They are simplified. The complete timing requirements among various signals can be very involved and complicated. Readers are encouraged to get the data sheet of the part and read it carefully. Now, let's see how we can write a VHDL code to model the DRAM basic functionality. The VHDL code is shown below. Lines 1 to 6 reference library and packages including textio packages. Lines 10 to 11 declare generics ASIZE and DSIZE to specify the width of the address and data bus respectively. Lines 12 to 20 declare generics for the timing parameters. Line 21 declares a generic CHECK_ON to enable or to disable the timing check. Lines 22 to 28 declare the ports for the DRAM.

```
1    library IEEE;
2    use IEEE.std_logic_1164.all;
3    use IEEE.std_logic_unsigned.all;
4    use IEEE.std_logic_arith.all;
5    use STD.textio.all;
6    use IEEE.std_logic_textio.all;
7    ----------------------------------------------------------------
8    entity DRAM is
9      generic(
10       ASIZE    : integer   :=  9    ; -- address width
11       DSIZE    : integer   := 16    ; -- data width
12       T_ASR    : time      :=  0 ns ; -- row address setup
13       T_RAH    : time      := 10 ns ; -- row address hold
14       T_ASC    : time      :=  0 ns ; -- column address setup
15       T_CAH    : time      := 10 ns ; -- column address hold
16       T_OE     : time      := 15 ns ; -- output enable time
17       T_OD     : time      :=  3 ns ; -- output disable time
18       T_DS     : time      :=  0 ns ; -- write data setup time
19       T_DH     : time      :=  8 ns ; -- write data hold  time
20       T_OFF    : time      :=  3 ns ; -- data to Z  time
21       CHECK_ON : boolean   := false); -- check timing or not
22     port (
23       A        : in    std_logic_vector(ASIZE-1 downto 0);
24       DQ       : inout std_logic_vector(DSIZE-1 downto 0);
25       RASn     : in    std_logic;
26       CASn     : in    std_logic;
27       WEn      : in    std_logic;
28       OEn      : in    std_logic);
29     end DRAM;
```

Line 31 declares a constant ASIZE2 to be the double of ASIZE. In practice, this can be any number greater than ASIZE. Line 32 declares PSIZE as the number of bits to specify the number of pages. Lines 33 and 34 define subtype and type to be used as a page of data. Line 35 declares an access type that is a pointer to the page_type. The page can then be dynamically allocated when the data in the page is required. This dynamic memory allocation reduces the run time memory requirement when just a small portion of the memory is accessed. Lines 37 to 39 declare an access type to an array of page pointers.

```
30    architecture BEH of DRAM is
31      constant ASIZE2 : integer := 2 * ASIZE; -- memory a size
32      constant PSIZE  : integer := (ASIZE2-ASIZE); -- # of page
33      subtype word_type is std_logic_vector(DSIZE-1 downto 0);
34      type    page_type is array(0 to 2**ASIZE-1) of word_type ;
35      type page_ptr_type is access page_type; -- pointer to page
36      constant XVAL    : word_type := (others=>'X');
37      type page_tab_type is
38        array(0 to 2**PSIZE-1) of page_ptr_type;
39      type page_tab_ptr_type is access page_tab_type;
40
41    ----------------------------------------------------------------
```

Lines 43 to 112 implement a process statement. Lines 44 to 52 declare variables. Lines 54 to 58 get the row and column addresses into variable MA. Line 59 checks whether the address is valid. Lines 61 to 62 set the variable M_WR to indicate a write access. Lines 62 to 63 set the variable M_OE to indicate a read access. Line 64 sets the variable M_OFF to indicate that there is no memory access. Lines 65 to 71 set the variable PAGE_INDEX and WORD_INDEX to the page table and the word index inside the page.

```
42    begin
43      memory : process (A, RASn, CASn, DQ, OEn, WEn)
44        variable PAGE_TABLE : page_tab_ptr_type := null;
45        variable PAGE       : page_ptr_type;
46        variable MA : std_logic_vector(ASIZE2-1 downto 0);
47        variable PAGE_INDEX : integer range 0 to 2**PSIZE-1;
48        variable WORD_INDEX : integer range 0 to 2**ASIZE-1;
49        variable MA_X  : boolean;
50        variable M_WR  : boolean;
51        variable M_OE  : boolean;
52        variable M_OFF : boolean;
53      begin
54        if (RASn = '0' and RASn'event) then
55          MA(ASIZE2-1 downto ASIZE ) := A; -- get row address
56        elsif (CASn'event and CASn = '0') then
57          MA(ASIZE-1 downto 0 ) := A;
58        end if;
59        MA_X  := IS_X(MA);
60        M_WR  := (RASn = '0') and (CASn  = '0') and
61                 (WEn  = '0') and (OEn  /= '0')  ;
62        M_OE  := (RASn = '0') and (CASn  = '0') and
63                 (WEn  = '1') and (OEn  = '0');
64        M_OFF := (RASn = '1') or ((WEn/='0') and (OEn/='0'));
65        if (MA_X) then
66          PAGE_INDEX := 0;
67          WORD_INDEX := 0;
68        else
69          PAGE_INDEX := conv_integer(MA(ASIZE2-1 downto ASIZE));
70          WORD_INDEX := conv_integer(MA(ASIZE -1 downto 0    ));
71        end if;
```

For the write cycle, line 74 checks whether this is the first time access and whether the page table has been allocated. After the page table is allocated in line 75, lines 76 to 78 initialize all pointers to null. Line 80 checks whether the page has been allocated. Lines 81 to 85 allocate a page if the page has not been allocated and initialize all data to invalid values (with every bit as 'X'). The data is written in lines 87 to 94 when the data is valid.

```
72              --------- write ----------------------------------------
73              if (M_WR and not MA_X ) then
74                if (PAGE_TABLE = null) then -- page table empty
75                  PAGE_TABLE := new page_tab_type; -- allocate table
76                  for j in page_tab_type'range loop
77                    PAGE_TABLE(j) := null; -- invalidate all pointers
78                  end loop;
79                end if;
80                if (PAGE_TABLE(PAGE_INDEX) = null) then
81                  PAGE := new page_type ; -- allocate a new page.
82                  PAGE_TABLE(PAGE_INDEX) := PAGE ; -- link the page
83                  for j in page_type'range loop
84                    PAGE(j) := XVAL ; -- invalidate all data
85                  end loop;
86                end if;
87                PAGE := PAGE_TABLE(PAGE_INDEX); -- point to page
88                if (IS_X(DQ)) then
89                  assert FALSE report "write invalid data"
90                    severity WARNING;
91                  PAGE(WORD_INDEX) := XVAL;
92                else
93                  PAGE(WORD_INDEX) := DQ; -- write new data
94                end if;
95              end if;
```

Lines 97 to 111 implement the read access. Line 97 checks whether the memory is idle. Line 98 releases DQ by setting DQ to all 'Z's. Line 99 checks whether the address is valid during the read cycle. Lines 102 and 103 check whether the page table exists before it is read. Line 104 puts the data to be read in the DQ bus after T_OE.

```
96              --------- read -----------------------------------------
97              if (M_OFF) then
98                DQ <= (others=>'Z') after T_OFF;
99              elsif (M_OE and MA_X) then -- undefined address
100               DQ <= XVAL after T_OE;
101             elsif (M_OE) then -- read
102               if ((PAGE_TABLE /= null) and
103                  (PAGE_TABLE(PAGE_INDEX) /= null)) then
104                 PAGE := PAGE_TABLE(PAGE_INDEX); -- get the page
105                 DQ   <= PAGE(WORD_INDEX) after T_OE; -- read data
106               else
107                 DQ   <= XVAL after T_OE ; -- page never written
108                 assert FALSE report "read uninitialized page"
109                   severity NOTE;
110               end if ;
111             end if;
112         end process;
```

The row address setup-timing check is implemented in lines 114 to 119. The row address hold time checking is done in lines 120 to 128. Similarly, the column address setup and hold time check is implemented in lines 129 to 143. The write data setup and hold time check is done in lines 145 to 161. After the checking of the row address hold time, line 127 waits until the CASn is asserted. Is this necessary? The answer is left as Exercise 9.12.

```
113    ---------- address timing check -------------------------
114    ra_setup : process
115    begin
116      wait until CHECK_ON and (RASn'event and RASn = '0');
117      assert (A'stable(T_ASR) or T_ASR = 0 ns) report
118        "row address setup error" severity NOTE;
119    end process;
120    ra_hold : process
121    begin
122      wait until CHECK_ON and
123        (A'event and RASn = '0' and CASn /= '0');
124        assert (RASn'last_event > T_RAH) or
125               (RASn'last_event = 0 ns) report
126          "row address hold error" severity NOTE;
127      wait until CASn'event and CASn = '0'; -- skip A glitch
128    end process;
129    ca_setup : process
130    begin
131      wait until CHECK_ON and (CASn'event and CASn = '0');
132        assert (A'stable(T_ASC) or T_ASC = 0 ns) report
133          "column address setup error" severity NOTE;
134      wait until CASn'event and CASn = '1'; -- skip A glitch
135    end process;
136    ca_hold : process
137    begin
138      wait until CHECK_ON and (A'event and CASn = '0');
139        assert (CASn'last_event > T_CAH) or
140               (CASn'last_event = 0 ns) report
141          "column address hold error" severity NOTE;
142      wait until CASn'event and CASn = '1'; -- skip A glitch
143    end process;
144    ---------- data timing check ----------------------------
145    d_setup : process
146    begin
147      wait until CHECK_ON and
148        (CASn'event and WEn='0' and CASn='0');
149        assert (DQ'stable(T_DS) or T_DS = 0 ns) report
150          "write data setup-time error" severity NOTE;
151      wait until CASn'event and CASn = '1'; -- skip A glitch
152    end process;
153    d_hold : process
154    begin
```

```
155        wait until CHECK_ON and
156          (DQ'event and WEn = '0' and CASn = '0');
157            assert (DQ'last_event > T_DH) or
158                  (DQ'last_event = 0 ns) report
159              "write data hold-time error" severity NOTE;
160          wait until CASn'event and CASn = '1'; -- skip A glitch
161      end process;
162    end BEH;
```

To test the DRAM VHDL code, the following test bench can be used. Lines 7 to 28 declare component DRAM. Constants and signals are declared in lines 29 to 35.

```
1      library IEEE;
2      use IEEE.std_logic_1164.all;
3      use IEEE.std_logic_unsigned.all;
4      entity TBDRAM is
5      end TBDRAM;
6      architecture BEH of TBDRAM is
7        component DRAM
8        generic(
9          ASIZE    : integer   := 9    ; -- address width
10         DSIZE    : integer   := 16   ; -- data width
11         T_ASR    : time      :=  0 ns ; -- row address setup
12         T_RAH    : time      := 10 ns ; -- row address hold
13         T_ASC    : time      :=  0 ns ; -- column address setup
14         T_CAH    : time      := 10 ns ; -- column address hold
15         T_OE     : time      := 15 ns ; -- output enable time
16         T_OD     : time      :=  3 ns ; -- output disable time
17         T_DS     : time      :=  0 ns ; -- write data setup time
18         T_DH     : time      :=  8 ns ; -- write data hold  time
19         T_OFF    : time      :=  3 ns ; -- data to Z   time
20         CHECK_ON : boolean   := false); -- check timing or not
21       port (
22         A        : in    std_logic_vector(ASIZE-1 downto 0);
23         DQ       : inout std_logic_vector(DSIZE-1 downto 0);
24         RASn     : in    std_logic;
25         CASn     : in    std_logic;
26         WEn      : in    std_logic;
27         OEn      : in    std_logic);
28     end component;
29     constant  PERIOD2 : time := 20 ns;
30     constant  ASIZE   : integer   := 9 ;
31     constant  DSIZE   : integer   := 8 ;
32     constant  CHECK_ON : boolean   := true;
33     signal A : std_logic_vector(ASIZE-1 downto 0);
34     signal DQ        : std_logic_vector(DSIZE-1 downto 0);
35     signal RASn, CASn, WEn, OEn, CLK   : std_logic;
```

Lines 36 to 67 declare a procedure to do one read or write data by asserting related signals such as A, CASn, WEn, and OEn. In line 38, the timing delay for each signal is passed in to the procedure so that each signal can be individually controlled by an "after clause". The signal timing relationship can then be skewed. For example, in line 47, column address and CASn signals are assigned. Depending on values T_A and T_CASn, signals A can be changed before, after, or at the same time when CASn is asserted. This allows the column address setup timing requirement to be checked.

```
36     procedure DO_ONE (
37        constant RW  : in  integer;
38        constant T_A, T_DQ,T_RASn, T_CASn, T_WEn, T_OEn:in time;
39        constant DATA : in  std_logic_vector(DSIZE-1 downto 0);
40        constant CA : in  std_logic_vector(ASIZE-1 downto 0);
41        signal CLK    : in  std_logic;
42        signal A      : out std_logic_vector(ASIZE-1 downto 0);
43        signal DQ     : out std_logic_vector(DSIZE-1 downto 0);
44        signal RASn, CASn, WEn, OEn : out std_logic) is
45     begin
46        wait until CLK'event and CLK = '1';
47        A   <= CA  after T_A;   CASn <= '0' after T_CASn;
48        WEn <= '1' after T_WEn; OEn  <= '1' after T_OEn;
49        DQ  <= (others => 'Z') after T_DQ;
50        if (RW = 0) then -- read
51          OEn <= '0'  after T_OEn;
52        elsif (RW = 1) then -- write
53          WEn  <= '0'  after T_WEn; DQ  <= DATA after T_DQ;
54        else -- read and then write
55          OEn  <= '0'  after T_OEn;
56          wait until CLK'event and CLK = '1';
57          A    <= (others => 'Z') after T_A;
58          OEn  <= '1'  after T_OEn;
59          wait until CLK'event and CLK = '1';
60          WEn  <= '0'  after T_WEn; DQ  <= DATA after T_DQ;
61        end if;
62        wait until CLK'event and CLK = '1';
63        CASn <= '1' after T_CASn;
64        WEn  <= '1' after T_WEn; OEn  <= '1' after T_OEn;
65        DQ   <= (others => 'Z') after T_DQ;
66        A    <= (others => 'Z') after T_A;
67     end DO_ONE;
```

Lines 68 to 99 declare a procedure RWCYCLE to do a complete read or write cycle. It asserts RASn with the row address, and then calls procedure DO_ONE in lines 93 to 94. Lines 85 to 90 assign timing delay values for each signal based on various bits of variable VD to get the random effect.

```
68     procedure RWCYCLE (
69        constant RW  : in  integer; -- R=0, W=1, RW=2
70        constant WORD_CNT : in  integer;
```

```
71        constant VA : std_logic_vector(ASIZE-1 downto 0);
72        constant VD : std_logic_vector(DSIZE-1 downto 0);
73        signal CLK    : in  std_logic;
74        signal A      : out std_logic_vector(ASIZE-1 downto 0);
75        signal DQ     : out std_logic_vector(DSIZE-1 downto 0);
76        signal RASn, CASn, WEn, OEn : out std_logic) is
77        variable T_A, T_DQ, T_RASn, T_CASn : time := 0 ns;
78        variable T_WEn, T_OEn              : time := 0 ns;
79        variable CA : std_logic_vector(ASIZE-1 downto 0);
80        variable CD : std_logic_vector(DSIZE-1 downto 0);
81     begin
82        CA := VA; CD := VD;
83        wait until CLK'event and CLK = '1';
84        for j in 1 to WORD_CNT loop
85          T_A    := conv_integer(VD(2 downto 0)) * 1 ns;
86          T_DQ   := conv_integer(VD(3 downto 1)) * 1 ns;
87          T_RASn := conv_integer(VD(4 downto 2)) * 1 ns;
88          T_CASn := conv_integer(VD(5 downto 3)) * 1 ns;
89          T_WEn  := conv_integer(VD(6 downto 4)) * 1 ns;
90          T_OEn  := conv_integer(VD(7 downto 5)) * 1 ns;
91          A  <= VA after T_A; RASn <= '0' after T_RASn;
92          CA := (not CA(0) xor CA(1)) & CA(ASIZE-1 downto 1);
93          DO_ONE(RW, T_A, T_DQ, T_RASn, T_CASn, T_WEn, T_OEn,
94                 CD, CA, CLK, A, DQ, RASn, CASn, WEn, OEn);
95          CD := (not CD(0) xor CD(1)) & CD(DSIZE-1 downto 1);
96        end loop;
97        RASn <= '1' after T_RASn;
98        wait until CLK'event and CLK = '1';
99     end RWCYCLE;
100    function LFB (constant D : std_logic_vector) return
101      std_logic_vector is
102      alias AD : std_logic_vector(D'length-1 downto 0) is D;
103    begin
104      return AD(AD'length-2 downto 0) &
105             (AD(AD'length-1) xor not AD(0));
106    end LFB;
```

Lines 100 to 106 declare a simple linear feedback function to randomly generate variable values for VA and VD. They are used to generate various timing delays for address and data signals. Line 118 reads a data, line 119 writes a data to the same address, and line 120 reads the data back. Line 122 writes 3 data and line 123 reads 3 data back.

```
107    begin
108    ----------------------------------------------------------------
109    p0 : process
110      variable VA : std_logic_vector(ASIZE-1 downto 0)
111                 := (others => '0');
112      variable VD : std_logic_vector(DSIZE-1 downto 0)
```

```
113                         := (others => '0');
114    begin
115      WEn    <= '1'; OEn <= '1'; RASn <= '1'; CASn <= '1';
116      A      <= (others => 'Z');
117      DQ     <= (others => 'Z');
118      RWCYCLE(0, 1, VA, VD, CLK,A,DQ, RASn, CASn, WEn, OEn);
119      RWCYCLE(1, 1, VA, VD, CLK,A,DQ, RASn, CASn, WEn, OEn);
120      RWCYCLE(0, 1, VA, VD, CLK,A,DQ, RASn, CASn, WEn, OEn);
121      VA := LFB(VA); VD := LFB(VD);
122      RWCYCLE(1, 3, VA, VD, CLK,A,DQ, RASn, CASn, WEn, OEn);
123      RWCYCLE(0, 3, VA, VD, CLK,A,DQ, RASn, CASn, WEn, OEn);
124      VA := LFB(VA); VD := LFB(VD);
125      RWCYCLE(2, 1, VA, VD, CLK,A,DQ, RASn, CASn, WEn, OEn);
126      RWCYCLE(0, 1, VA, VD, CLK,A,DQ, RASn, CASn, WEn, OEn);
127      VA := LFB(VA); VD := LFB(VD);
128      RWCYCLE(2, 3, VA, VD, CLK,A,DQ, RASn, CASn, WEn, OEn);
129      RWCYCLE(0, 3, VA, VD, CLK,A,DQ, RASn, CASn, WEn, OEn);
130      VA := LFB(VA); VD := LFB(VD);
131    end process;
132    clkgen : process
133    begin
134      CLK <= '0'; wait for PERIOD2;
135      CLK <= '1'; wait for PERIOD2;
136    end process;
137    dram0 : DRAM
138    generic map (
139      ASIZE  => ASIZE, DSIZE => DSIZE, CHECK_ON => true)
140    port map (
141      A    => A,    DQ  => DQ,  RASn    => RASn,
142      CASn => CASn, WEn => WEn, OEn     => OEn);
143    end BEH;
144    configuration CFG_TBDRAM of TBDRAM is
145      for BEH
146      end for;
147    end CFG_TBDRAM;
```

Lines 132 to 136 generate the clock to be used as the timing reference. Lines 137 to 142 instantiate component DRAM. Lines 144 to 147 configure the test bench. Figure 9-15 shows a simulation waveform with read, write, and read cycles. The first read cycle gets XX in the DQ bus since the data has not been written, and the page has not been allocated. The second cycle is a write cycle that writes 00 into memory. The third cycle is a read cycle that reads the same address with 00 in DQ bus. Note that data 00 appears after OEn is asserted with a delay of T_OE. Valid data 00 is changed to ZZ after OEn is deasserted with a delay of T_OD.

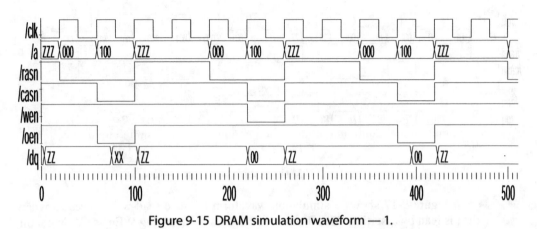

Figure 9-15 DRAM simulation waveform — 1.

Figure 9-16 shows a write cycle of three data and a read cycle of three data. Due to the timing violations as the following messages indicate, data is not written into memory.

```
1    #     Time: 501 ns   Iteration: 0   Instance:/dram0
2    # ** Note: column address hold error
3    #     Time: 541 ns   Iteration: 0   Instance:/dram0
4    # ** Note: column address hold error
5    #     Time: 621 ns   Iteration: 0   Instance:/dram0
6    # ** Note: column address hold error
7    #     Time: 701 ns   Iteration: 0   Instance:/dram0
8    # ** Note: row address hold error
9    #     Time: 821 ns   Iteration: 0   Instance:/dram0
10   # ** Note: column address hold error
11   #     Time: 861 ns   Iteration: 0   Instance:/dram0
12   # ** Note: column address hold error
13   #     Time: 941 ns   Iteration: 0   Instance:/dram0
14   # ** Note: column address hold error
15   #     Time: 1021 ns  Iteration: 0   Instance:/dram0
16   # ** Note: row address hold error
17   #     Time: 1142 ns  Iteration: 0   Instance:/dram0
18   # ** Note: column address hold error
19   #     Time: 1182 ns  Iteration: 0   Instance:/dram0
```

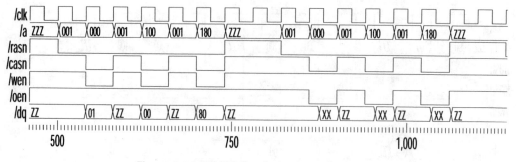

Figure 9-16 DRAM simulation waveform — 2.

Figure 9-17 shows a simulation waveform of three read-write accesses where data is read by asserting OEn. Then it writes back by asserting WEn as CASn continues to be asserted. More simulations are left as Exercises 9.13 and 9.14.

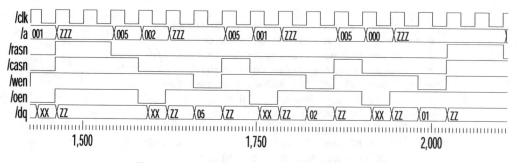

Figure 9-17 DRAM simulation waveform — 3.

9.5 EXERCISES

9.1   Refer to TBDUALRAM and DUALRAM VHDL code. Run VHDL simulation to observe the waveform. Verify that DUALRAM VHDL code is correct. Does the test bench verify all possible combinations of the AWR and BWR timing relationship? Justify your conclusions.

9.2   Refer to TBDUALRAM and DUALRAM VHDL code. What improvements can you suggest to the VHDL models?

9.3   Refer to TBDUALRAM and DUALRAM VHDL code. The DUALRAM model assumes that address is valid before the write pulse is asserted. What would happen if the address changes after the write pulse is asserted? Does the model still work? Modify the test bench and run the simulation to verify your conclusions.

9.4 Refer to IDT part number IDT7025S high-speed 8Kx16 dual-port static RAM data sheets. Write a behavioral functional VHDL model for the part. Develop a test bench to verify the model.

9.5 Refer to TBFIFOSYN and FIFOSYN VHDL code. Run more simulations to verify the FIFO operations. What improvements can you suggest for the test bench?

9.6 Refer to FIFOSYN VHDL code. It can be synthesized. Synthesize the VHDL code. How many flip-flops will you get?

9.7 Refer to FIFOWXN VHDL code and its simulation waveforms. Develop a test bench to verify the FIFOWXN functions.

9.8 Refer to FIFOWXN and its related VHDL code. They are used for simulation, not for synthesis. How can the code be modified so that the whole asynchronous FIFO circuit is synthesized? How can the delay be implemented? How can the synthesized schematic be verified?

9.9 Refer to FIFOWXN VHDL code and its simulation waveforms. Design a controller to write data to and read data from the asynchronous FIFO FIFOWXN so that the FIFO can be reset. How can the full and empty flags be used in the FIFO controller?

9.10 Refer to FIFOWXN VHDL code. Some FIFO circuits also bring the read and write pointer values to the outputs. How can we use these pointer outputs to determine the number of data in the FIFO which has not been read?

9.11 Refer to the Cypress CY7C401 FIFO data sheet. Develop a behavioral VHDL model for this part.

9.12 Refer to DRAM VHDL code line 127. Is line 127 necessary? What is its purpose?

9.13 Refer to DRAM and TBDRAM VHDL code. Run a VHDL simulator. Does the DRAM VHDL code model the DRAM function correctly? How can the TBDRAM VHDL code be improved?

9.14 Refer to the Micron MT4C16257 data sheet. There are many timing parameters and timing relationships. Develop a VHDL code to model that part.

Chapter 10

A Design Case Study: Finite Impulse Response Filter ASIC Design

A Finite Impulse Response (FIR) filter is common in digital signal processing. In this chapter, a FIR filter is used as an illustration to go through the design process of design description, VHDL coding, verification, synthesis, place/route, and postlayout verification.

10.1 DESIGN DESCRIPTION

A FIR filter usually has many stages. Each stage is referred to as a TAP. The organization of a TAP is shown in Figure 10-1. Each stage has a 16-bit coefficient register. Input signal I, 16-bits wide is going through two-stage registers. The first stage of the register is multiplied with the coefficient. The product (32 bits) is registered. The registered product is sign extended to add with the 64-bit input PI. The sum is again registered in a 64-bit register. Both the second-stage registered input I and the registered sum outputs are fed to the next stage of the TAP. Figure 10-2 Shows the connection of two adjacent TAPs.

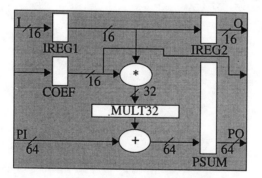

Figure 10-1 Organization of a TAP.

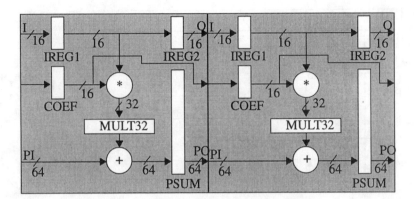

Figure 10-2 Connection of adjacent TAPS.

From Figure 10-2, it is easy to see that many TAPs can be cascaded in series. Here, we assume that the application-specific integrated circuit (ASIC) would have 32 stages. From the system point of view, the design is better if it can cascade the same ASIC in a chain to form a FIR with 64, 96, or more stages. Another feature is the ability to bypass any ASIC in the chain to allow that ASIC to be ignored if it fails. This supports a fault-tolerant architecture. The outputs O and PO can also be three-stated so that the ASIC can be connected in parallel. Figure 10-3 shows the organization of a 32-TAP FIR with capabilities of bypass and three-state outputs. Figure 10-4 illustrates a FIR system with six FIR ASICs. They are configured as three parallel FIR ASICs in series. Many combinations of FIR organizations can be formed by controlling three-state input OEn and bypass input BYPASS.

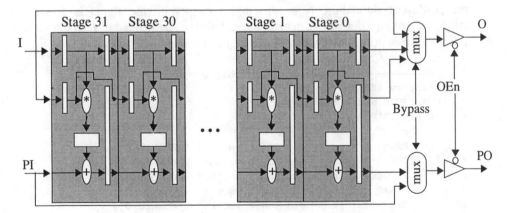

Figure 10-3 A 32-stage FIR organization.

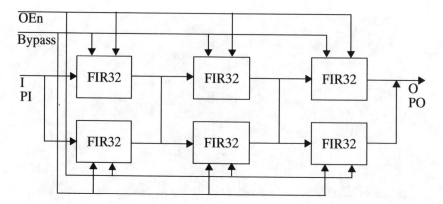

Figure 10-4 An FIR system with six FIR ASICs.

Referring to Figure 10-3, the FIR ASIC may have the following entity declaration. Before the FIR starts to perform the filtering, coefficient registers should be loaded. Input LDCOEF (line 8) is asserted during the loading of the coefficient registers. Input signal START (line 9) tells the FIR ASIC to start the filtering process. Input signal OEn (line 10) is a low-active signal that controls the three-state buffers for outputs O and PO. Input signal BYPASS (line 11) is a high-active signal controlling whether the data will be bypassed. Line 12 declares input I to be multiplied with the coefficient and shifted to the next stage of the TAP. Line 13 declares the partial sum input. Lines 14 to 15 declare outputs O and PO as output data and the sum output. Lines 16 and 17 declare power and ground pads. This is usually not necessary. These are added here due to the place and route tool to be used in this chapter to generate postlayout netlist with the power and ground pads. Including these allows the same test bench to be used without any modifications in both the functional simulation and the postlayout simulation. Note also that a single-bit-wide signal is declared as std_logic_vector(0 downto 0), not std_logic. This is also included to accommodate the interface to the primitive layout tool used for this design.

```
1        --------------------- file firchip_e.vhd
2        library IEEE;
3        use IEEE.std_logic_1164.all;
4        entity FIRCHIP is
5          port(
6            RSTn     : in    std_logic_vector( 0 downto 0);
7            CLK      : in    std_logic_vector( 0 downto 0);
8            LDCOEF   : in    std_logic_vector( 0 downto 0);
9            START    : in    std_logic_vector( 0 downto 0);
10           OEn      : in    std_logic_vector( 0 downto 0);
11           BYPASS   : in    std_logic_vector( 0 downto 0);
```

```
12        I      : in    std_logic_vector(15 downto 0);
13        PI     : in    std_logic_vector(63 downto 0);
14        O      : out   std_logic_vector(15 downto 0);
15        PO     : out   std_logic_vector(63 downto 0);
16        VDDS   : inout std_logic_vector(20 downto 0);
17        GNDS   : inout std_logic_vector(20 downto 0));
18    end FIRCHIP;
```

As another point, there are 166 signal pins for the FIRCHIP. The design requires a package. The package will need to have more than 166 pins to accommodate power and ground pins. A 180-pin package will allow 14 extra pins for the power and ground pins. As a rule of thumb, there are a pair of power and ground pins for every 48 milli-amperes (mA) of output pins. The power and ground may require more than 14 pins. The next larger package has 208 pins. This provides 42 pins for power and ground: 21 for the power pins, and 21 for the ground pins.

The FIRCHIP design can be behaviorally described with the following VHDL code. Line 6 declares constant K to be 32, indicating 32 stages of TAPs. By changing only this constant, a different number of stages can be described without modifying other VHDL codes. Lines 7 to 14 declare subtypes and types for the signals defined in lines 16 to 22.

```
1     ---------------------- file firchip_beh.vhd
2     library IEEE;
3     use IEEE.std_logic_1164.all;
4     use IEEE.std_logic_arith.all;
5     architecture BEH of FIRCHIP is
6        constant K : integer := 32;
7        subtype size16 is std_logic_vector(15 downto 0);
8        subtype size32 is std_logic_vector(31 downto 0);
9        subtype size64 is std_logic_vector(63 downto 0);
10       type    kx16   is array (K-1 downto 0) of size16;
11       type    k1x16  is array (K   downto 0) of size16;
12       type    kx32   is array (K-1 downto 0) of size32;
13       type    kx64   is array (K-1 downto 0) of size64;
14       type    k1x64  is array (K   downto 0) of size64;
15
16       signal IREG1, IREG2           : k1x16;
17       signal COEF                   : k1x16;
18       signal MULT32                 : kx32;
19       signal MULT64                 : kx64;
20       signal PSUM                   : k1x64;
21       signal O_S                    : size16;
22       signal PO_S                   : size64;
23       signal RSTn_F1, RSTn_S, CLOCK : std_logic;
```

Lines 16 declare signals IREG1 and IREG2 for the two stages of registers for input I. Line 17 declares the COEF signal for the coefficient registers. Line 18 declares the MULT32 signal for the multiplier product registers. Line 19 declares the

MULT64 signal for the MULT32 signal to be sign extended to 64 bits. Line 20 declares the PSUM signal for the sum registers.

Line 25 assigns the CLOCK signal. Lines 26 and 27 initialize the IREG2(K) signal. It is set to input I when the START input signal is asserted '1', otherwise, it is set to '0'. Lines 28 to 30 initialize the IREG1(K), PSUM(K), and COEF(K) signals. Note that the generate statement in line 37 has a range from K-1 downto 0. The purpose of lines 26 to 30 is to initialize the signals to be used inside the generate statement. For example, in line 47, the IREG1 register is getting the previous stage IREG2 register value. If the IREG2(K) signal is not initialized, the assignment IREG1(K-1) <= IREG2(K) will be wrong. The IREG2 as defined in line 16 has K+1 16-bits. There are only K stages. By declaring one more than the number of stages and initializing the first one, there is no need to use the **if generate** statement inside the generate statement to test the boundary conditions. This reduces the amount of VHDL coding at the expense of more signals.

Lines 33 to 35 synchronize the reset input signal RSTn. Lines 37 to 60 use a generate statement to create 32 stages of TAPs. Each TAP has IREG1, IREG2, COEF, MULT32, and PSUM registers. They are asynchronously reset in lines 40 to 45. IREG1 register gets the IREG2 register value from the previous stage as shown in line 47. The IREG2 register gets the IREG1 register value from the same stage as shown in line 48.

```
24   begin
25     CLOCK      <= CLK(0);
26     IREG2(K)   <= I when (START(0) = '1')
27                     else (IREG2(K)'range => '0');
28     IREG1(K)   <= (IREG1(K)'range => '0');
29     PSUM(K)    <= PI;
30     COEF(K)    <= I;
31     sync : process
32     begin
33       wait until CLOCK'event and CLOCK = '1';
34       RSTn_F1 <= RSTn(0);
35       RSTn_S  <= RSTn_F1;
36     end process;
37     gen32 : for j in K-1 downto 0 generate
38       ffs : process (RSTn_S, CLOCK)
39       begin
40         if (RSTn_S = '0') then
41            IREG1(j)  <= (IREG1(j)'range  => '0');
42            IREG2(j)  <= (IREG2(j)'range  => '0');
43            COEF(j)   <= (COEF(j)'range   => '0');
44            MULT32(j) <= (MULT32(j)'range => '0');
45            PSUM(j)   <= (PSUM(j)'range   => '0');
46         elsif (CLOCK'event and CLOCK = '1') then
47            IREG1(j)  <= IREG2(j+1);
48            IREG2(j)  <= IREG1(j);
49            if (LDCOEF(0) = '1') then
50               COEF(j) <= COEF(j+1);
```

```
51              end if;
52            MULT32(j) <= signed(IREG1(j))  * signed(COEF(j));
53            PSUM(j)   <= signed(MULT64(j)) + signed(PSUM(j+1));
54          end if;
55        end process;
56        signext : for m in 63 downto 32 generate
57          MULT64(j)(m) <= MULT32(j)(31);
58        end generate;
59        MULT64(j)(31 downto 0) <= MULT32(j);
60      end generate;
61      O_S  <= I        when BYPASS(0) = '1' else
62              IREG2(0) when LDCOEF(0) = '0' else
63              COEF(0);
64      PO_S <= PI   when BYPASS(0) = '1' else PSUM(0);
65      O    <= O_S  when OEn(0)    = '0' else (O'range => 'Z');
66      PO   <= PO_S when OEn(0)    = '0' else (PO'range => 'Z');
67    end BEH;
```

Lines 49 to 51 shift the coefficient registers when the LDCOEF input signal is asserted. Lines 52 and 53 perform the multiplication and addition. Lines 56 to 59 form the 64-bit MULT64 signal by sign extension of the MULT32 register. Lines 61 to 63 model a multiplexer function. Signal O_S gets the I input when the chip is to be bypassed. It gets the IREG2(0) value when the LDCOEF signal is not asserted. Otherwise, O_S gets the value of the COEF(0) so that the coefficient can be passed to the next chip. Line 64 assigns PO_S signal to be PI input when the chip is bypassed. Otherwise, PO_S gets the PSUM(0) value. Lines 65 and 66 model the three-state behavior.

10.2 DESIGN PARTITION

The above FIRCHIP architecture, BEH VHDL code, is actually synthesizable. However, there are 32, 16x16 multipliers and 32, 64-bit adders. As a result, it will take a long time to synthesize the whole circuit. Also, the three-state behavior is usually included as part of three-state output pads. There is no indication of the pads used. The clock and reset signals are connected to every flip-flop. There are no clock and reset buffer trees.

To improve the synthesis run time, a multiplier block is created. The following shows a simple VHDL code that uses the multiplication operator "*". The Synopsys synthesis tool would synthesize a DesignWare multiplier. Note that the operands are converted to signed type to get the 2's complement multiplication.

```
1    library IEEE;
2    use IEEE.std_logic_1164.all;
3    use IEEE.std_logic_arith.all;
4    entity MULT is
5      port (
```

```
6        DATA  : in  std_logic_vector(15 downto 0);
7        COEF  : in  std_logic_vector(15 downto 0);
8        PROD  : out std_logic_vector(31 downto 0));
9    end MULT;
10   architecture RTL of MULT is
11   begin
12     PROD <= signed(DATA) * signed(COEF);
13   end RTL;
```

The adder is also isolated as a block. The following shows the VHDL code. Note that input signal B is sign extended before it is added with A, since the multiplier product has only 32 bits.

```
1    library IEEE;
2    use IEEE.std_logic_1164.all;
3    use IEEE.std_logic_unsigned.all;
4    entity ADD64 is
5      port (
6        A : in  std_logic_vector(63 downto 0);
7        B : in  std_logic_vector(31 downto 0);
8        S : out std_logic_vector(63 downto 0));
9    end ADD64;
10   architecture RTL of ADD64 is
11     signal BEXT : std_logic_vector(63 downto 0);
12   begin
13     signext : for m in 63 downto 32 generate
14       BEXT(m) <= B(31);
15     end generate;
16     BEXT(31 downto 0) <= B;
17     S  <= A + BEXT;
18   end RTL;
```

The coefficient register block is shown below. When the COEFEN signal is asserted, the coefficient register is loaded with input data D.

```
1    library IEEE;
2    use IEEE.std_logic_1164.all;
3    entity COEFREG is
4      port(
5        RSTn     : in  std_logic;
6        CLK      : in  std_logic;
7        D        : in  std_logic_vector(15 downto 0);
8        COEFEN   : in  std_logic;
9        COEF_FF  : out std_logic_vector(15 downto 0));
10   end COEFREG;
11   architecture RTL of COEFREG is
12   begin
13     ffs : process(RSTn, CLK)
14     begin
```

```
15          if (RSTn = '0') then
16            COEF_FF <= (COEF_FF'range => '0');
17          elsif (CLK'event and CLK = '1') then
18            if (COEFEN = '1') then
19              COEF_FF <= D;
20            end if;
21          end if;
22        end process;
23      end RTL;
```

The bypass function can be grouped as a block as shown below.

```
1       library IEEE;
2       use IEEE.std_logic_1164.all;
3       entity MUX is
4         port(
5           BYPASS     : in  std_logic;
6           LDCOEF     : in  std_logic;
7           I          : in  std_logic_vector(15 downto 0);
8           O_in       : in  std_logic_vector(15 downto 0);
9           COEF_in    : in  std_logic_vector(15 downto 0);
10          O_out      : out std_logic_vector(15 downto 0);
11          PI         : in  std_logic_vector(63 downto 0);
12          PO_in      : in  std_logic_vector(63 downto 0);
13          PO_out     : out std_logic_vector(63 downto 0));
14      end MUX;
15      architecture RTL of MUX is
16      begin
17        O_out   <= I      when BYPASS = '1' else
18                   O_in   when LDCOEF = '0' else
19                   COEF_in;
20        PO_out  <= PI  when BYPASS = '1' else PO_in;
21      end RTL;
```

There are many registers in each TAP. The following VHDL code shows a parameterized register block. This block can be used to implement the IREG1, IREG2, MULT, and PSUM registers.

```
1       library IEEE;
2       use IEEE.std_logic_1164.all;
3       entity REGS is
4         generic ( N : integer := 16);
5         port (
6           RSTn : in  std_logic;
7           CLK  : in  std_logic;
8           DIN  : in  std_logic_vector(N-1 downto 0);
9           DFF  : out std_logic_vector(N-1 downto 0));
10      end REGS;
11      architecture RTL of REGS is
```

```
12    begin
13      ffs : process(CLK, RSTn)
14      begin
15        if (RSTn = '0') then
16          DFF  <= (DFF'range => '0');
17        elsif (CLK = '1' and CLK'event) then
18          DFF  <= DIN;
19        end if;
20      end process;
21    end RTL;
```

There are other miscellaneous functions such as the synchronization of the input signal RSTn and the disabling of input data that goes to IREG1 when input signal START is not asserted. These functions are grouped as a GLUE logic block.

```
1     library IEEE;
2     use IEEE.std_logic_1164.all;
3     entity GLUE is
4       port(
5       RSTn        : in  std_logic;
6       RSTno       : out std_logic;
7       CLK         : in  std_logic;
8       START       : in  std_logic;
9       I           : in  std_logic_vector(15 downto 0);
10      I_out       : out std_logic_vector(15 downto 0));
11    end GLUE;
12    architecture RTL of GLUE is
13      signal RSTn_FF1 : std_logic;
14    begin
15      rstsync : process
16      begin
17        wait until (CLK'event and CLK = '1') ;
18          RSTn_FF1 <= RSTn;
19          RSTno    <= RSTn_FF1;
20      end process;
21      I_out   <= I   when (START = '1')
22                      else (I_out'range => '0');
23    end RTL;
```

Now, we have all the building blocks. We can put them all together by instantiating the GLUE and MUX blocks, and using a generate statement to instantiate REGS, COEFREG, MULT, and ADD64 blocks for all 32 stages of TAPs. Another way to accomplish this is to create another level of hierarchy that encompasses one TAP stage. The TAP stage is then instantiated 32 times.

Examining the organization of one TAP stage, we see that IREG1 registers feed IREG2 registers directly. IREG2 registers feed directly to IREG1 registers in the next TAP stage. The coefficient COEFREG registers feed to the next TAP stage COEFREG directly. The clock skew, as discussed in Chapter 8, may introduce setup and hold time violations. The minimum paths that may cause setup and hold time violations should be corrected. Commands **set_fix_hold** and "**compile - prioritize_min_paths -only_design_rule**" can be used to add more delays to the paths that are less than a certain value. These two commands can otherwise be executed in a higher level block to add delay for all timing paths in or among the lower level blocks. This takes more time compared to executing the same two commands in a smaller block.

Another partition consideration is the place and route tool. The place and route tool is more efficient for the smaller blocks. Smaller blocks can be placed and routed first. They are then hierarchically placed and routed for the whole design.

Based on the above two considerations, we will illustrate a partition such that the setup and hold time issues and the place and route tool can be completed much faster. The goal is to be able to use the **set_fix_hold** command in a smaller lower-level block. The **set_fix_hold** does not need to be done in a higher-level (which is bigger) block that demands that the whole design have a design margin to avoid any setup and hold time problems. If the set_fix_hold command applies only to each stage, the minimum paths between adjacent TAPs from IREG2 to IREG1 and between COEFREG registers need to be corrected. Consider the second TAP stage. We can add a register IREG1D before IREG1 register. IREG1D is implemented with negatively triggered flip-flops. The first stage IREG2 output can be registered with IREG1D at the falling edge of the clock. The IREG1D register output is then registered with IREG1. There will be no minimum path problems. Similarly, another register COEFD is added for the COEFREG register. The following shows VHDL code for the TAP stages which are not the first TAP stage. The IREG1D register is implemented in lines 52 to 54. The COEFD register is implemented in lines 62 to 64.

```
1     library IEEE;
2     use IEEE.std_logic_1164.all;
3     entity FIRTAPA is
4       port(
5         CLK_S    : in  std_logic_vector(8 downto 0);
6         CLKn_S   : in  std_logic_vector(1 downto 0);
7         RSTn_S   : in  std_logic_vector(10 downto 0);
8         LDCOEF   : in  std_logic;
9         COEFi    : in  std_logic_vector(15 downto 0);
10        COEFo    : out std_logic_vector(15 downto 0);
11        I        : in  std_logic_vector(15 downto 0);
12        O        : out std_logic_vector(15 downto 0);
13        PI       : in  std_logic_vector(63 downto 0);
14        PO       : out std_logic_vector(63 downto 0));
15    end FIRTAPA;
16    architecture RTL of FIRTAPA is
17      component REGS
18      generic ( N : integer := 16);
19      port (
20        RSTn : in  std_logic;
21        CLK  : in  std_logic;
22        DIN  : in  std_logic_vector(N-1 downto 0);
23        DFF  : out std_logic_vector(N-1 downto 0));
```

```
24      end component;
25      component MULT
26      port (
27        DATA  : in  std_logic_vector(15 downto 0);
28        COEF  : in  std_logic_vector(15 downto 0);
29        PROD  : out std_logic_vector(31 downto 0));
30      end component;
31      component ADD64
32      port (
33        A : in  std_logic_vector(63 downto 0);
34        B : in  std_logic_vector(31 downto 0);
35        S : out std_logic_vector(63 downto 0));
36      end component;
37      component COEFREG
38      port(
39        RSTn      : in  std_logic;
40        CLK       : in  std_logic;
41        D         : in  std_logic_vector(15 downto 0);
42        COEFEN    : in  std_logic;
43        COEF_FF   : out std_logic_vector(15 downto 0));
44      end component;
45      subtype size16 is std_logic_vector(15 downto 0);
46      subtype size32 is std_logic_vector(31 downto 0);
47      subtype size64 is std_logic_vector(63 downto 0);
48      signal IREG1n, COEFin, COEF, IREG1, IREG2 : size16;
49      signal MULT32, MULT32_D                   : size32;
50      signal PSUM                               : size64;
51    begin
52      ireg1d : REGS generic map ( N => 16) port map (
53        RSTn => RSTn_S(9),    CLK => CLKn_S(0),
54        DIN  => I,            DFF => IREG1n);
55      uireg1 : REGS generic map ( N => 16) port map (
56        RSTn => RSTn_S(0),    CLK => CLK_S(0),
57        DIN  => IREG1n,       DFF => IREG1);
58      uireg2 : REGS generic map ( N => 16) port map (
59        RSTn => RSTn_S(1),    CLK => CLK_S(1),
60        DIN  => IREG1,        DFF => O    );
61
62      coefd  : REGS generic map ( N => 16) port map (
63        RSTn => RSTn_S(10),     CLK => CLKn_S(1),
64        DIN  => COEFi,          DFF => COEFin);
65      ucoef  : COEFREG port map (
66        RSTn => RSTn_S(2), CLK => CLK_S(2), D => COEFin,
67        COEFEN => LDCOEF,  COEF_FF => COEF);
68      COEFo <= COEF;
69
70      umult  : MULT port map (
71        DATA =>IREG1, COEF=>COEF, PROD=>MULT32_D);
72      umultreg0 : REGS generic map ( N => 16) port map (
73        RSTn => RSTn_S(3),  CLK => CLK_S(3),
```

```
74          DIN  => MULT32_D(15 downto 0),
75          DFF  => MULT32(15 downto 0));
76      umultreg1 : REGS generic map ( N => 16) port map (
77          RSTn => RSTn_S(4),  CLK => CLK_S(4),
78          DIN  => MULT32_D(31 downto 16),
79          DFF  => MULT32(31 downto 16));
80
81      uadd64 : ADD64 port map (
82          A => PI, B => MULT32, S => PSUM);
83      addreg0 : REGS generic map ( N => 16) port map (
84          RSTn => RSTn_S(5),  CLK => CLK_S(5),
85          DIN  => PSUM(15 downto 0),
86          DFF  => PO(15 downto 0));
87      addreg1 : REGS generic map ( N => 16) port map (
88          RSTn => RSTn_S(6),  CLK => CLK_S(6),
89          DIN  => PSUM(31 downto 16),
90          DFF  => PO(31 downto 16));
91      addreg2 : REGS generic map ( N => 16) port map (
92          RSTn => RSTn_S(7),  CLK => CLK_S(7),
93          DIN  => PSUM(47 downto 32),
94          DFF  => PO(47 downto 32));
95      addreg3 : REGS generic map ( N => 16) port map (
96          RSTn => RSTn_S(8),  CLK => CLK_S(8),
97          DIN  => PSUM(63 downto 48),
98          DFF  => PO(63 downto 48));
99      end RTL;
```

Note that each clock and reset input signal drives 16 flip-flops. Therefore, the PSUM register is implemented by instantiating REGS 4 times as shown in lines 83 to 98. If the place and route tool has the clock synthesis capability, this can be implemented with one REGS instantiation by mapping the generic N to 64. The same clock and reset signal can be used for MULT32, IREG1, and other registers. We will see how IREG1D and COEFD can help in the minimum path problems in the later synthesis section. The first TAP stage VHDL code is shown below. It is almost the same as FIRTAPA, except that there are no IREG1D and COEFD registers.

```
1       library IEEE;
2       use IEEE.std_logic_1164.all;
3       entity FIRTAPB is
4         port(
5           CLK_S   : in  std_logic_vector( 8 downto 0);
6           RSTn_S  : in  std_logic_vector( 8 downto 0);
7           LDCOEF  : in  std_logic;
8           COEFi   : in  std_logic_vector(15 downto 0);
9           COEFo   : out std_logic_vector(15 downto 0);
10          I       : in  std_logic_vector(15 downto 0);
11          O       : out std_logic_vector(15 downto 0);
12          PI      : in  std_logic_vector(63 downto 0);
13          PO      : out std_logic_vector(63 downto 0));
```

```
14    end FIRTAPB;
15    architecture RTL of FIRTAPB is
16      component REGS
17      generic ( N : integer := 16);
18      port (
19        RSTn : in  std_logic;
20        CLK  : in  std_logic;
21        DIN  : in  std_logic_vector(N-1 downto 0);
22        DFF  : out std_logic_vector(N-1 downto 0));
23      end component;
24      component MULT
25      port (
26        DATA : in  std_logic_vector(15 downto 0);
27        COEF : in  std_logic_vector(15 downto 0);
28        PROD : out std_logic_vector(31 downto 0));
29      end component;
30      component ADD64
31      port (
32        A : in  std_logic_vector(63 downto 0);
33        B : in  std_logic_vector(31 downto 0);
34        S : out std_logic_vector(63 downto 0));
35      end component;
36      component COEFREG
37      port(
38        RSTn     : in  std_logic;
39        CLK      : in  std_logic;
40        D        : in  std_logic_vector(15 downto 0);
41        COEFEN   : in  std_logic;
42        COEF_FF  : out std_logic_vector(15 downto 0));
43      end component;
44      subtype size16 is std_logic_vector(15 downto 0);
45      subtype size32 is std_logic_vector(31 downto 0);
46      subtype size64 is std_logic_vector(63 downto 0);
47      signal COEF, IREG1, IREG2 : size16;
48      signal MULT32, MULT32_D  : size32;
49      signal PSUM              : size64;
50    begin
51      uireg1 : REGS generic map ( N => 16) port map (
52        RSTn => RSTn_S(0),    CLK => CLK_S(0),
53        DIN  => I,            DFF => IREG1);
54      uireg2 : REGS generic map ( N => 16) port map (
55        RSTn => RSTn_S(1),    CLK => CLK_S(1),
56        DIN  => IREG1,        DFF => O     );
57
58      ucoef  : COEFREG port map (
59        RSTn => RSTn_S(2), CLK => CLK_S(2), D => COEFi,
60        COEFEN => LDCOEF,  COEF_FF => COEF);
61      COEFo <= COEF;
62
63      umult  : MULT port map (
```

```
64        DATA =>IREG1, COEF=>COEF, PROD=>MULT32_D);
65    umultreg0 : REGS generic map ( N => 16) port map (
66        RSTn => RSTn_S(3), CLK => CLK_S(3),
67        DIN  => MULT32_D(15 downto 0),
68        DFF  => MULT32(15 downto 0));
69    umultreg1 : REGS generic map ( N => 16) port map (
70        RSTn => RSTn_S(4), CLK => CLK_S(4),
71        DIN  => MULT32_D(31 downto 16),
72        DFF  => MULT32(31 downto 16));
73
74    uadd64 : ADD64 port map (
75        A => PI, B => MULT32, S => PSUM);
76    addreg0 : REGS generic map ( N => 16) port map (
77        RSTn => RSTn_S(5), CLK => CLK_S(5),
78        DIN  => PSUM(15 downto 0),
79        DFF  => PO(15 downto 0));
80    addreg1 : REGS generic map ( N => 16) port map (
81        RSTn => RSTn_S(6), CLK => CLK_S(6),
82        DIN  => PSUM(31 downto 16),
83        DFF  => PO(31 downto 16));
84    addreg2 : REGS generic map ( N => 16) port map (
85        RSTn => RSTn_S(7), CLK => CLK_S(7),
86        DIN  => PSUM(47 downto 32),
87        DFF  => PO(47 downto 32));
88    addreg3 : REGS generic map ( N => 16) port map (
89        RSTn => RSTn_S(8), CLK => CLK_S(8),
90        DIN  => PSUM(63 downto 48),
91        DFF  => PO(63 downto 48));
92  end RTL;
```

The following VHDL code combines all the blocks we have discussed. Line 4 describes generic K with a default value of 32. Note that clock and reset signals are declared with generic K for parameterizing the array sizes as shown in lines 8 to 10. CLKn_S is the inversion of CLK_S. For each stage of FIRTAPA, there are nine clock signals, two inverted clock signals, and nine reset signals. The first stage FIRTAPB has nine clock signals and nine reset signals.

```
1   library IEEE;
2   use IEEE.std_logic_1164.all;
3   entity FIRSYN is
4     generic (K : integer := 32);
5     port(
6       RSTn    : in  std_logic;
7       RSTno   : out std_logic;
8       CLK_S   : in  std_logic_vector(9*K-1 downto 0);
9       CLKn_S  : in  std_logic_vector(2*K-3 downto 0);
10      RSTn_S  : in  std_logic_vector(11*K-3 downto 0);
11      LDCOEF  : in  std_logic; -- enable loading coef
12      START   : in  std_logic; -- start filtering
13      BYPASS  : in  std_logic; -- bypass the chip
```

```
14        I          : in  std_logic_vector(15 downto 0); -- data
15        PI         : in  std_logic_vector(63 downto 0); -- sum
16        O          : out std_logic_vector(15 downto 0); -- data
17        PO         : out std_logic_vector(63 downto 0));
18  end FIRSYN;
19  architecture RTL of FIRSYN is
20    component GLUE
21    port(
22      RSTn      : in  std_logic;
23      RSTno     : out std_logic;
24      CLK       : in  std_logic;
25      START     : in  std_logic;
26      I         : in  std_logic_vector(15 downto 0);
27      I_out     : out std_logic_vector(15 downto 0));
28    end component;
29    component MUX
30    port(
31      BYPASS    : in  std_logic;
32      LDCOEF    : in  std_logic;
33      I         : in  std_logic_vector(15 downto 0);
34      O_in      : in  std_logic_vector(15 downto 0);
35      COEF_in   : in  std_logic_vector(15 downto 0);
36      O_out     : out std_logic_vector(15 downto 0);
37      PI        : in  std_logic_vector(63 downto 0);
38      PO_in     : in  std_logic_vector(63 downto 0);
39      PO_out    : out std_logic_vector(63 downto 0));
40    end component;
41    component FIRTAPA
42    port(
43      CLK_S  : in  std_logic_vector(8 downto 0);
44      CLKn_S : in  std_logic_vector(1 downto 0);
45      RSTn_S : in  std_logic_vector(10 downto 0);
46      LDCOEF : in  std_logic;
47      COEFi  : in  std_logic_vector(15 downto 0);
48      COEFo  : out std_logic_vector(15 downto 0);
49      I      : in  std_logic_vector(15 downto 0);
50      O      : out std_logic_vector(15 downto 0);
51      PI     : in  std_logic_vector(63 downto 0);
52      PO     : out std_logic_vector(63 downto 0));
53    end component;
54    component FIRTAPB
55    port(
56      CLK_S  : in  std_logic_vector( 8 downto 0);
57      RSTn_S : in  std_logic_vector( 8 downto 0);
58      LDCOEF : in  std_logic;
59      COEFi  : in  std_logic_vector(15 downto 0);
60      COEFo  : out std_logic_vector(15 downto 0);
61      I      : in  std_logic_vector(15 downto 0);
62      O      : out std_logic_vector(15 downto 0);
63      PI     : in  std_logic_vector(63 downto 0);
64      PO     : out std_logic_vector(63 downto 0));
```

```
65       end component;
66       subtype size16 is std_logic_vector(15 downto 0);
67       subtype size64 is std_logic_vector(63 downto 0);
68       type    k1x16  is array (K   downto 0) of size16;
69       type    k1x64  is array (K   downto 0) of size64;
70       signal IREG, COEF : k1x16;
71       signal PSUM        : k1x64;
72     begin
73       PSUM(K)   <= PI; COEF(K)   <= I;
74       glue0 : GLUE port map (
75         RSTn       => RSTn,           RSTno      => RSTno,
76         CLK        => CLK_S(0),       START      => START,
77         I          => I,             I_out      => IREG(K));
78       mux0 : MUX port map (
79         BYPASS     => BYPASS,         LDCOEF     => LDCOEF,
80         I          => I,             O_in       => IREG(0),
81         COEF_in    => COEF(0),        O_out      => O,
82         PI         => PI,            PO_in      => PSUM(0),
83         PO_out     => PO);
84       gen32 : for j in K-1 downto 0 generate
85         tapb : if (j = K-1) generate
86           firtapb0 : FIRTAPB port map (
87             CLK_S    => CLK_S(j*9+8 downto j*9),
88             RSTn_S   => RSTn_S(j*11+8 downto j*11),
89             LDCOEF   => LDCOEF,
90             COEFi    => COEF(j+1),
91             COEFo    => COEF(j),
92             I        => IREG(j+1),
93             O        => IREG(j),
94             PI       => PSUM(j+1),
95             PO       => PSUM(j));
96         end generate;
97         tapa : if (j /= K-1) generate
98           firtapa0 : FIRTAPA port map (
99             CLK_S    => CLK_S(j*9+8 downto j*9),
100            CLKn_S   => CLKn_S(j*2+1 downto j*2),
101            RSTn_S   => RSTn_S(j*11+10 downto j*11),
102            LDCOEF   => LDCOEF,
103            COEFi    => COEF(j+1),
104            COEFo    => COEF(j),
105            I        => IREG(j+1),
106            O        => IREG(j),
107            PI       => PSUM(j+1),
108            PO       => PSUM(j));
109        end generate;
110      end generate;
111    end RTL;
```

The following VHDL code can be used to generate a two-stage tree of buffers. In line 6, generic WIDE1 is used to specify the number of buffers in the first level of the tree buffers. In line 7, generic BUF is used to specify the number of buffers in the second level of the tree buffers. The inverted buffer component stdinv_60x is used which has a buffer size of 60. The stdinv_60x component is declared in package GATECOMP as referenced in line 3. Lines 16 to 19 instantiate component stdinv_60x WIDE1 times. Lines 20 to 23 instantiate the second level of buffers BUF times. Note that these buffers are almost evenly distributed to connect to the first level of the buffers with (j mod (WIDE1)) as the index as shown in line 22. Also note that TREE components can be used for both the clock and reset tree buffers.

```
1    library IEEE;
2    use IEEE.std_logic_1164.all;
3    use work.GATECOMP.all;
4    entity TREE is
5      generic(
6        WIDE1  : integer range 1 to 20 := 4;
7        BUF    : integer range 1 to 400 := 20);
8      port(
9        CLKIN  : in  std_logic; -- input clock
10       CLK    : out std_logic_vector(BUF-1 downto 0));
11     end TREE;
12   architecture RTL of TREE is
13     signal CLKtemp1 : std_logic_vector(WIDE1-1 downto 0);
14     signal CLK1     : std_logic_vector(BUF-1 downto 0);
15   begin
16     genclk1 : for j in 0 to WIDE1-1 generate
17       u0 : stdinv_60x
18         port map (IN0 => CLKIN, Y => CLKtemp1(j));
19     end generate;
20     genclk4 : for j in 0 to BUF-1 generate
21       u2 : stdinv_60x
22         port map (IN0=>CLKtemp1(j mod (WIDE1)), Y=>CLK1(j));
23     end generate;
24     CLK <= CLK1;
25   end RTL;
```

The following FIRCORE VHDL code includes all of the chip logic excluding the I/O pads of the chip. Components TREE and FIRSYN are declared in lines 20 to 43. Line 44 declares constant K with value 32. K is used in lines 46 to 48 to declare signals and in line 64 for the generic mapping. Lines 53 to 56 instantiate inverted buffers for the inverted clock signals.

```
1    --------------------- file fircore.vhd
2    library IEEE;
3    use IEEE.std_logic_1164.all;
4    use work.GATECOMP.all;
5    entity FIRCORE is
6      port(
7        RSTn   : in  std_logic;
8        CLK    : in  std_logic;
9        LDCOEF : in  std_logic; -- enable loading coef
10       START  : in  std_logic; -- start filtering
11       OEn    : in  std_logic; -- enable tristate pad
```

```
12        BYPASS  : in  std_logic; -- bypass the chip
13        I       : in  std_logic_vector(15 downto 0); -- data
14        PI      : in  std_logic_vector(63 downto 0); -- sum
15        O       : out std_logic_vector(15 downto 0); -- data
16        PO      : out std_logic_vector(63 downto 0);
17        OEn_S   : out std_logic_vector(79 downto 0)); -- OEn
18    end FIRCORE;
19    architecture RTL of FIRCORE is
20      component TREE
21      generic(
22        WIDE1  : integer range 1 to 20  := 4;
23        BUF    : integer range 1 to 400 := 20);
24      port(
25        CLKIN  : in  std_logic; -- input clock
26        CLK    : out std_logic_vector(BUF-1 downto 0));
27      end component;
28      component FIRSYN
29      generic (K : integer := 32);
30      port(
31        RSTn    : in  std_logic;
32        RSTno   : out std_logic;
33        CLK_S   : in  std_logic_vector(9*K-1 downto 0);
34        CLKn_S  : in  std_logic_vector(2*K-3 downto 0);
35        RSTn_S  : in  std_logic_vector(11*K-3 downto 0);
36        LDCOEF  : in  std_logic; -- enable loading coef
37        START   : in  std_logic; -- start filtering
38        BYPASS  : in  std_logic; -- bypass the chip
39        I       : in  std_logic_vector(15 downto 0); -- data
40        PI      : in  std_logic_vector(63 downto 0); -- sum
41        O       : out std_logic_vector(15 downto 0); -- data
42        PO      : out std_logic_vector(63 downto 0));
43      end component;
44      constant K : integer := 32;
45      signal RSTno : std_logic;
46      signal CLK_S   : std_logic_vector(9*K downto 0);
47      signal CLKn_S  : std_logic_vector(2*K-3 downto 0);
48      signal RSTn_S  : std_logic_vector(11*K-3 downto 0);
49    begin
50      clktree : TREE
51        generic map (WIDE1 => 16,    BUF => 9*K+1)
52        port map    (CLKIN => CLK,   CLK => CLK_S);
53      clkngen : for j in CLKn_S'range generate
54      u0 : stdinv_60x
55        port map (IN0 => CLK_S(9*K), Y => CLKn_S(j));
56      end generate;
57      rstntree : TREE
58        generic map (WIDE1 => 16,    BUF => 11*K-2)
59        port map    (CLKIN => RSTno, CLK => RSTn_S);
60      oentree : TREE
61        generic map (WIDE1 => 8,     BUF => 80)
62        port map    (CLKIN => OEn,   CLK => OEn_S);
```

```
63      firsyn0 : FIRSYN
64      generic map (K => K)
65      port map (
66        RSTn      => RSTn,
67        RSTno     => RSTno,
68        CLK_S     => CLK_S(9*K-1 downto 0),
69        CLKn_S    => CLKn_S,
70        RSTn_S    => RSTn_S,
71        LDCOEF    => LDCOEF,
72        START     => START,
73        BYPASS    => BYPASS,
74        I         => I,
75        PI        => PI,
76        O         => O,
77        PO        => PO);
78    end RTL;
```

Figure 10-5 shows the hierarchy of the FIRCHIP VHDL code. The blocks to be synthesized are all inside the FIRSYN block. In the FIRCORE block, all blocks are instantiated so that there is no need to synthesize. The output enable signals are simply connected from the input signal OEn to a tree of buffers. In common practice, the output enables may be controlled by an internal block such as a finite-state machine. In that case, the output enable signals and related circuit would be inside the FIRSYN block. Also, it is important to know whether the three-state pad output enable is low- or high-active in the technology library. The output enable signal can be generated at the right polarity with the synthesis inside the synthesized block. Otherwise, the signal may need to be inverted before connecting to the pad. The clock and reset tree are separated from the FIRSYN block so that they can be easily changed when another process technology is used and the buffer names are not the same. It also helps expedite the analysis process to specify the clock with flexibility when FIRSYN is analyzed.

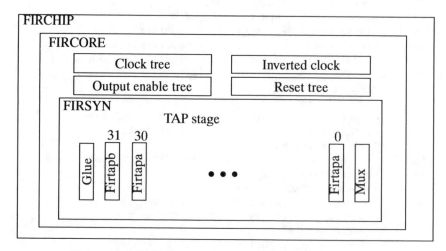

Figure 10-5 FIRCHIP VHDL hierarchical organization.

10.3 DESIGN VERIFICATION

How do we verify this design? We have presented the FIRCHIP with two architectures: BEH and RTL. Architecture BEH is self-contained which does not have lower level blocks. It is not intended for synthesis. Architecture RTL is intended for synthesis; that's why the design is partitioned hierarchically and functionally. This helps the synthesis process since the design is smaller. It also allows the blocks such as the multiplier MULT and the adder ADD64 to be easily substituted with different architectures. For example, ADD64 component can be implemented with the CLA64 as discussed in Chapter 5, rather than using the operator "+".

In the last section, we have shown the VHDL code for all the blocks. That remains an issue. The stdinv_60x component was used, but the VHDL code was not shown. The following VHDL code shows the stdinv_60x VHDL code. It is inside a text file named gates.vhd. The idea is to combine all components that are to be instantiated in the synthesis process together. When the design is to be ported to another process technology, equivalent components are then replaced in this file.

```
1    library IEEE;
2    use IEEE.std_logic_1164.all;
3    entity STDINV_60x is
4      port (
5        IN0 : in  std_logic;
6        Y   : out std_logic);
7    end STDINV_60x;
8    architecture RTL of STDINV_60x is
9    begin
10     Y <= not IN0;
11   end RTL;
```

The following VHDL code shows a package containing the stdinv_60x component. This package is referenced in the TREE VHDL code.

```
1    library IEEE;
2    use IEEE.std_logic_1164.all;
3    package GATECOMP is
4      component STDINV_60x
5      port (
6        IN0 : in  std_logic;
7        Y   : out std_logic);
8      end component;
9    end GATECOMP;
```

The following VHDL code shows the configuration for the FIRCHIP RTL architecture. From the following configuration, it is easy to see the VHDL hierarchy as referenced in Figure 10-5. Line 8 shows that there is a CFG_TREE configuration. The VHDL code is shown below.

```
1      ---------------------- file firchip_cfg.vhd
2      configuration CFG_FIRCHIP of FIRCHIP is
3        for RTL
4          for fircore0 : FIRCORE
5            use entity work.FIRCORE(RTL);
6            for RTL
7              for all : TREE
8                use configuration work.CFG_TREE;
9              end for;
10             for clkngen
11               for all : stdinv_60x
12                 use entity work.stdinv_60x(RTL);
13               end for;
14             end for;
15             for all : FIRSYN
16               use entity work.FIRSYN(RTL);
17               for RTL
18                 for glue0 : GLUE
19                   use entity work.GLUE(RTL);
20                 end for;
21                 for mux0 : MUX
22                   use entity work.MUX(RTL);
23                 end for;
24                 for gen32
25                   for tapb
26                     for all : FIRTAPB
27                       use entity work.FIRTAPB(RTL);
28                       for RTL
29                         for all : COEFREG
30                           use entity work.COEFREG(RTL);
31                         end for;
32                         for all : REGS
33                           use entity work.REGS(RTL);
34                         end for;
35                         for all : MULT
36                           use entity work.MULT(RTL);
37                         end for;
38                         for all : ADD64
39                           use entity work.ADD64(RTL);
40                         end for;
41                       end for;
42                     end for;
43                   end for;
44                   for tapa
45                     for all : FIRTAPA
46                       use entity work.FIRTAPA(RTL);
47                       for RTL
48                         for all : COEFREG
49                           use entity work.COEFREG(RTL);
50                         end for;
```

```
51                           for all : REGS
52                               use entity work.REGS(RTL);
53                             end for;
54                           for all : MULT
55                               use entity work.MULT(RTL);
56                             end for;
57                           for all : ADD64
58                               use entity work.ADD64(RTL);
59                             end for;
60                         end for;
61                       end for;
62                     end for;
63                   end for;
64                 end for;
65               end for;
66             end for;
67           end for;
68       end for;
69   end CFG_FIRCHIP;


1    configuration CFG_TREE of TREE is
2      for RTL
3        for genclk1
4          for all : STDINV_60x
5              use entity WORK.STDINV_60x(RTL);
6          end for;
7        end for;
8        for genclk4
9          for all : STDINV_60x
10             use entity WORK.STDINV_60x(RTL);
11         end for;
12       end for;
13     end for;
14   end CFG_TREE;
```

Now we bring the whole design together using the CFG_FIRCHIP configuration. How do we verify the design? First of all, we need to determine how the coefficient registers are being loaded and how input data I and PI are driven to the design. The following VHDL code shows an example of how to drive the LDCOEF, I, and PI signals. Lines 3 to 6 reference packages. Lines 9 to 11 declare the delay for input signals relative to the rising edge of the clock. Line 12 declares generic RSTn2COEF to specify the number of clock between the deassertion of RSTn and the loading of the first coefficient. Line 13 declares generic COEF2DATA to specify the number of clock cycles between coefficient load completion and the time to start driving I and PI inputs. Lines 14 to 16 declare the three files containing values of I, PI, and COEF which can be read in as text files.

```
1     -------------------- file drive.vhd
2     library IEEE;
3     use IEEE.std_logic_1164.all;
4     use IEEE.std_logic_arith.all;
5     use IEEE.std_logic_textio.all;
6     use STD.TEXTIO.all;
7     entity DRIVE is
8       generic (
9         LDCOEF_DLY   : time    := 15 ns;
10        I_DLY        : time    := 15 ns;
11        PI_DLY       : time    := 15 ns;
12        RSTn2COEF    : integer := 10; -- from rstn to coef
13        COEF2DATA    : integer := 10; -- from coef to data
14        coeffile     : string  := "coef.in";
15        idatafile    : string  := "idata.in";
16        pidatafile   : string  := "pidata.in");
17      port (
18        RSTn      : in  std_logic;
19        CLK       : in  std_logic;
20        LDCOEF    : out std_logic;
21        I         : out std_logic_vector(15 downto 0);
22        PI        : out std_logic_vector(63 downto 0));
23      end DRIVE;
```

Lines 27 to 29 declare three file pointers. Lines 31 to 34 declare variables to be used. Lines 36 to 38 first initialize signals LDCOEF, I, and PI. Lines 39 to 40 wait for the RSTn to transition from '0' to '1'. Line 41 waits until the next rising edge of the clock. Lines 42 to 44 wait for RSTn2COEF+1 clock cycles.

```
24    architecture BEH of DRIVE is
25    begin
26      r0 : process
27        file coeffp      : text is in coeffile;
28        file idatafp     : text is in idatafile;
29        file pidatafp    : text is in pidatafile;
30        variable lin     : line;
31        variable DATA16  : std_logic_vector(15 downto 0);
32        variable DATA64  : std_logic_vector(63 downto 0);
33        variable COEFNUM : integer;
34        variable DATANUM : integer;
35      begin
36        LDCOEF   <= '0';
37        I        <= (I'range  => '0');
38        PI       <= (PI'range => '0');
39        wait until RSTn'event and RSTn = '1' and
40                   RSTn'last_value = '0';
41        wait until CLK'event and CLK = '1';
42        for j in 0 to RSTn2COEF loop
43          wait until CLK'event and CLK = '1';
44        end loop;
```

Lines 46 and 47 read the number of coefficients from the coefficient text file. Lines 48 to 54 use a loop statement to repeat reading the specified number of coefficients. LDCOEF signal is asserted '1' in line 52. Signal I is assigned in line 51. Both of them are assigned with delays I_DLY and LDCOEF_DLY, respectively. Line 55 deasserts LDCEOF to '0'. Otherwise, the coefficient registers will be shifted when LDCOEF is asserted. It is important to assert LDCOEF signal the number of clock cycles exactly the same as the number of TAP stages. Line 56 assigns all '0' to signal I. Lines 57 to 60 wait another two clock cycles. Lines 61 to 63 wait for COEF2DATA clock cycles. Line 65 to 66 read the number of I and PI data. The loop statement is approximately the same as before. Signals I and PI are assigned in lines 70 and 73, respectively. Lines 76 to 77 assign all '0' to signals I and PI. Line 78 waits for one clock cycle. Line 79 waits forever.

```
45          -- start to write to coef registers
46          readline (coeffp, lin);
47          read (lin, COEFNUM);
48          for j in 0 to COEFNUM-1 loop
49            readline (coeffp, lin);
50            hread (lin, DATA16);
51            I     <= DATA16 after I_DLY;
52            LDCOEF <= '1' after LDCOEF_DLY;
53            wait until CLK'event and CLK = '1';
54          end loop;
55          LDCOEF <= '0' after LDCOEF_DLY;
56          I     <= (I'range  => '0') after I_DLY;
57          wait until CLK'event and CLK = '1';
58          -----------------------------------------------
59            -- start to drive data
60          wait until CLK'event and CLK = '1';
61          for j in 1 to COEF2DATA loop
62            wait until CLK'event and CLK = '1';
63          end loop;
64          wait until CLK'event and CLK = '1';
65          readline (idatafp, lin);
66          read (lin, DATANUM);
67          for j in 1 to DATANUM loop
68            readline (idatafp, lin);
69            hread (lin, DATA16);
70            I   <= DATA16 after I_DLY;
71            readline (pidatafp, lin);
72            hread (lin, DATA64);
73            PI  <= DATA64 after PI_DLY;
74            wait until CLK'event and CLK = '1';
75          end loop;
76          I   <= (I'range  => '0') after I_DLY;
77          PI  <= (PI'range => '0') after PI_DLY;
78          wait until CLK'event and CLK = '1';
79          wait;
```

```
80      end process;
81    end BEH;
```

The following simple CONTROL VHDL code generates the clock and reset signals.

```
1     library IEEE;
2     use IEEE.std_logic_1164.all;
3     entity CONTROL is
4       generic (
5         PERIOD   : time := 30 ns;
6         RSTn_DLY : time := 15 ns;
7         RSTNUM   : integer := 10);
8       port (
9         CLK      : out std_logic := '1';
10        RSTn     : out std_logic := '0');
11    end CONTROL;
12    architecture BEH of CONTROL is
13      constant CLK_P2  : time := PERIOD / 2;
14    begin
15      resetn_gen : process
16      begin
17        RSTn <= '0' after RSTn_DLY; wait for RSTNUM * PERIOD;
18        RSTn <= '1' after RSTn_DLY; wait;
19      end process;
20      clk_gen : process
21      begin
22        CLK   <= '1'; wait for CLK_P2;
23        CLK   <= '0'; wait for CLK_P2;
24      end process;
25    end BEH;
```

The following test bench VHDL code combines CONTROL, DRIVE, and FIRCHIP. You are encouraged to pay attention to how generics and signals are mapped. The idea is to be able to change the generics at the very top level so that the whole test setup can be changed from a single point of entry. For example, the clock period can be changed to 1000 ns in line 6 so that the clock period is 1000 ns long. Some fabrication vendors may require the released test vector in either a 1000-ns or 2000-ns cycle. The input delay for each signal can also be changed. Note that signals OEn, START, and BYPASS are not driven by DRIVE VHDL code. They will be assigned as VHDL simulator commands which will be shown below.

```
1     -------------------- file tbfirchip.vhd
2     library IEEE;
3     use IEEE.std_logic_1164.all;
4     entity TBFIRCHIP is
5       generic (
6         PERIOD     : time    := 30 ns;
7         RSTn_DLY   : time    := 15 ns;
```

```
 8        RSTNUM      : integer := 10;
 9        LDCOEF_DLY : time     := 15 ns;
10        START_DLY  : time     := 15 ns;
11        I_DLY      : time     := 15 ns;
12        PI_DLY     : time     := 15 ns;
13        RSTn2COEF  : integer := 10; -- from rstn to coef setup
14        COEF2DATA  : integer := 10; -- from coef to data setup
15        coeffile   : string   := "coef.in";
16        idatafile  : string   := "idata.in";
17        pidatafile : string   := "pidata.in");
18    end TBFIRCHIP;
19    architecture BEH of TBFIRCHIP is
20      component FIRCHIP
21      port(
22        RSTn     : in    std_logic_vector( 0 downto 0);
23        CLK      : in    std_logic_vector( 0 downto 0);
24        LDCOEF   : in    std_logic_vector( 0 downto 0);
25        START    : in    std_logic_vector( 0 downto 0);
26        OEn      : in    std_logic_vector( 0 downto 0);
27        BYPASS   : in    std_logic_vector( 0 downto 0);
28        I        : in    std_logic_vector(15 downto 0);
29        PI       : in    std_logic_vector(63 downto 0);
30        O        : out   std_logic_vector(15 downto 0);
31        PO       : out   std_logic_vector(63 downto 0);
32        VDDS     : inout std_logic_vector(20 downto 0);
33        GNDS     : inout std_logic_vector(20 downto 0));
34      end component;
35      component CONTROL
36      generic (
37        PERIOD   : time := 30 ns;
38        RSTn_DLY : time := 15 ns;
39        RSTNUM   : integer := 10);
40      port (
41        CLK      : out std_logic := '1';
42        RSTn     : out std_logic := '0');
43      end component;
44      component DRIVE
45      generic (
46        LDCOEF_DLY : time     := 15 ns;
47        I_DLY      : time     := 15 ns;
48        PI_DLY     : time     := 15 ns;
49        RSTn2COEF  : integer := 10; -- from rstn to coef setup
50        COEF2DATA  : integer := 10; -- from coef to data setup
51        coeffile   : string   := "coef.in";
52        idatafile  : string   := "idata.in";
53        pidatafile : string   := "pidata.in");
54      port (
55        RSTn   : in  std_logic;
56        CLK    : in  std_logic;
57        LDCOEF : out std_logic;
```

```
58        I      : out std_logic_vector(15 downto 0);
59        PI     : out std_logic_vector(63 downto 0));
60     end component;
61     signal RSTn    : std_logic_vector( 0 downto 0);
62     signal CLK     : std_logic_vector( 0 downto 0);
63     signal LDCOEF  : std_logic_vector( 0 downto 0);
64     signal START   : std_logic_vector( 0 downto 0);
65     signal OEn     : std_logic_vector( 0 downto 0);
66     signal BYPASS  : std_logic_vector( 0 downto 0);
67     signal I       : std_logic_vector(15 downto 0);
68     signal PI      : std_logic_vector(63 downto 0);
69     signal O       : std_logic_vector(15 downto 0);
70     signal PO      : std_logic_vector(63 downto 0);
71     signal VDDS    : std_logic_vector(20 downto 0);
72     signal GNDS    : std_logic_vector(20 downto 0);
73   begin
74     firchip0 : FIRCHIP
75     port map (
76       RSTn       => RSTn,    CLK       => CLK,
77       LDCOEF     => LDCOEF,  START     => START,
78       OEn        => OEn,     BYPASS    => BYPASS,
79       I          => I,      PI        => PI,
80       O          => O,      PO        => PO,
81       VDDS       => VDDS,    GNDS      => GNDS);
82     control0 : CONTROL
83     generic map (
84       PERIOD=>PERIOD, RSTn_DLY=>RSTn_DLY, RSTNUM=>RSTNUM)
85     port map ( CLK => CLK(0), RSTn => RSTn(0));
86
87     drive0 : DRIVE
88     generic map (
89       LDCOEF_DLY   => LDCOEF_DLY,
90       I_DLY        => I_DLY,
91       PI_DLY       => PI_DLY,
92       RSTn2COEF    => RSTn2COEF, -- from rstn to coef setup
93       COEF2DATA    => COEF2DATA, -- from coef to data setup
94       coeffile     => coeffile,
95       idatafile    => idatafile,
96       pidatafile   => pidatafile)
97     port map (
98       RSTn       => RSTn(0),    CLK       => CLK(0),
99       LDCOEF     => LDCOEF(0),  I         => I,
100      PI         => PI);
101  end BEH;
```

The test bench can be configured with the following VHDL code. Note that line 5 configures the FIRCHIP to use architecture BEH. Line 4, which is commented out, can be used to configure the FIRCHIP with architecture RTL. By toggling lines 4 and 5, either BEH or RTL architecture can be selected. This requires that we reanalyze the

following configuration VHDL code. Another approach is to have two separate configurations, one with line 4 and the other with line 5.

```
1    configuration CFG_TBFIRCHIP of TBFIRCHIP is
2      for BEH
3        for firchip0 : FIRCHIP
4          -- use configuration work.CFG_FIRCHIP;
5          use entity work.FIRCHIP(BEH);
6        end for;
7        for drive0 : DRIVE
8          use entity work.DRIVE(BEH);
9        end for;
10       for control0 : CONTROL
11         use entity work.CONTROL(BEH);
12       end for;
13     end for;
14   end CFG_TBFIRCHIP;
```

All the VHDL code can be analyzed with the following command file for Synopsys VSS VHDL simulator. The command file for other VHDL simulators can be
written in a similar manner.

```
1        echo "../vhdl/drive.vhd"
2    vhdlan -c -nc ../vhdl/drive.vhd
3        echo "../vhdl/control.vhd"
4    vhdlan -c -nc ../vhdl/control.vhd
5        echo "../vhdl/gates.vhd"
6    vhdlan -c -nc ../vhdl/gates.vhd
7        echo "../vhdl/gatecomp.vhd"
8    vhdlan -c -nc ../vhdl/gatecomp.vhd
9        echo "../vhdl/tree.vhd"
10   vhdlan -c -nc ../vhdl/tree.vhd
11       echo "../vhdl/tree_cfg.vhd"
12   vhdlan -c -nc ../vhdl/tree_cfg.vhd
13       echo "../vhdl/regs.vhd"
14   vhdlan -c -nc ../vhdl/regs.vhd
15       echo "../vhdl/coefreg.vhd"
16   vhdlan -c -nc ../vhdl/coefreg.vhd
17       echo "../vhdl/glue.vhd"
18   vhdlan -c -nc ../vhdl/glue.vhd
19       echo "../vhdl/mux.vhd"
20   vhdlan -c -nc ../vhdl/mux.vhd
21       echo "../vhdl/mult.vhd"
22   vhdlan -c -nc ../vhdl/mult.vhd
23       echo "../vhdl/add64.vhd"
24   vhdlan -c -nc ../vhdl/add64.vhd
25       echo "../vhdl/firtapa.vhd"
26   vhdlan -c -nc ../vhdl/firtapa.vhd
27       echo "../vhdl/firtapb.vhd"
```

```
28   vhdlan -c -nc ../vhdl/firtapb.vhd
29      echo "../vhdl/firsyn.vhd"
30   vhdlan -c -nc ../vhdl/firsyn.vhd
31      echo "../vhdl/fircore.vhd"
32   vhdlan -c -nc ../vhdl/fircore.vhd
33      echo "../vhdl/firchip_e.vhd"
34   vhdlan -c -nc ../vhdl/firchip_e.vhd
35      echo "../vhdl/firchip_beh.vhd"
36   vhdlan -c -nc ../vhdl/firchip_beh.vhd
37      echo "../vhdl/firchip_rtl.vhd"
38   vhdlan -c -nc ../vhdl/firchip_rtl.vhd
39      echo "../vhdl/firchip_cfg.vhd"
40   vhdlan -c -nc ../vhdl/firchip_cfg.vhd
41      echo "../vhdl/tbfirchip.vhd"
42   vhdlan -c -nc ../vhdl/tbfirchip.vhd
43      echo "../vhdl/tbfirchip_cfg.vhd"
44   vhdlan -c -nc ../vhdl/tbfirchip_cfg.vhd
```

Now, we discover whether the design works as expected. First, let's try to verify architecture BEH. Figure 10-6 shows a simulation waveform in a "global picture." The LDCOEF signal is asserted exactly 32 clock cycles. At that period of time, the input signal I is set to 1 so that the coefficient register of each stage is loaded with value 1. Start input signal is asserted a couple of clock cycles later to enable the FIR function. A couple of clock cycles later, input data I comes in to go through the TAP stages. Output data PO comes out last.

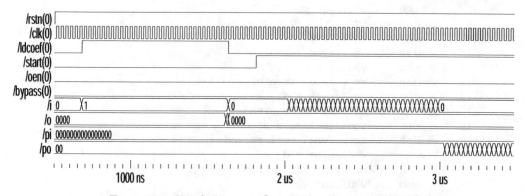

Figure 10-6 Simulation waveform test1.1 for FIRCHIP BEH.

Figure 10-7 shows an enlargement of the point in Figure 10-6 where PO starts to change. PO output values are 1, 3, 6, 10, 15, 21, 28, and so on, in decimal. The input data I is 1, 2, 3, . . . , 30, 31, 32 in decimal. Since all the coefficients are 1, the resulting sequence is just the accumulated sum of the input data I sequence, 1, 1+2, 1+2+3,

1+2+3+4, and so on. The design is "somewhat" working. This is the first major accomplishment — we have verified that the test bench is working!

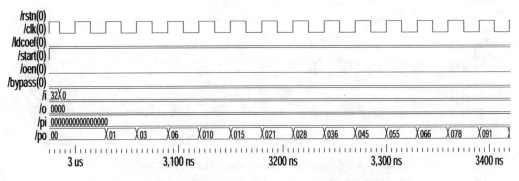

Figure 10-7 Simulation waveform test1.2 for FIRCHIP BEH.

Figure 10-8 shows an enlargement of the point in Figure 10-6 where output O starts to change values. They are 1, 2, 3, and so on, which is the exact sequence of the input data I. This verifies the shifting from I, to each stage of IREG1 and IREG2, and to output O is working.

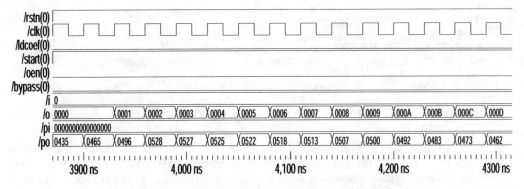

Figure 10-8 Simulation waveform test1.3 for FIRCHIP BEH.

Figure 10-9 shows another test with coefficients loaded as 1, 2, 3, and so on. Coefficient 1 will be loaded to the stage that is closest to the output. Coefficient 32 will be loaded to the stage that is closest to the input. The input data sequence is still 1, 2, 3, and so on. Output PO has the values 32, 95, 182, 310, 460, and so on. They are exactly $1*32 = 32$, $1*31 + 2*32 = 95$, $1*30 + 2*31 + 3*32 = 188$, $1*29 + 2*30 + 3*31 + 4*32 = 310$, and so on.

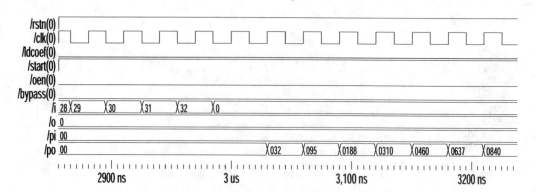

Figure 10-9 Simulation waveform test2 for FIRCHIP BEH.

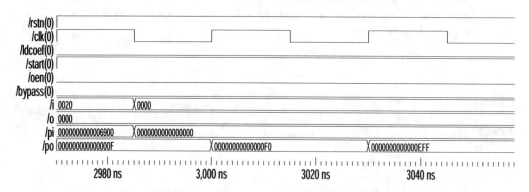

Figure 10-10 Simulation waveform test3 for FIRCHIP BEH.

Figure 10-10 shows another simulation waveform with coefficients loaded with 1, −1, 1, −1 and so on. The input signal PI has values F, F0, F00, and so on in hexadecimal. Input signal I follows the sequence 1, 2, 3, and so on. Output PO has the value F, F0, EFF, and so on. The first PO output values F and F0 are coming from PI since input I needs to go in to IREG1, and then to the MULT register, which requires two clock cycles before the value is added with PI. The third PO value is F00 (from PI) + 1*(−1), which is EFF. This verifies that negative coefficients work.

Figure 10-11 shows the simulation waveform for verifying the OEn and BYPASS input signals. When OEn is deasserted to '1', outputs O and PO are threestated with all 'Z' values. When BYPASS is asserted, outputs O and PO are the same as I and PI, respectively. There are many other tests that can be run. Let's consider how these tests are set up.

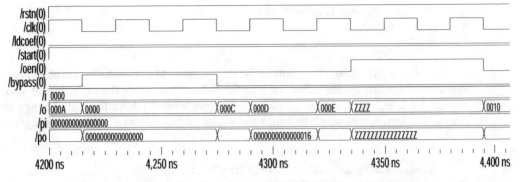

Figure 10-11 Simulation waveform test3.2 for FIRCHIP BEH.

The following file sim.cmd shows how a test is run. Lines 1 and 14 print out the time stamp so that we know how long each test takes. Lines 2 and 3 set up the directory names. The text files for the coefficient, I and PI, are stored in the $TEST_DIR directory which is mapped to ../infile. Lines 5 to 8 copy the four input files to the directory where the VHDL simulation is run. Lines 9 to 10 invoke Synopsys the VSS VHDL simulator. The simulation log and the waveform are saved in the $RESULT_DIR (./_result) directory. The simulation is controlled by the simulation command file sim.do as shown below.

```
1    date
2    export TEST_DIR=../infile
3    export RESULT_DIR=./_result
4       echo "Simulating $1..."
5       cp -f $TEST_DIR/$1.coef ./coef.in
6       cp -f $TEST_DIR/$1.idata ./idata.in
7       cp -f $TEST_DIR/$1.pidata ./pidata.in
8       cp -f $TEST_DIR/$1.vss ./sim.do
9       vhdlsim -pro -o $RESULT_DIR/$1.log -w $RESULT_DIR/$1.ow \
10          -nc -i sim.do cfg_tbfirchip
11      echo "Done!"
12      echo
13      ;
14   date
```

Lines 1 to 3 assign values to BYPASS, OEn, and START signals in the test bench that is not driven by the DRIVE VHDL code. Line 4 includes a trace file as shown below. Line 5 runs 1,815,000 picoseconds. Line 6 asserts the START signal. Line 7 runs for a while. Line 8 quits the simulator.

```
1    assign "0" /tbfirchip/BYPASS
2    assign "0" /tbfirchip/OEn
3    assign "0" /tbfirchip/START
4    include trace.chip
5    run 1815000
6    assign "1" /tbfirchip/START
7    run 3185000
8    quit
```

All the input/output (I/O) pins of the design should be traced. The enable signals that control the three-state output pads or the bidirectional pads should be traced also. These enable signals are used so the tester knows when to drive the signal, when to release the signal, and when to strobe the signal and to compare. The enable signals are traced in line 6.

```
1    trace /tbfirchip/RSTn(0) /tbfirchip/CLK(0)
2    trace /tbfirchip/LDCOEF(0) /tbfirchip/START(0)
3    trace /tbfirchip/OEn(0)
4    trace /tbfirchip/BYPASS(0) /tbfirchip/I
5    trace /tbfirchip/O /tbfirchip/PI
6    trace /tbfirchip/PO /tbfirchip/FIRCHIP0/OEn_S
```

To run all the above three tests, the following command file batchsim.cmd is used.

```
1    sim.cmd test1
2    sim.cmd test2
3    sim.cmd test3
```

After the above command is run and completed, the simulation waveform files are saved as test1.ow, test2.ow, and test3.ow files in _result directory. The following command file ow2vec.cmd can be used to extract values from the waveform file. It takes a format file vec.fmt as shown below. The vec.fmt specifies the order of the signals to be extracted. It is strobed every 30,000 picoseconds with an offset of 28,000 picoseconds. All signal values are extracted 2 ns before the rising edge of the clock. Three lines from the vector output file are also shown below. First, it has a time stamp, followed by a long sequence of 0's and 1's.

```
1    wif2tab -f vec.fmt -nc -nh -strobe 30000 -offset 28000
_result/$1.ow > _result/$1.vec
```

```
1    /tbfirchip/RSTn(0);
2    /tbfirchip/CLK(0);
3    /tbfirchip/LDCOEF(0);
4    /tbfirchip/START(0);
5    /tbfirchip/OEn(0);
6    /tbfirchip/BYPASS(0);
```

```
7    /tbfirchip/I(15 downto 0);
8    /tbfirchip/O(15 downto 0);
9    /tbfirchip/PI(63 downto 0);
10   /tbfirchip/PO(63 downto 0);


1       4378000
10010000000000000000000000000000001111000000000000000000000000000000
00000000000000000000000000000000000000000000000000000000000000000000
00000000000000000000000000110110101
2       4408000
10010000000000000000000000000000001000000000000000000000000000000000
00000000000000000000000000000000000000000000000000000000000000000000
00000000000000000000000000110100111
3       4438000
10010000000000000000000000000000001000100000000000000000000000000000
00000000000000000000000000000000000000000000000000000000000000000000
00000000000000000000000000110011000
```

The above vector file can be processed to prepare a foundry release test vector. The test vector is used by the tester to test the ASIC after it is fabricated. Each foundry may require a different format for the release vector. For example, the test cycle may be 1000 ns. The input timing and output strobe timing of signals are defined. The details of this process are not covered here. However, this vector file is the starting point for the formatting process.

Another good use of this vector file is for comparison. For example, in this design, the same test bench can be configured with FIRCHIP architecture RTL. The same process can be run with results saved in different directories. A simple Unix **diff** command can be used to compare the results of architectures BEH and RTL automatically. They did compare with the following command file. This means that the implementation of architecture RTL is functionally the same as architecture BEH as far as the tests we have run. For a designer, ensuring the tests to cover all requirements and functions is a major challenge. It has often been said that, "if it is not tested, it is not going to work as expected." The same configuration can be changed to simulate the postlayout netlist with back-annotated timing delays. They can also be compared automatically.

```
1    diff _result/test1.vec _beh/test1.vec
2    diff _result/test2.vec _beh/test2.vec
3    diff _result/test3.vec _beh/test3.vec
```

We have spent a great deal of time in the test bench and the test setup. We have also illustrated a process and methodology designed to run many tests as a batch. The results can be automatically compared to reduce human error in simply viewing the waveforms. This allows more tests to be run as a batch job, taking advantage of the over-night run capability. The daytime is well spent in troubleshooting and updating the design.

10.4 DESIGN SYNTHESIS

By now, you must be an expert in individual block synthesis. Therefore, individual block synthesis is not presented here. Figure 10-12 shows the synthesized schematic for the GLUE block. Note that there are two flip-flops in series to synchronize the RSTn input.

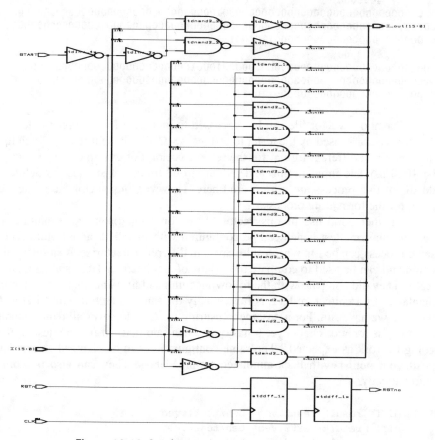

Figure 10-12 Synthesized schematic for the GLUE block.

Figure 10-13 shows part of the synthesized schematic for the COEFREG block. Note that there is a multiplexer before the flip-flop to enable the loading of the coefficient register.

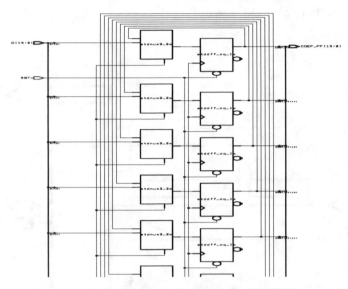

Figure 10-13 Synthesized schematic for COEFREG.

Assume that we have all the lowest-level blocks synthesized and the results are saved. The following synthesis script file compile.fircore is used to combine all blocks inside the FIRCORE block. Line 2 sets the **hdlin_auto_save_templates** variable to TRUE so that the block with generic is saved with a template. Lines 3 to 6 read the saved synthesized results for the lower-level blocks. Note that mult.clean.db and add64.clean.db are modified from the original synthesis result by flattening the original Synopsys DesignWare components and recompiling. This removes any inputs that are tied to ground and open outputs. Line 7 reads the gatecomp.vhd package so that it can be used in the TREE block. Note that the file gates.vhd is not read. The name "stdinv_60x" matches the exact component in the synthesis library. The synthesis tool just instantiates it without synthesis. Line 9 reads in firsyn.vhd. Line 10 reads in fircore.vhd. Figure 10-14 shows the FIRCORE schematic. Note that the inverted buffers are grouped in **clkngen** with the command in line 11. The name **clkngen** is the label for the generate statement from line 53 of FIRCORE VHDL code. Figure 10-15 shows part of the FIRSYN schematic.

```
1    design = fircore
2    hdlin_auto_save_templates = TRUE
3    read -f db ../db/glue.db
4    read -f db ../db/mux.db
5    read -f db ../db/firtapa.db
6    read -f db ../db/firtapb.db
7    read -f vhdl VHDLDIR + "gatecomp.vhd"
```

```
8    read -f vhdl VHDLDIR + "tree.vhd"
9    read -f vhdl VHDLDIR + "firsyn.vhd"
10   read -f vhdl VHDLDIR + design + ".vhd"
11   group -hdl_block clkngen
12   include compile.common
```

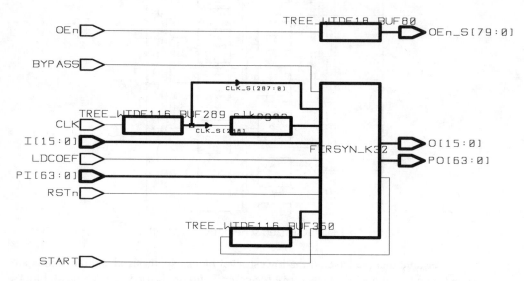

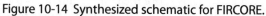

Figure 10-14 Synthesized schematic for FIRCORE.

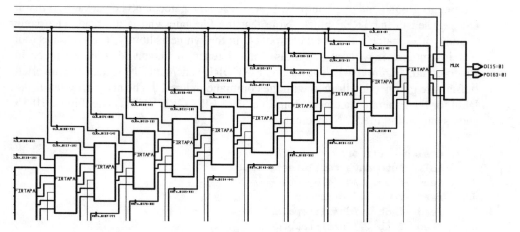

Figure 10-15 Synthesized schematic for FIRSYN.

Figure 10-16 shows part of the synthesized TREE block for the OEn tree. There are eight inverted buffers in the first level of the buffer tree.

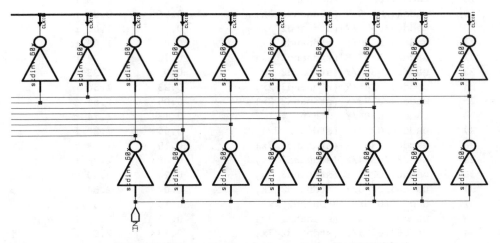

Figure 10-16 Synthesized schematic for the TREE block.

10.5 WORST-CASE TIMING ANALYSIS

Note that the design name for the FIRSYN block is now changed to FIRSYN_K32. That's because Synopsys appends the generic K and generic mapped value 32 to the design name of the block synthesized with a generic. After setting up clock operating conditions, and wire_load model, the **report_timing** command can be used to analyze the timing between flip-flops, from chip input to flip-flops, D inputs, from flip-flops, outputs to chip outputs, and from chip inputs to chip outputs. For example, the timing path for the multiplier is about 16 ns. The input timing from the PI input, going through the adder, and reaching the flip-flops' D inputs is shown below.

```
1       Startpoint: PI[1] (input port)
2       Endpoint: addreg3/DFF_reg[15]
3                 (rising edge-triggered flip-flop clocked by clk)
4       Path Group: clk
5       Path Type: max
6
7       Point                                    Incr        Path
8       ----------------------------------------------------------------
9       clock (input port clock) (rise edge)     0.00        0.00
10      input external delay                     0.00        0.00 r
11      PI[1] (in)                               0.00        0.00 r
12      uadd64/A[1] (ADD64_21)                   0.21        0.21 r
```

13	uadd64/U347/Y (stdinv_5x)	0.11	0.32	f
14	uadd64/U195/Y (stdnor2_2x)	0.14	0.45	r
15	uadd64/U194/Y (stdinv_2x)	0.05	0.50	f
16	uadd64/U351/Y (stdnand2_3x)	0.15	0.65	r
17	uadd64/U353/Y (stdnand2_3x)	0.06	0.71	f
18	uadd64/U354/Y (stdnand2_3x)	0.12	0.83	r
19	uadd64/U356/Y (stdnand2_3x)	0.06	0.89	f
20	uadd64/U357/Y (stdnand2_3x)	0.12	1.01	r
21	uadd64/U359/Y (stdnand2_3x)	0.07	1.08	f
22	uadd64/U297/Y (stdnand2_2x)	0.16	1.24	r
23	uadd64/U362/Y (stdnand2_3x)	0.09	1.33	f
24	uadd64/U241/Y (stdnand3_2x)	0.23	1.56	r
25	uadd64/U185/Y (stdnand2_2x)	0.05	1.61	f
26	uadd64/U184/Y (stdinv_4x)	0.13	1.74	r
27	uadd64/U299/Y (stdnor2_2x)	0.06	1.80	f
28	uadd64/U363/Y (stdnand2_3x)	0.19	1.98	r
29	uadd64/U364/Y (stdnand2_3x)	0.08	2.06	f
30	uadd64/U203/Y (stdnand3_2x)	0.24	2.30	r
31	uadd64/U243/Y (stdnand2_2x)	0.04	2.34	f
32	uadd64/U56/Y (stdnand2_2x)	0.17	2.50	r
33	uadd64/U373/Y (stdnand2_3x)	0.05	2.55	f
34	uadd64/U374/Y (stdnand2_3x)	0.13	2.68	r
35	uadd64/U375/Y (stdnand2_3x)	0.05	2.73	f
36	uadd64/U301/Y (stdnand2_2x)	0.15	2.88	r
37	uadd64/U181/Y (stdnand2_2x)	0.05	2.93	f
38	uadd64/U376/Y (stdnand3_2x)	0.26	3.19	r
39	uadd64/U215/Y (stdinv_4x)	0.10	3.29	f
40	uadd64/U378/Y (stdnor2_3x)	0.14	3.42	r
41	uadd64/U227/Y (stdnor2_2x)	0.07	3.50	f
42	uadd64/U379/Y (stdnand2_3x)	0.16	3.66	r
43	uadd64/U380/Y (stdnand2_3x)	0.06	3.72	f
44	uadd64/U381/Y (stdnand2_3x)	0.12	3.84	r
45	uadd64/U246/Y (stdnand2_3x)	0.04	3.89	f
46	uadd64/U248/Y (stdnand2_3x)	0.11	4.00	r
47	uadd64/U382/Y (stdnand2_3x)	0.04	4.04	f
48	uadd64/U383/Y (stdnand2_3x)	0.16	4.20	r
49	uadd64/U212/Y (stdnand2_2x)	0.04	4.24	f
50	uadd64/U303/Y (stdnand2_2x)	0.15	4.38	r
51	uadd64/U217/Y (stdnand2_2x)	0.05	4.43	f
52	uadd64/U384/Y (stdnand3_2x)	0.26	4.69	r
53	uadd64/U216/Y (stdinv_4x)	0.10	4.79	f
54	uadd64/U386/Y (stdnor2_3x)	0.14	4.93	r
55	uadd64/U229/Y (stdnor2_2x)	0.07	5.00	f
56	uadd64/U387/Y (stdnand2_3x)	0.16	5.16	r
57	uadd64/U388/Y (stdnand2_3x)	0.06	5.22	f
58	uadd64/U389/Y (stdnand2_3x)	0.12	5.34	r
59	uadd64/U390/Y (stdnand2_3x)	0.08	5.42	f
60	uadd64/U187/Y (stdnand2_3x)	0.12	5.54	r
61	uadd64/U188/Y (stdnand2_2x)	0.05	5.59	f
62	uadd64/U178/Y (stdinv_3x)	0.14	5.73	r

63	uadd64/U391/Y (stdnor2_3x)	0.09	5.81 f
64	uadd64/U266/Y (stdnand2_2x)	0.12	5.94 r
65	uadd64/U265/Y (stdnand2_2x)	0.07	6.01 f
66	uadd64/U284/Y (stdnand2_2x)	0.14	6.15 r
67	uadd64/U393/Y (stdoai21_1x)	0.09	6.24 f
68	uadd64/U287/Y (stdnand2_2x)	0.18	6.42 r
69	uadd64/U193/Y (stdinv_4x)	0.07	6.49 f
70	uadd64/U277/Y (stdnand2_2x)	0.13	6.62 r
71	uadd64/U200/Y (stdnand2_3x)	0.07	6.69 f
72	uadd64/U288/Y (stdnand2_2x)	0.14	6.83 r
73	uadd64/U394/Y (stdoai21_1x)	0.09	6.92 f
74	uadd64/U219/Y (stdnand2_2x)	0.20	7.13 r
75	uadd64/U395/Y (stdinv_6x)	0.08	7.20 f
76	uadd64/U396/Y (stdnor2_3x)	0.12	7.33 r
77	uadd64/U80/Y (stdinv_3x)	0.07	7.39 f
78	uadd64/U290/Y (stdnand2_2x)	0.14	7.53 r
79	uadd64/U398/Y (stdoai21_1x)	0.09	7.62 f
80	uadd64/U250/Y (stdnand2_2x)	0.17	7.79 r
81	uadd64/U186/Y (stdnand2_2x)	0.04	7.83 f
82	uadd64/U399/Y (stdaoai211_1x)	0.24	8.06 r
83	uadd64/U221/Y (stdnand2_2x)	0.06	8.13 f
84	uadd64/U180/Y (stdnand2_3x)	0.14	8.27 r
85	uadd64/U179/Y (stdinv_4x)	0.06	8.33 f
86	uadd64/U83/Y (stdnand2_2x)	0.10	8.43 r
87	uadd64/U401/Y (stdao21_3x)	0.29	8.72 r
88	uadd64/U402/Y (stdnand2_3x)	0.06	8.78 f
89	uadd64/U403/Y (stdinv_4x)	0.10	8.88 r
90	uadd64/U404/Y (stdnand2_3x)	0.03	8.91 f
91	uadd64/U405/Y (stdaoai211_1x)	0.23	9.13 r
92	uadd64/U406/Y (stdaoai211_1x)	0.20	9.34 f
93	uadd64/U196/Y (stdnand2_2x)	0.22	9.55 r
94	uadd64/U407/Y (stdaoai211_1x)	0.15	9.70 f
95	uadd64/U100/Y (stdnand2_2x)	0.25	9.95 r
96	uadd64/U408/Y (stdoai21_1x)	0.09	10.05 f
97	uadd64/U214/Y (stdnand2_2x)	0.19	10.24 r
98	uadd64/U410/Y (stdnand2_3x)	0.07	10.31 f
99	uadd64/U411/Y (stdnand2_3x)	0.13	10.43 r
100	uadd64/U413/Y (stdnand2_3x)	0.06	10.49 f
101	uadd64/U414/Y (stdnand2_3x)	0.12	10.61 r
102	uadd64/U416/Y (stdnor2_3x)	0.06	10.67 f
103	uadd64/U417/Y (stdnand2_3x)	0.12	10.79 r
104	uadd64/U418/Y (stdnand2_3x)	0.06	10.85 f
105	uadd64/U419/Y (stdnand2_3x)	0.12	10.97 r
106	uadd64/U421/Y (stdnand2_3x)	0.06	11.03 f
107	uadd64/U422/Y (stdnand2_3x)	0.12	11.15 r
108	uadd64/U424/Y (stdnand2_3x)	0.06	11.21 f
109	uadd64/U425/Y (stdnand2_3x)	0.12	11.33 r
110	uadd64/U427/Y (stdnand2_3x)	0.06	11.39 f
111	uadd64/U428/Y (stdnand2_3x)	0.12	11.51 r
112	uadd64/U430/Y (stdnand2_3x)	0.06	11.57 f

113	uadd64/U431/Y (stdnand2_3x)	0.12	11.69	r
114	uadd64/U433/Y (stdnand2_3x)	0.06	11.75	f
115	uadd64/U434/Y (stdnand2_3x)	0.12	11.87	r
116	uadd64/U436/Y (stdnand2_3x)	0.06	11.93	f
117	uadd64/U437/Y (stdnand2_3x)	0.12	12.05	r
118	uadd64/U439/Y (stdnand2_3x)	0.06	12.11	f
119	uadd64/U440/Y (stdnand2_3x)	0.12	12.23	r
120	uadd64/U442/Y (stdnand2_3x)	0.06	12.28	f
121	uadd64/U443/Y (stdnand2_3x)	0.12	12.40	r
122	uadd64/U445/Y (stdnand2_3x)	0.06	12.46	f
123	uadd64/U446/Y (stdnand2_3x)	0.16	12.62	r
124	uadd64/U447/Y (stdnand2_3x)	0.07	12.69	f
125	uadd64/U448/Y (stdnand2_3x)	0.12	12.81	r
126	uadd64/U450/Y (stdnand2_3x)	0.05	12.86	f
127	uadd64/U315/Y (stdxnor2_1x)	0.30	13.17	f
128	uadd64/S[63] (ADD64_21)	0.00	13.17	f
129	addreg3/DIN[15] (REGS_242)	0.00	13.17	f
130	addreg3/DFF_reg[15]/D (stddff_cq_2x)	0.00	13.17	f
131	data arrival time		13.17	
132				
133	clock clk (rise edge)	30.00	30.00	
134	clock network delay (ideal)	0.00	30.00	
135	clock uncertainty	-1.00	29.00	
136	addreg3/DFF_reg[15]/CLK (stddff_cq_2x)	0.00	29.00	r
137	library setup time	-0.10	28.90	
138	data required time		28.90	
139	---			
140	data required time		28.90	
141	data arrival time		-13.17	
142	---			
143	slack (MET)		15.73	

Since the pads are not included in this primitive synthesis library, there is no timing analysis from the very top level which includes the I/O pads. In practice, input and output timing should be carefully analyzed so that the timing characteristics of the ASIC are fully understood. When it is integrated with other systems, there should be no surprises. This input to output timing is important since the design may be cascaded. If the input to output delay is longer than expected, the cascading of the same ASIC in series may not work. For example, if PO output data available time is 10 ns after the rising edge of the clock, it takes 13.17 ns to go to the next FIRCHIP ASIC. This leaves an input pad delay of 6.83 ns for the next FIRCHIP ASIC in order to run at a 30-ns clock period (33 MHz). Of course, this analysis is further complicated by a clock delay from the clock input to the pad, and a further delay caused by the data subsequently going through the clock tree buffers. The longest timing path between flip-flops is about 16 ns.

10.6 BEST-CASE TIMING ANALYSIS

As shown in Figure 10-12, there are flip-flops in series. There may be setup and hold time violations, especially when there is a clock skew between these two flip-flops. To accommodate this, the best-case timing is performed. Based on the best-case timing, delays can be inserted between these flip-flops to ensure that there is a design margin for clock skew, and no setup and hold timing problems. This process is called fix_hold. The following synthesis script file fixhold.scr shows an example. Line 1 specifies the design name. Line 2 sets up variables, wire_load model, and the worst-case operating condition. The operating condition is overwritten by line 3 so that the best-case timing is used. Lines 6 and 7 declare the clocks. Line 11 declares the clock skew to be 1 ns. This skew is determined by how well the place and route tool can control the clock skew. Line 12 attaches the CLKSKEW to the clock clk. Line 14 specifies that the timing paths dependent on the clock clk will be examined for the fix_hold process. Line 15 tells the synthesis tool not to touch ADD64 and MULT blocks. Line 16 uniquifies all blocks since some blocks are instantiated many times. Line 17 updates the schematic to fix the paths for which the timing delay between flip-flops is less than a value. Similar synthesis script can be used for FIRTAPB and GLUE blocks.

```
1      current_design FIRTAPA
2      include compile.common
3      set_operating_conditions CDA_BEST
4      set_dont_touch_network {CLK_S CLKn_S RSTn_S}
5      set_drive 0            {CLK_S CLKn_S RSTn_S}
6      create_clock "CLK_S*" -name clk -period CLKPERIOD -waveform
{0.0 HALFCLK}
7      create_clock "CLKn_S*" -name clkn -period CLKPERIOD -waveform
{HALFCLK CLKPERIOD}
8      set_dont_touch_network all_outputs()
9      set_input_delay 0 -clock clk {PI I COEFi}
10     set_input_delay CLKPER1 -clock clk {LDCOEF}
11     CLKSKEW = 1.0
12     set_clock_skew -ideal -uncertainty CLKSKEW clk
13     update_timing
14     set_fix_hold clk
15     set_dont_touch {ADD64 MULT}
16     uniquify
17     compile -prioritize_min_paths -only_design_rule >
fix_hold_firtapa.trans
18     write -f db -hier -o ../db/firtapa_fixhold.db
```

After running the above synthesis script, part of the modified schematic for FIR-TAPA is shown in Figure 10-17. Note that many buffers are inserted. No buffer is inserted between IREG1D and IREG1.

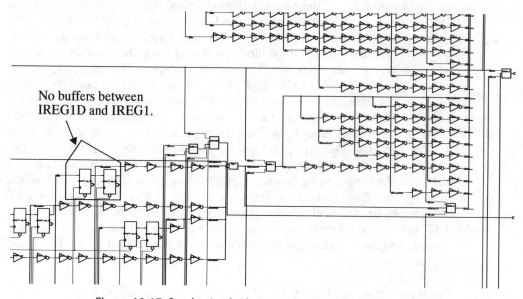

No buffers between
IREG1D and IREG1.

Figure 10-17 Synthesized schematic for FIRTAPA after fix_hold.

Many blocks such as FIRTAPA block are modified. These added buffers introduce more timing delays for timing paths that are short. It is important to analyze the best case minimum delay paths. The following shows that the minimum timing path is between the two synchronizing flip-flops in the GLUE block. The timing path is at least the clock skew (1 ns) and the library hold time value (0.12 ns).

```
1.     Startpoint: glue0/RSTn_FF1_reg
2                  (rising edge-triggered flip-flop clocked by clk)
3      Endpoint: glue0/RSTno_reg
4                  (rising edge-triggered flip-flop clocked by clk)
5      Path Group: clk
6      Path Type: min
7
8      Point                                      Incr        Path
9      -----------------------------------------------------------
10     clock clk (rise edge)                      0.00        0.00
11     clock network delay (ideal)                0.00        0.00
12     glue0/RSTn_FF1_reg/CLK (stddff_1x)         0.00        0.00 r
13     glue0/RSTn_FF1_reg/Q (stddff_1x)           0.24        0.24 f
14     glue0/U45/Y (stdbuf_1x)                    0.09        0.33 f
15     glue0/U44/Y (stdbuf_1x)                    0.09        0.42 f
16     glue0/U43/Y (stdbuf_1x)                    0.09        0.51 f
17     glue0/U42/Y (stdbuf_1x)                    0.09        0.60 f
18     glue0/U41/Y (stdbuf_1x)                    0.09        0.69 f
```

19	glue0/U40/Y (stdbuf_1x)	0.09	0.78 f
20	glue0/U38/Y (stdbuf_5x)	0.15	0.92 f
21	glue0/U39/Y (stdinv_1x)	0.09	1.02 r
22	glue0/U37/Y (stdinv_7x)	0.11	1.12 f
23	glue0/RSTno_reg/D (stddff_1x)	0.00	1.12 f
24	data arrival time		1.12
25			
26	clock clk (rise edge)	0.00	0.00
27	clock network delay (ideal)	0.00	0.00
28	clock uncertainty	1.00	1.00
29	glue0/RSTno_reg/CLK (stddff_1x)	1.00	1.00 r
30	library hold time	0.12	1.12
31	data required time		1.12
32	--		
33	data required time		1.12
34	data arrival time		-1.12
35	--		
36	slack (MET)		0.00

10.7 NETLIST GENERATION

The following script file can be used to generate a VHDL netlist. Line 1 specifies the design FIRCORE to be written. Lines 2 to 5 specify variables used when generating the VHDL netlist. The names are descriptive. Line 3 requests that the top configuration VHDL code not be generated. Line 4 and 5 request the types used in ports be maintained as the original. If they are not specified, the bus type would be changed from std_logic_vector(15 downto 0) to std_logic_vector(0 to 15). In general, this presents no problem since VHDL maps signal left to left, right to right, without considering the index value for the bus as long as the bus widths are the same. However, in the chip boundary when the package pin location is specified, the pads may be placed in positions that are not intended. Maintaining consistency by using lines 4 and 5 removes any ambiguity.

```
1    current_design FIRCORE
2    vhdlout_top_configuration_arch_name = "structure"
3    vhdlout_write_top_configuration = "FALSE"
4    vhdlout_single_bit = "USER"
5    vhdlout_preserve_hierarchical_types = "USER"
6    vhdlout_use_packages = { \
7       IEEE.STD_LOGIC_1164.ALL \
8       EPOCH_LIB.COMPONENTS.ALL \
9       EPOCH_LIB.STANDARD_COMPONENTS.ALL \
10      }
11   write -f vhdl -hier -o ../strvhdl/fircore.str.vhd
```

Lines 6 to 10 tell the netlist generation to use the library and packages in the netlist. The following shows part of the VHDL code in the file fircore.str.vhd. Lines 1 to 6 are generated based on lines 7 to 9 of the above script file. The netlist has more than 15,000 lines of VHDL code. It will not be shown here!

```
1    library IEEE;
2    library EPOCH_LIB;
3
4    use IEEE.STD_LOGIC_1164.ALL;
5    use EPOCH_LIB.COMPONENTS.ALL;
6    use EPOCH_LIB.STANDARD_COMPONENTS.ALL;
7
8    use work.CONV_PACK_FIRCORE.all;
9
10   entity COEFREG_0 is
11
12       port( RSTn, CLK : in std_logic;  D : in std_logic_vector (15
downto 0);
13              COEFEN : in std_logic;  COEF_FF : out std_logic_vector
(15 downto 0));
14
15   end COEFREG_0;
```

To complete the netlist, pads are required. The following architecture EPK shows part of the VHDL code to instantiate pads. Lines 3 and 4 are used to reference the pad components to be instantiated. The component FIRCORE is declared and instantiated the same way as other architectures.

```
1    library IEEE;
2    use IEEE.std_logic_1164.all;
3    library EPOCH_LIB;
4    use EPOCH_LIB.components.all;
5    architecture EPK of FIRCHIP is
6      component FIRCORE
7      port(
8        RSTn    : in  std_logic;
9        CLK     : in  std_logic;
10       LDCOEF  : in  std_logic; -- enable loading coef
11       START   : in  std_logic; -- start filtering
12       OEn     : in  std_logic; -- enable tristate pad
13       BYPASS  : in  std_logic; -- bypass the chip
14       I       : in  std_logic_vector(15 downto 0); -- data
15       PI      : in  std_logic_vector(63 downto 0); -- sum
16       O       : out std_logic_vector(15 downto 0); -- data
17       PO      : out std_logic_vector(63 downto 0);
18       OEn_S   : out std_logic_vector(79 downto 0)); -- OEn
19     end component;
20     signal RSTni   : std_logic;
21     signal CLKi    : std_logic;
22     signal LDCOEFi : std_logic; -- enable loading coef
23     signal STARTi  : std_logic; -- start filtering
```

```
24      signal OEni    : std_logic; -- enable tristate pad
25      signal BYPASSi : std_logic; -- bypass the chip
26      signal Ii      : std_logic_vector(15 downto 0); -- data
27      signal PIi     : std_logic_vector(63 downto 0); -- sum
28      signal Oo      : std_logic_vector(15 downto 0); -- data
29      signal POo     : std_logic_vector(63 downto 0);
30      signal OEn_S   : std_logic_vector(79 downto 0); -- OEn
```

Lines 20 to 30 declare signals to be used to connect the pads. Lines 39 to 45 are comments used to describe the 208-pin package. The package pin number is numbered counterclockwise from the bottom of the right side. Various power and ground pads, input pads, and three-state pads are shown. Note that pads are specified with some generics. For example, in line 55, the pad has a 4-mA output drive. It is pin number 206. This file is to be done carefully so that signals are mapped correctly. Package pins are mapped with the correct generic values and package numbers. The whole file has more than 800 lines. Therefore only part of it is shown here.

```
31    begin
32      fircore0 : FIRCORE
33      port map (
34        RSTn  => RSTni,  CLK    => CLKi, LDCOEF  => LDCOEFi,
35        START => STARTi, OEn    => OEni, BYPASS  => BYPASSi,
36        I     => Ii,     PI     => PIi,  O       => Oo,
37        PO    => POo,    OEn_S  => OEn_S);
38      ----------------------------------------------------
39      -- pin numbers top view 208, counter clockwise,
40      --           104                          53
41      --           105                            52
42      --
43      --           156                            1
44      --           157                          208
45      ----------------------------------------------------
46      gndpad0 : padgnd
47        generic map (N=>208,SLIM_FLAG=>0, PTYPE=>"REGULAR",
48                     YPITCH=>"4MA")
49        port map (GNDS(0));
50      vddpad0 : padvdd
51        generic map (N=>207,SLIM_FLAG=>0, PTYPE=>"REGULAR",
52                     YPITCH=>"4MA")
53        port map (VDDS(0));
54      o15pad  : padout_tri
55        generic map (N=>206,M=>206,SLIM_FLAG=>0,OUTDRIVE=>"4MA")
56        port map (OEn_S(15 downto 15), Oo(15 downto 15),
57                  O(15 downto 15));
58      ---
59      pi63pad : padin_buf
60       generic map(N=>186,M=>186,SLIM_FLAG=>0,LEVEL_SHIFTING=>0,
61         SCHMITT_TRIGGER=>0, PULL_TYPE =>"None", YPITCH=>"4MA")
62        port map (PI(63 downto 63), PIi(63 downto 63));
```

```
63    ---
64    clkpad  : padin_buf
65     generic map(N=>86,M=>86,SLIM_FLAG=>0,LEVEL_SHIFTING=>0,
66        SCHMITT_TRIGGER=>0, PULL_TYPE =>"None", YPITCH=>"4MA")
67     port map (CLK, CLKi);
68    ldcoefpad  : padin_buf
69     generic map(N=>85,M=>85,SLIM_FLAG=>0,LEVEL_SHIFTING=>0,
70        SCHMITT_TRIGGER=>0, PULL_TYPE =>"None", YPITCH=>"4MA")
71     port map (LDCOEF, LDCOEFi);
72   end EPK;
```

By using the following command, a netlist file is generated which combines the FIRCHIP entity, architecture EPK, and the FIRCORE netlist. The purpose of the netlist is to describe the placing and routing of the design. The picture shows the final plot. The FIRTAPA and FIRTAPB blocks are first placed and routed. They are then combined with the rest of the circuit by the automatic place and route tool. It is easy to see that the input and output signals of FIRTAPA and FIRTAPB can be lined up and optimized so that signals between the TAP stages can be connected more easily. This will reduce the routing area to save on silicon. However, the placing and routing design process is outside the scope of this book. The whole design has 82,466 standard cells which have 700,026 transistors. The die size is 270x296 mils (1 mil = 25.4 microns).

```
1    cat ../vhdl/firchip_e.vhd ../vhdl/firchip_epk.vhd
fircore.str.vhd > firchip.str.vhd
```

10.8 POSTLAYOUT VERIFICATION

After the placing and routing process is done, a back-annotated timing file can be generated. The timing is usually in the industry standard Standard Delay Format (SDF). The following shows part of the SDF file. The whole file has more than 732K lines! Line 17 shows the delay of an inverter for an instance shown in line 15. The delay is shown with the rise time and fall time. Each has the best-case, typical case, and the worst-case process. The temperature and process for the best-case, typical case, and the worst-case processes are shown in lines 9 and 11.

```
1    (DELAYFILE
2     (SDFVERSION  "1.7.3")
3     (DESIGN      "firchip")
4     (DATE        "Sun Mar 15 18:18:25 1998")
5     (VENDOR      "XXXXXXX Design Automation Corp.")
6     (PROGRAM     "vhdlout")
7     (VERSION     "3.2")
8     (DIVIDER     .)
9     (VOLTAGE     3.600:3.300:3.000)
10    (PROCESS     "min:typ:max, NMOS=TYP, PMOS=TYP")
```

```
11        (TEMPERATURE  0.000:25.000:70.000)
12        (TIMESCALE    1ps)
13
14        (CELL (CELLTYPE "cda_inv")
15          (INSTANCE firchip_corea0.bypasspad_iobuff83.inst1)
16          (DELAY (ABSOLUTE
17            (IOPATH  in0    out0    (190:190:190) (115:115:115))
18        )))
19        (CELL (CELLTYPE "cda_inv")
20          (INSTANCE firchip_corea0.bypasspad_sigcon83.inst1)
21          (DELAY (ABSOLUTE
22            (IOPATH  in0    out0    (174:174:174) (168:168:168))
```

The place and route tool also generates a VHDL netlist so that the instance names in the SDF match these in the VHDL netlist. Part of the VHDL netlist (it has more than 41K lines) is shown below. Note that the component instantiation label "bypasspad_iobuff83" in line 10 is the same as shown in line 15 of the SDF.

```
1        for all:  ist_stdtribuf_pad_3x
2            use entity WORK.ist_stdtribuf_pad_3x(cda_structural);
3        for all:  ist_stdtribuf_pad_10x
4            use entity WORK.ist_stdtribuf_pad_10x(cda_structural);
5        for all:  ist_stdtribuf_pad_11x
6            use entity WORK.ist_stdtribuf_pad_11x(cda_structural);
7    begin
8
9
10       bypasspad_iobuff83 :  std_stdinv_1x
11           port map (bypasspad_sigcon_out(83), bypassi);
```

The place and route tool generates the following FIRCHIP entity. Other than lines 1 to 6, the entity ports are exactly the same as we have shown (with the single-bit signal declared as a std_logic_vector 1 bit wide). This requires no changes to be made in the test bench.

```
1    library IEEE;
2    use IEEE.std_logic_1164.all;
3    library EPOCH_VHDLOUT;
4    use EPOCH_VHDLOUT.PRIMTYPES.all;
5    use EPOCH_VHDLOUT.PRIMCOMP.all;
6    use IEEE.Vital_Timing.all;
7
8    entity firchip is
9      port (
10       rstn            :in     std_logic_vector(0 downto 0);
11       clk             :in     std_logic_vector(0 downto 0);
12       ldcoef          :in     std_logic_vector(0 downto 0);
13       start           :in     std_logic_vector(0 downto 0);
14       oen             :in     std_logic_vector(0 downto 0);
```

```
15        bypass             :in     std_logic_vector(0 downto 0);
16        i                  :in     std_logic_vector(15 downto 0);
17        pi                 :in     std_logic_vector(63 downto 0);
18        o                  :out    std_logic_vector(15 downto 0);
19        po                 :out    std_logic_vector(63 downto 0);
20        vdds               :inout  std_logic_vector(20 downto 0);
21        gnds               :inout  std_logic_vector(20 downto 0));
22   end firchip;
```

The following shows the command script used to analyze the VHDL code for the postlayout simulation. A new configuration is used as shown in line 10. The configuration VHDL code is shown below. Note that this configuration file is in the **gatesim** directory.

```
1        echo "../vhdl/drive.vhd"
2    vhdlan -c -nc ../vhdl/drive.vhd
3        echo "../vhdl/control.vhd"
4    vhdlan -c -nc ../vhdl/control.vhd
5        echo "../layout1/firchip.vhd"
6    vhdlan -c -nc ../layout1/firchip.vhd
7        echo "../vhdl/tbfirchip.vhd"
8    vhdlan -c -nc ../vhdl/tbfirchip.vhd
9        echo "../gatesim/tbfirchip_cfg.vhd"
10   vhdlan -c -nc ../gatesim/tbfirchip_cfg.vhd
```

```
1    configuration CFG_TBFIRCHIP of TBFIRCHIP is
2      for BEH
3        for firchip0 : FIRCHIP
4          -- use configuration work.CFG_FIRCHIP;
5          use entity work.FIRCHIP(cda_structural);
6        end for;
7        for drive0 : DRIVE
8          use entity work.DRIVE(BEH);
9        end for;
10       for control0 : CONTROL
11         use entity work.CONTROL(BEH);
12       end for;
13     end for;
14   end CFG_TBFIRCHIP;
```

The sim.cmd file is changed as shown below. Note that the SDF file is overlaid in lines 10 and 11. The rest of the lines remain the same as above.

```
1    date
2    export TEST_DIR=../infile
3    export RESULT_DIR=./_result
4        echo "Simulating $1..."
5        cp -f $TEST_DIR/$1.coef ./coef.in
6        cp -f $TEST_DIR/$1.idata ./idata.in
```

```
7          cp -f $TEST_DIR/$1.pidata ./pidata.in
8          cp -f $TEST_DIR/$1.vss ./sim.do
9          vhdlsim -pro -o $RESULT_DIR/$1.log -w $RESULT_DIR/$1.ow \
10            -sdf_af -sdf_max -sdf_top /tbfirchip/firchip0 \
11            -sdf ../layout1/firchip.sdf \
12            -nc -i sim.do cfg_tbfirchip
13         echo "Done!"
14         echo
15         ;
16    date
```

The same command file batchsim.cmd can be run again. The command file batchvec.cmd can also be run. A comparison of the vector output file with the command "diff _result/test1.vec ../rtlsim/_result/test1.vec" shows that the first two vectors are not the same. This is caused by the reset signal that is synchronized. In practice, they can be ignored. This completes the postlayout timing simulation which matches the RTL simulation.

```
1    1,2c1,2
2    <      28000
0000000000000000000000UUUUUUUUUUUUUUUUU00000000000000000000000000000
00000000000000000000000000000000000UUUUUUUUUUUUUUUUUUUUUUUUUUUUUUUUUU
UUUUUUUUUUUUUUUUUUUUUUUUUUUUUUUUUUUU
3    <      58000
0000000000000000000000UUUUUUUUUUUUUUUUU00000000000000000000000000000
00000000000000000000000000000000000UUUUUUUUUUUUUUUUUUUUUUUUUUUUUUUUUU
UUUUUUUUUUUUUUUUUUUUUUUUUUUUUUUUUUUU
4    ---
5    >      28000
0000000000000000000000UUUUUUUUUUUUUUUUU00000000000000000000000000000
00000000000000000000000000000000000XXXXXXXXXXXXXXXXXXXXXXXXXXXXXXXXXX
XXXXXXXXXXXXXXXXXXXXXXXXXXXXXXXXXXXX
6    >      58000
000000000000000000000000000000000000000000000000000000000000000000000
000000000000000000000000000000000000000000000000000000000000000000000
00000000000000000000000000000000000
```

10.9 DESIGN MANAGEMENT

The following shows that the design directory has the following subdirectories: bin, db, gatesim, gatesim.work, infile, layout1, qsim, qsim.work, report, rtlsim, rtlsim.work, strvhdl, syn, syn.work, vhdl, and vhdlsave. Directory vhdl contains all the VHDL codes. The directory vhdlsave saves a copy of the VHDL codes in the vhdl directory. The directory rtlsim is used for running RTL simulation. The associated VHDL working library is mapped to the rtlsim.work directory. The directory gatesim is used for running postlayout simulation with the VHDL working library mapped to gatesim.work directory. The directory syn is used for running the synthesis with synthesis script files. The VHDL working library is mapped to the syn.work directory.

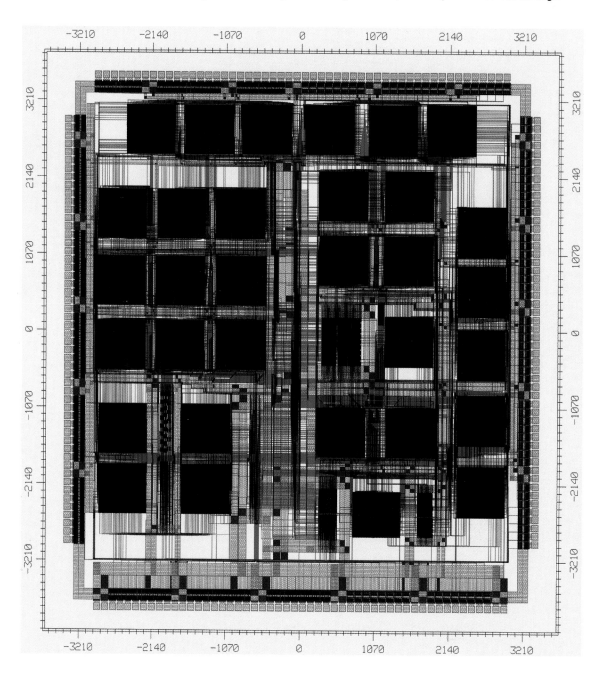

The directory strvhdl contains the VHDL netlist generated from the synthesis tool. The directory db saves the synthesized database. The directory report contains synthesis report files. The directory infile contains the input files for the test bench. The directory layout1 is used to perform the place and route functions. The directories qsim and qsim.work are similar to rtlsim and rtlsim.work, except that qsim is used for running the QuickVHDL while rtlsim is used for running Synopsys VSS. The directory bin contains special command files.

Note also that the directory references are better used as relative positions as shown throughout this chapter. This allows the directory to be moved to another place without changing the command script files. In practice, the design is usually archived when the design is completed. When the design is reloaded, it may not be in the same place. The command script should have minimal dependency on the tool versions, synthesis library, and VHDL simulation libraries.

It is also important to appreciate the size of each file and directory. The VHDL working library is usually big. The postlayout VHDL netlist and SDF files are huge. This problem will become worse when multiple versions are used.

We have come a long way from starting the design specification to the point where the design can be shipped for fabrication. There may be a couple of issues that we have not covered. The following exercises will allow you to dig in further.

10.10 EXERCISES

10.1 Assume you are fortunate enough to have access to a state-of-the-art place and route tool. The tool can take 82,466 standard cells and complete the layout within a week. The hierarchy of using FIRTAPA and FIRTAPB may not be necessary. How would you partition the design inside the FIRSYN block? How would the synthesis process be changed?

10.2 If your place and route tool has the clock synthesis capability, how would you change the VHDL code to take advantage of it?

10.3 The TREE block is implemented with two levels of inverted buffers. Develop a synthesis script so that the TREE block would be synthesized without any buffers. The CLK input is connected to all the flip-flops directly.

10.4 Refer to FIRTAPA VHDL code. Registers IREG1D and COEFD are added that are clocked by the falling edge of the input clock. Does their use guarantee that there will be no setup and hold time from IREG2 to IREG1 registers and between COEFREG registers? What are the design trade-offs involving using such additional flip-flops versus using the series of buffers?

10.5 Refer to the FIRTAPA schematic in Figure 10-17. Many buffers are inserted to increase timing delays for short timing paths. How does the synthesis tool determine the number of buffers that are inserted in each timing path? What factors should you consider for controlling the number of buffers inserted? What are the design trade-offs?

10.6 Refer to the FIRCHIP plot. Many improvements can be made to make the die size smaller. Based on the design construction, suggest several improvements that can reduce the die size.

10.7 The postlayout timing simulation was done on FIRCHIP. However, the timing simulation waveform was not shown. After the layout, the static timing analyzer is usually run to ensure that the timing paths are still within the requirement. The timing analyzer reports that the worst-case timing at 3 V, 125 degrees C is 26.23 ns. This is much more than the synthesis static timing analyzer which reports 16 ns. The timing discrepancy amounts to 64 percent. What are the reasons for this timing discrepancy?

10.8 For any commercial integraged circuit (IC), the timing characteristics of all I/O pins are usually described. This information is used by the designers to better control the interface with ICs or components. How do you suggest that the I/O timing characteristics for FIRCHIP IC be shown? How do you control these I/O timing characteristics in the VHDL design, synthesis process, and the place and route process? How can the test bench verify the I/O timing characteristics of the FIRCHIP?

10.9 Refer to the FIRCHIP plot. I/O pads are placed on the boundary. The positions of these I/O pads are certain to relate to the position of the package pins. What factors should you consider in selecting a package for your design?

10.10 Refer to the FIRCHIP plot. The I/O pad positions are determined in the VHDL code. What are the advantages and disadvantages for including pad positions in the VHDL code? Can you suggest another method?

10.11 Based on your synthesis tools and the place and route (or FPGA) tools capability, change the design so that it has three TAP stages. Complete the design process of VHDL coding, simulation, synthesis, placing and routing, and postlayout timing simulation.

10.12 The functional vectors for the fabrication vendor are required to test the part after it has been fabricated. This process is not shown here, even though the wif2tab command to generate vectors is illustrated. Contact a vendor for the release functional vector and research it. Develop a process to incorporate that format into your design environment.

10.13 Refer to the FIRCHIP test bench. Three separate tests were performed. In general, they were not enough. Develop more tests to verify the design. How do you know that you have tested enough for a given design?

Chapter 11

A Design Case Study: A Microprogram Controller Design

In this chapter we will discuss and describe the design, architecture, and features of an AM2910 equivalent microcontroller. The design is captured with VHDL, simulated, and synthesized. We will describe the backend place and route interface process, and provide a test bench. We will also discuss and illustrate how manufacturing test vectors can be generated and verified.

11.1 MICROPROGRAM CONTROLLER

A microprogram controller is also called as a microcontroller. It has a small set of instructions to select, store, retrieve, manipulate, and test addresses that control the execution sequence of instructions stored in a memory. AMD developed the first AM2910 microcontroller architecture with five words of stack. Later, AM2910A was developed with nine words of stack. Cypress Semiconductors produced CY7C910, which is the functional equivalent to AM2910 and AM2910A with 17 words of stack. In this chapter, we will design a functional equivalent ASIC to the CY7C910 part. For more detailed information on these parts, please refer to AMD and Cypress data books. Here, we will introduce the block diagram and instructions to illustrate the VHDL design process.

Figure 11-1 shows the CY7C910 block diagram. It consists of a 17-word by 12 bit stack which is accessed in last-in first-out manner. The stack is addressed by a stack pointer SP. There is a 12-bit register/counter. The microprogram counter and incrementer are also 12 bits wide. The multiplexer selects data from the register counter, stack, microprogram counter, D input, or zero. The instruction decoder decodes instructions to generate control signals.

The block diagram also shows input and output signals. They are described in the pin description table shown in Figure 11-2. D11, I3, Y11 are the most significant bits of D, I, and Y bus, respectively.

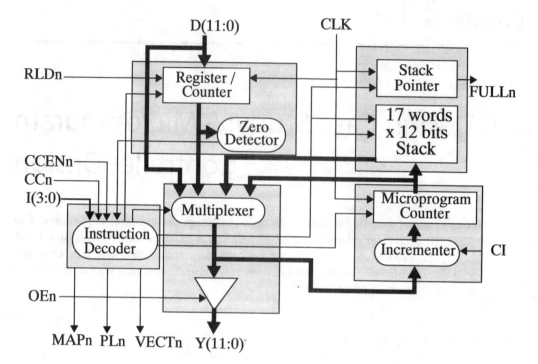

Figure 11-1 CY7C910 block diagram.

Pin name	I/O	Description
D11-D0	I	Direct input to register/counter and multiplexor
I3-I0	I	Select 1-of-16 instructions
CCn	I	Used as test criteria. Pass test is low on CCn
CCENn	I	Low active enable signal to control CCn
CI	I	Low order carry input to incrementer
RLDn	I	When low forces loading register/counter
OEn	I	Three-state low active control of Y output
CLK	I	Clock input for flip flops
Y11-Y0	O	Address to microprogram memory
FULLn	O	Low active signal to indicate the stack is full
PLn	O	Select #1 source (usually the pipeline register)
MAPn	O	Select #2 source (usually mapping PROM or PLA)
VECTn	O	Select #3 source (usually interrupt starting address)

Figure 11-2 CY7C910 pin description table. (PLA = Programmable Logic Array; PROM = Programmable Read Only Memory.)

The CY7C910 comes in two types of packages: a 40-pin DIP (Dual In-line Package) and a 44-pin flat pack package. Figure 11-3 shows the pin configuration on both packages. Each package has prefabricated footprints that connect the pins on the boundary of the package and the bonding fingers in the boundary of the cavity. The silicon die is then attached inside the package cavity. Bonding wires are used to connect the pads in the silicon die to the cavity bonding fingers. The same die is packaged with two different types of packages with 40 pins and 44 pins. To achieve this, the relative order of the pin configuration of both packages is the same so that the planar package footprint and bonding wires can be achieved. There are 4 no-connect (NC) pins in the 44-pin flat pack package (pins 6, 17, 32, and 36).

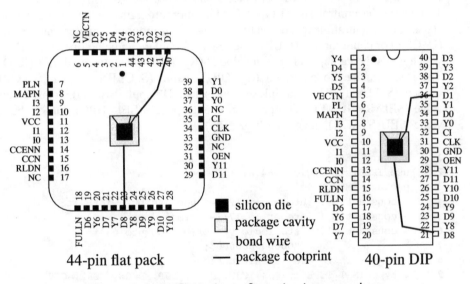

Figure 11-3 CY7C910 pin configuration in two packages.

CY7C910 has different parts with various performance grades. The fastest part can be run with a 40-ns clock period in commercial environments (0 to 70 degrees C, 5 V). The military environment (minus 55 to 125 degrees C, 5 V) runs with a 46-ns clock period. CY7C910 was fabricated on a 1.2 microns complementary metal oxide semiconductor (CMOS) process. In this chapter, our goal is to design an ASIC that is functionally equivalent to the CY7C910 using the Chip Express 0.8-micron Laser Programmable Gate Array (LPGA) technology. Doing this will take us through each step of the design process.

11.2 DESIGN DESCRIPTION AND PARTITION

Figure 11-1 gives us a good overall architecture description. As we approach this design, it would be beneficial to structure our VHDL code for reusability. For example, the number of words in the stack can be 5, 9, and 17 in AM2910, AM2910A, and CY7C910, respectively. The address and data input are 12 bits wide. It can be changed to 16 bits wide later. Also, for debugging purposes, we may want to know the instruction type directly shown on the simulation waveform rather than a 4-bit value. The control signals for stack, microprogram counter, register/counter, and the multiplexer can also be more descriptive. Therefore, the following package has been developed. Line 4 defines the address bus width as a constant N. Line 5 declares the depth of the stack as a constant M to indicate the number of words in the stack. Lines 6 and 7 declare a constant of 12-bit zero. Lines 9 and 10 declare 16 instructions. Lines 11 to 15 declare the control signal types for the microprogram counter, register/counter, stack, and the multiplexer, respectively. The microprogram counter can be cleared (U_CLR) or counting up (U_UP). The register/counter can be no operation (R_NOP) which holds the value, loaded with value D (R_LOAD), decremented by 1 (R_DEC). The stack can be no operation (S_NOP), cleared (S_CLR), popped (S_POP), or pushed (S_PUSH). The multiplexer can select D input (SEL_D), register/counter value (SEL_REGCNT), microprogram counter value (SEL_UPC), top of the stack value (SEL_STACK), or zero (SEL_ZERO).

```
1    library IEEE;
2    use IEEE.std_logic_1164.all;
3    package AMPACK is
4       constant N : integer := 12; -- address bus width
5       constant M : integer := 17; -- stack depth
6       constant Z12 : std_logic_vector(N-1 downto 0) :=
7                     "000000000000";
8          -- There are 16 instructions
9       type INST_TYPE is (JZ, CJS, JMAP, CJP, PUSH, JSRP, CJV,
10         JRP, RFCT, RPCT, CRTN, CJPP, LDCT, LOP, CONT, TWB);
11      type UPC_OPS    is (U_CLR, U_UP);
12      type REGCNT_OPS is (R_NOP, R_LOAD, R_DEC);
13      type STACK_OPS  is (S_NOP, S_CLR, S_POP, S_PUSH);
14      type MUX_OPS    is (SEL_D, SEL_REGCNT, SEL_UPC,
15                         SEL_STACK, SEL_ZERO);
16    end AMPACK;
```

The microprogram counter block is named UPC. The VHDL code is shown below. Package AMPACK is referenced in line 5. CLK is the clock signal. CI is the carry input to be added in the incrementer. Line 10 declares input signal UPC_CNTL that is in UPC_OPS type to control the microprogram counter operations. This UPC_CNTL will be from the instruction decoder. Line 11 declares input signal MUX-OUT which is the output of the multiplexer. Line 12 is the output signal UPC_DOUT for the content of the microprgram counter. UPC_DOUT is going to both multiplexer and stack blocks. Note that lines 11 and 12 are declared as N-bits wide.

```
1          --------------------- file upc.vhd
2     library IEEE;
3     use IEEE.std_logic_1164.all;
4     use IEEE.std_logic_unsigned.all;
5     use work.AMPACK.all;
6     entity UPC is
7       port (
8         CLK           : in  std_logic;
9         CI            : in  std_logic;
10        UPC_CNTL      : in  UPC_OPS;
11        MUXOUT        : in  std_logic_vector(N-1 downto 0);
12        UPC_DOUT      : out std_logic_vector(N-1 downto 0));
13    end UPC;
14    architecture RTL of UPC is
15    begin
16      process (CLK)
17      begin
18        if (CLK'event and CLK = '1') then
19          if (UPC_CNTL = U_UP) THEN
20            UPC_DOUT <= MUXOUT + CI;
21          else
22            UPC_DOUT <= Z12;
23          end if;
24        end if;
25      end process;
26    end RTL;
```

The microprogram counter can be cleared to zero in line 22 when UPC_CNTL is not equal to U_UP. When UPC_CNTL is equal to U_UP, the microprogram counter gets the value of MUXOUT + CI as shown in line 20. Note that CI can be '0' or '1'. Therefore, the microprogram counter can be incremented by 1 or can retain its old value. Line 3 references package std_logic_unsigned for the operator "+" used in line 20.

The register counter block is named REGCNT. The VHDL code is shown below. The REGCNT block has an input D as declared in line 8. Line 9 declares REGCNT_CNTL of REGCNT_OPS type to control the register/counter operations. REGCNT_CNTL is generated by the instruction decoder. Line 10 declares output REGCNT_DOUT for the content of the register/counter to go to the multiplexer. Line 11 declares another output signal REGCNT_ZERO to indicate that the register/counter value is zero or not. When it is asserted '1', the register counter value is zero.

```
1     library IEEE;
2     use IEEE.std_logic_1164.all;
3     use IEEE.std_logic_unsigned.all;
4     use work.AMPACK.all;
5     entity REGCNT is
6       port (
7         CLK           : in  std_logic;
8         D             : in  std_logic_vector(N-1 downto 0);
```

```
9            REGCNT_CNTL   : in  REGCNT_OPS;
10           REGCNT_DOUT   : out std_logic_vector(N-1 downto 0);
11           REGCNT_ZERO   : out std_logic);
12    end REGCNT;
13    architecture RTL of REGCNT is
14      signal REGCNT_FF : std_logic_vector(N-1 downto 0);
15    begin
16      process (CLK)
17      begin
18        if (CLK'event and CLK = '1') then
19          if REGCNT_CNTL = R_LOAD then
20            REGCNT_FF <= D;
21          elsif REGCNT_CNTL = R_DEC then
22            REGCNT_FF <= REGCNT_FF - 1;
23          end if;
24        end if;
25      end process;
26      REGCNT_DOUT <= REGCNT_FF;
27      REGCNT_ZERO <= '1' when REGCNT_FF = Z12 else '0';
28    end RTL;
```

The register counter can be loaded with input D when REGCNT_CNTL is equal to R_LOAD as shown in lines 19 to 20. When REGCNT_CNTL is equal to R_DEC, the register/counter is decremented by 1 as shown in lines 21 to 22. Otherwise, the register/counter retains its old value. Line 27 implemented the zero detector function to generate REGCNT_ZERO output to the instruction decoder. Line 26 connects the register/counter flip-flops to output.

The multiplexer block is named MUX. The VHDL code is shown below. Four data inputs are declared in lines 7 to 10. Input MUX_CNTL is from the instruction decoder. Line 11 declares the result of the multiplexer output.

```
1     library IEEE;
2     use IEEE.std_logic_1164.all;
3     use work.AMPACK.all;
4     entity MUX is
5       port (
6         MUX_CNTL      : in  MUX_OPS;
7         D             : in  std_logic_vector(N-1 downto 0);
8         REGCNT_DOUT   : in  std_logic_vector(N-1 downto 0);
9         STACK_DOUT    : in  std_logic_vector(N-1 downto 0);
10        UPC_DOUT      : in  std_logic_vector(N-1 downto 0);
11        MUXOUT        : out std_logic_vector(N-1 downto 0));
12    end MUX;
13    architecture RTL of MUX is
14    begin
15      with MUX_CNTL select
16        MUXOUT <=
17          D             when SEL_D,
18          REGCNT_DOUT when SEL_REGCNT,
19          UPC_DOUT    when SEL_UPC,
```

```
20              STACK_DOUT  when SEL_STACK,
21              Z12         when SEL_ZERO;
22      end RTL;
```

Based on the value of MUX_CNTL, MUXOUT value is selected from the value of D, REGCNT_DOUT, UPC_DOUT, STACK_DOUT, and zero.

The stack block is named STACK. The VHDL code is shown below. Line 7 declares STACK_CNTL input from the instruction decoder to control the stack operations. Line 8 declares the input to the stack to be pushed into the top of the stack. STACK_DIN will be connected to the UPC_DOUT. Line 9 declares STACK_DOUT which is the content of the word in the top of the stack. Line 10 declares an output signal FULLn to indicate the stack is full. It is a low-active signal.

```
1       library IEEE;
2       use IEEE.std_logic_1164.all;
3       use work.AMPACK.all;
4       entity STACK is
5         port (
6           CLK            : in  std_logic;
7           STACK_CNTL     : in  STACK_OPS;
8           STACK_DIN      : in  std_logic_vector(N-1 downto 0);
9           STACK_DOUT     : out std_logic_vector(N-1 downto 0);
10          FULLn          : out std_logic );
11      end STACK;
```

Lines 13 to 15 declare subtype and type so that the M-word by N-bit stack can be easily declared. Line 16 declares signal MEM to contain the M-word by N-bit stack. A stack pointer is required to keep track of the stack. Line 18 declares signals FULLn_FF and EMPTYn_FF to indicate the stack is full or empty, respectively. There are many ways to construct a stack. For example, the stack can start on a lower address or a higher address. When a word is pushed, the stack pointer can be incremented or decremented. The stack pointer can either point to the empty slot or the first valid word at the top of the stack. The stack pointer can be incremented (or decremented) before or after the word is pushed or popped.

In this design, the stack pointer type is from M-1 down to 0. The stack can be cleared initially. The stack pointer SP is cleared to be 0. All words in MEM are cleared to zero also. Since the value at the top of the stack should always be visible to the multiplexer, SP will point to the top of the stack with a valid value, except when the stack is empty. This implies that SP should be incremented before the word is pushed, except when the stack is empty, the word is pushed and SP is not updated.

```
12      architecture RTL of STACK is
13        subtype nbit is std_logic_vector(N-1 downto 0);
14        subtype sptype is integer range M-1 downto 0;
15        type mem_type is array (sptype) of nbit;
16        signal MEM : mem_type;
17        signal SP  : sptype;
18        signal FULLn_FF, EMPTYn_FF : std_logic;
```

The stack MEM, stack pointer SP, signals FULLn_FF and EMPTYn_FF will be implemented as flip-flops. Clock signal CLK is used in line 20 as the process sensitivity list and the clock rising edge check is done in line 23. Based on the STACK_CNTL value, the stack operations can be described in a **case statement** starting in line 24. Lines 25 to 26 implement the stack function when STACK_CNTL is equal to S_NOP. Nothing needs to be done so a **null statement** is used in line 26. Lines 27 to 33 implement the operation when the stack is cleared. EMPTYn_FF is asserted '0' in line 28 to indicate that the stack is actually empty. In line 29, FULLn_FF is deasserted to indicate that the stack is not full. In line 30, stack pointer SP is initialized to its lowest value which is 0. Lines 31 to 33 clear every word in the stack.

```
19    begin
20      p0 : PROCESS (CLK)
21        variable SP_PUSH : sptype;
22      begin
23        if (CLK'event and CLK = '1') then
24          case STACK_CNTL is
25            when S_NOP =>
26              NULL;
27            when S_CLR =>
28              EMPTYn_FF  <= '0';
29              FULLn_FF   <= '1';
30              SP         <= sptype'low;
31              for j in sptype loop
32                MEM(j) <= Z12;
33              end loop;
34            when S_PUSH =>
35              EMPTYn_FF <= '1';
36              if (FULLn_FF = '0') or (EMPTYn_FF = '0') then
37                MEM(SP) <= STACK_DIN;
38              else
39                SP_PUSH      := SP + 1;
40                MEM(SP_PUSH) <= STACK_DIN;
41                SP           <= SP_PUSH;
42                if (SP_PUSH = sptype'high) then
43                  FULLn_FF <= '0';
44                else
45                  FULLn_FF <= '1';
46                end if;
47              end if;
48            when S_POP =>
49              FULLn_FF <= '1';
50              if (SP = sptype'low) then
51                EMPTYn_FF <= '0';
52              else
53                SP <= SP - 1;
54              end if;
55          end case;
```

```
56          end if;
57        end process;
58        STACK_DOUT <= MEM(SP);
59        FULLn       <= FULLn_FF;
60      end RTL;
```

Lines 34 to 47 implement the stack push function. The stack is only pushed when the stack is not full. When the stack is full, the push operation will overwrite the top of the stack with the stack pointer unmodified. Lines 36 to 37 push the word into the stack without modifying the stack pointer when the stack is either empty or full. Line 39 increases the temporary stack pointer implemented with a variable SP_PUSH. It is used to push the word into the stack as shown in line 40. Line 41 updates the stack pointer. Lines 42 to 46 set the FULLn_FF signal to '0' or '1'.

Lines 48 to 54 implement the stack pop function. FULLn_FF signal is deasserted in line 49 to indicate that the stack cannot be full when it is popped. When the stack pointer SP is equal to the lowest index, EMPTYn_FF signal is asserted '0' while the stack pointer is not changed. Otherwise, the stack pointer is deceased by 1 as shown in line 53. Line 58 sends the value indicated by the stack pointer to output STACK_DOUT to go to the multiplexer. Line 59 connects the FULLn_FF signal to the output FULLn.

The instruction decoder is named CNTL and the VHDL code is shown below. The instruction input I is declared in line 10. Condition code low-active signal input CCn is declared in line 7. Condition code enable low-active signal input CCENn is declared in line 8. In line 9, input RLDn is used to control the loading of the register/counter. Line 11 declares REGCNT_ZERO input indicating whether the register/counter is zero. Line 12 declares OEn low-active three-state control signal. It is used to generate the high-active three-state control signal OE (in line 20) since the three-state pads control signal is high active in the Chip Express QYH500 library. Lines 13 to 15 declare three output signals. Lines 16 to 19 declare control signals to MUX, UPC, STACK, and REGCNT blocks.

```
1       ------------------------- file cntl.vhd
2       library IEEE;
3       use IEEE.std_logic_1164.all;
4       use work.AMPACK.all;
5       entity CNTL is
6         port (
7           CCn           : in  std_logic;
8           CCENn         : in  std_logic;
9           RLDn          : in  std_logic;
10          I             : in  std_logic_vector(3 downto 0);
11          REGCNT_ZERO   : in  std_logic;
12          OEn           : in  std_logic;
13          VECTn         : out std_logic;
14          MAPn          : out std_logic;
15          PLn           : out std_logic;
16          MUX_CNTL      : out MUX_OPS;
17          REGCNT_CNTL   : out REGCNT_OPS;
```

```
18          STACK_CNTL    : out STACK_OPS;
19          UPC_CNTL      : out UPC_OPS;
20          OE            : out std_logic_vector(N-1 downto 0));
21      end CNTL;
```

To design the instruction decoder, more detailed information about each instruction is required. Figure 11-4 shows the instruction table. There are 16 instructions. An instruction execution and output values may depend on whether the register/counter value is zero and the conditions indicated by CCENn and CCn.

I	Inst.	Name	Reg/cntr value	CCENn='0' and CCn = '1' Y	CCENn='0' and CCn = '1' Stack	CCENn='1' or CCn='0' Y	CCENn='1' or CCn='0' Stack	Reg/cntr	Enable
0	JZ	Jump Zero	X	0	Clear	0	Clear	Hold	PLn
1	CJS	Con JSB PL	X	PC	Hold	D	Push	Hold	PLn
2	JMAP	Jump Map	X	D	Hold	D	Hold	Hold	MAPn
3	CJP	Cond Jump PL	X	PC	Hold	D	Hold	Hold	PLn
4	PUSH	Push/Cond LD CNTR	X	PC	Push	PC	Push	*1	PLn
5	JSRP	Cond JSB R/PL	X	R	Push	D	Push	Hold	PLn
6	CJV	Cond Jump Vector	X	PC	Hold	D	Hold	Hold	VECTn
7	JRP	Cond Jump R/PL	X	R	Hold	D	Hold	Hold	PLn
8	RFCT	Repeat Loop, CNTR/= 0	/=0	F	Hold	F	Hold	Dec	PLn
8	RFCT	Repeat Loop, CNTR/= 0	=0	PC	Pop	PC	Pop	Hold	PLn
9	RPCT	Repeat PL, CNTR /= 0	/=0	D	Hold	D	Hold	Dec	PLn
9	RPCT	Repeat PL, CNTR /= 0	=0	PC	Hold	PC	Hold	Hold	PLn
A	CRTN	Cond RTN	X	PC	Hold	F	Pop	Hold	PLn
B	CJPP	Cond Jump PL & Pop	X	PC	Hold	D	Pop	Hold	PLn
C	LDCT	LD CNTR & Continue	X	PC	Hold	PC	Hold	Load	PLn
D	LOOP	Test End Loop	X	F	Hold	PC	Pop	Hold	PLn
E	CONT	Continue	X	PC	Hold	PC	Hold	Hold	PLn
F	TWB	Three-way Branch	/=0	F	Hold	PC	Pop	Dec	PLn
F	TWB	Three-way Branch	=0	D	Pop	PC	Pop	Hold	PLn

*1 if CCENn='0' and CCn='1', hold; else load;   X = Don't Care.

Figure 11-4 CY7C910 instruction summary table.

The output enable signal OE can be implemented in lines 24 to 26. Line 27 starts a process statement with I, CCn, CCEN, RLDn, REGCNT_ZERO as the process sensitivity list. By examining the above instruction table, outputs VECTn and MAPn are only asserted in one place while PLn is mostly asserted. This helps us to set the default values in lines 29 to 31 for PLn, MAPn, and VECTn to reduce the number of lines of VHDL code. Lines 32 to 34 set the default values for control signals UPC_CNTL, STACK_CNTL, and MUX_CNTL. Lines 35 to 38 set the default value for the REGCNT_CNTL.

```
22    architecture RTL of CNTL is
23    begin
24      invgen : for j in OE'range generate
25        OE(j) <= not OEn;
26      end generate;
27      p0 : process (I, CCn, CCENn, RLDn, REGCNT_ZERO)
28      begin
29        PLn          <= '0';
30        MAPn         <= '1';
31        VECTn        <= '1';
32        UPC_CNTL     <= U_UP;
33        STACK_CNTL <= S_NOP;
34        MUX_CNTL     <= SEL_UPC;
35        if (RLDn = '1') then
36          REGCNT_CNTL <= R_NOP;
37        else
38          REGCNT_CNTL <= R_LOAD;
39        end if;
```

Line 40 starts a **case statement** based on the input I value. Lines 41 to 44 implement the JZ instruction. The microprogram counter and the stack are cleared. The multiplexer output is set to zero so that the microprogram can start at address 0.

Lines 45 to 49 implement CJS instruction. Line 46 checks the condition. When the condition passes, the next instruction address is pushed into the stack. The next address is jumped to value D. Otherwise, the instruction executes normally and the output and control signals have been set in the default. There is no need to write additional VHDL code to cover the else clause for the **if statement** because default values are assigned in lines 29 to 34. This saves a couple of lines of VHDL code.

Lines 50 to 54 implement the JMAP instruction. The PLn and MAPn values are reversed from their default values in lines 51 and 52. Line 53 assigns the MUL_CNTL to select the D input.

```
40        case I is
41          when "0000" => -- JZ
42            UPC_CNTL     <= U_CLR;
43            STACK_CNTL <= S_CLR;
44            MUX_CNTL     <= SEL_ZERO;
45          when "0001" => -- CJS
46            if not (CCENn = '0' and CCn = '1') then
47              STACK_CNTL <= S_PUSH;
48              MUX_CNTL     <= SEL_D;
49            end if;
50          when "0010" => -- JMAP
51            PLn          <= '1';
52            MAPn         <= '0';
53            MUX_CNTL <= SEL_D;
```

Lines 54 to 57 implement the CJP instruction. If the condition passes, the MUX_CNTL is set to select the D input. Everything else is the same as the default value.

Lines 58 to 62 implement the PUSH instruction. The STACK_CNTL is assigned to the push operation. When the condition passes, the REGCNT_CNTL is set to load the D input. Everything else is the same as the default value.

Lines 63 to 69 implement JSRP instruction. It pushes the stack in line 64. When the condition fails, MUX_CNTL is set to select the register/counter in line 66. If the condition passes, MUX_CNTL is set to select the D input.

```
54          when "0011" => -- CJP
55            if not (CCENn = '0' and CCn = '1') then
56              MUX_CNTL <= SEL_D;
57            end if;
58          when "0100" => -- PUSH
59            STACK_CNTL <= S_PUSH;
60            if not (CCENn = '0' and CCn = '1') then
61              REGCNT_CNTL <= R_LOAD;
62            end if;
63          when "0101" => -- JSRP
64            STACK_CNTL <= S_PUSH;
65            if (CCENn = '0' and CCn = '1') then
66              MUX_CNTL <= SEL_REGCNT;
67            else
68              MUX_CNTL <= SEL_D;
69            end if;
```

Lines 70 to 75 implement the CJV instruction. Outputs PLn and VECTn values are reversed from their default values in lines 71 and 72. If the condition passes, the MUX_CNTL is set to select the D input.

Lines 76 to 81 implement the JRP instruction. If the condition fails, the MUX_CNTL is set to select the register/counter in line 78. If the condition passes, MUX_CNTL is set to select the D input.

```
70          when "0110" => -- CJV
71            PLn   <= '1';
72            VECTn <= '0';
73            if not (CCENn = '0' and CCn = '1') then
74              MUX_CNTL <= SEL_D;
75            END if;
76          when "0111" => -- JRP
77            if (CCENn = '0' and CCn = '1') then
78              MUX_CNTL <= SEL_REGCNT;
79            else
80              MUX_CNTL <= SEL_D;
81            end if;
```

Lines 82 to 88 implement the RFCT instruction. If the register/counter is not zero such that REGCNT_ZERO is equal to '0', MUX_CNTL is set to select the stack in line 84 and register/counter is decremented by setting the REGCNT_CNTL to R_DEC in line 85. If the register/counter value is 0, the STACK_CNTL is set to S_POP so that the stack can be popped. A subroutine is then returned.

Lines 89 to 93 implement the RPCT instruction. If the register/counter value is not zero, the MUX_CNTL is set to select the D input in line 91. The REGCNT_CNTL is set to R_DEC to reduce the register/counter by 1.

Lines 94 to 98 implement the CRTN instruction. If the condition passes, the MUX_CNTL is set to select the stack output and the stack is popped.

```
82          when "1000" => -- RFCT
83            if (REGCNT_ZERO = '0') then
84              MUX_CNTL    <= SEL_STACK;
85              REGCNT_CNTL <= R_DEC;
86            else
87              STACK_CNTL  <= S_POP;
88            end if;
89          when "1001" => -- RPCT
90            if (REGCNT_ZERO = '0') then
91              MUX_CNTL    <= SEL_D;
92              REGCNT_CNTL <= R_DEC;
93            end if;
94          when "1010" => -- CRTN
95            if not (CCENn = '0' and CCn = '1') then
96              MUX_CNTL    <= SEL_STACK;
97              STACK_CNTL <= S_POP;
98            end if;
```

Lines 99 to 103 implement the CJPP instruction. If the condition passes, MUX_CNTL is set to select the D input and the stack is popped.

Lines 104 to 105 implement the LDCT instruction. The register/counter is loaded with the D input by setting the REGCNT_CNTL to R_LOAD.

Lines 106 to 111 implement the LOOP instruction. If the condition fails, the MUX_CNTL is set to select the stack value so that the execution can go to the address which is last pushed. It is also the starting address for the instructions to be looped. If the condition passes, the looping is completed. The stack is popped. The execution goes to the instruction that follows the LOOP instruction.

```
99          when "1011" => -- CJPP
100           if not (CCENn = '0' and CCn = '1') then
101             MUX_CNTL    <= SEL_D;
102             STACK_CNTL <= S_POP;
103           end if;
104         when "1100" => -- LDCT
105           REGCNT_CNTL <= R_LOAD;
106         when "1101" => -- LOP
107           if (CCENn = '0' and CCn = '1') then
```

```
108                    MUX_CNTL    <= SEL_STACK;
109                else
110                    STACK_CNTL <= S_POP;
111                end if;
```

Lines 112 and 113 implement the CONT instruction. All is well by the default. A **null statement** is used.

Lines 114 to 127 implement the most complicated instruction, TWB. It is a three-way branch instruction. If the register/counter value is not zero, the REGCNT_CNTL is set to R_DEC to decrease the register/counter by 1 in line 116. If the condition fails, the MUX_CNTL is set to select the stack output so that the loop can be performed. If the condition passes, the stack is popped by setting the STACK_CNTL to S_POP in line 120. The execution follows the next instruction. If the register/counter is zero, the stack is popped by setting the STACK_CNTL to S_POP in line 123. If the condition fails, the MUX_CNTL is set to select D. The execution jumps to the address pointed by D input. If the condition passes, the normal execution sequence is performed. For a more detailed description of each instruction, you are referred to AMD or Cypress data books.

```
112          when "1110" => --   CONT
113             null;
114          when others => -- TWB
115             if (REGCNT_ZERO = '0') then -- REGCNT /= 0
116                REGCNT_CNTL <= R_DEC;
117                if (CCENn = '0' and CCn = '1') then
118                    MUX_CNTL <= SEL_STACK;
119                else
120                    STACK_CNTL <= S_POP;
121                end if;
122             else
123                STACK_CNTL <= S_POP;
124                if (CCENn = '0' and CCn = '1') then
125                    MUX_CNTL <= SEL_D;
126                end if;
127             end if;
128          end case;
129       end process;
130    end RTL;
```

The above blocks can be combined in the following VHDL code. Package AMPACK is referenced in line 4. The input and output signals in lines 7 to 20 are very close to the pin description. The differences are as follows:

• Signals such as D (I, Y) are combined as a bus signal with multiple bits.

• The OE (line 20) output signal is added to control the active high three-state enable pads.

```
1        ------------------- file core2910.vhd
2        library IEEE;
3        use IEEE.std_logic_1164.all;
4        use work.AMPACK.all;
5        entity CORE2910 is
6          port (
7            CLK     : in  std_logic;
8            CCn     : in  std_logic;
9            CCENn   : in  std_logic;
10           RLDn    : in  std_logic;
11           CI      : in  std_logic;
12           D       : in  std_logic_vector(N-1 downto 0);
13           I       : in  std_logic_vector(3 downto 0);
14           OEn     : in  std_logic;
15           FULLn   : out std_logic;
16           PLn     : out std_logic;
17           MAPn    : out std_logic;
18           VECTn   : out std_logic;
19           Y       : out std_logic_vector(N-1 downto 0);
20           OE      : out std_logic_vector(N-1 downto 0));
21       end CORE2910;
22       -----------------------------------------------------------------
23       architecture RTL of CORE2910 is
24         component MUX
25         port (
26           MUX_CNTL     : in  MUX_OPS;
27           D            : in  std_logic_vector(N-1 downto 0);
28           REGCNT_DOUT  : in  std_logic_vector(N-1 downto 0);
29           STACK_DOUT   : in  std_logic_vector(N-1 downto 0);
30           UPC_DOUT     : in  std_logic_vector(N-1 downto 0);
31           MUXOUT       : out std_logic_vector(N-1 downto 0));
32         end component;
33         component UPC
34         port (
35           CLK          : in  std_logic;
36           CI           : in  std_logic;
37           UPC_CNTL     : in  UPC_OPS;
38           MUXOUT       : in  std_logic_vector(N-1 downto 0);
39           UPC_DOUT     : out std_logic_vector(N-1 downto 0));
40         end component;
41         component CNTL
42         port (
43           CCn          : in  std_logic;
44           CCENn        : in  std_logic;
45           RLDn         : in  std_logic;
46           I            : in  std_logic_vector(3 downto 0);
47           REGCNT_ZERO  : in  std_logic;
48           OEn          : in  std_logic;
49           VECTn        : out std_logic;
50           MAPn         : out std_logic;
```

```
51        PLn             : out std_logic;
52        MUX_CNTL        : out MUX_OPS;
53        REGCNT_CNTL     : out REGCNT_OPS;
54        STACK_CNTL      : out STACK_OPS;
55        UPC_CNTL        : out UPC_OPS;
56        OE              : out std_logic_vector(N-1 downto 0));
57     end component;
58     component STACK
59     port (
60        CLK             : in  std_logic;
61        STACK_CNTL      : in  STACK_OPS;
62        STACK_DIN       : in  std_logic_vector(N-1 downto 0);
63        STACK_DOUT      : out std_logic_vector(N-1 downto 0);
64        FULLn           : out std_logic);
65     end component;
66     component REGCNT
67     port (
68        CLK             : in  std_logic;
69        D               : in  std_logic_vector(N-1 downto 0);
70        REGCNT_CNTL     : in  REGCNT_OPS;
71        REGCNT_DOUT     : out std_logic_vector(N-1 downto 0);
72        REGCNT_ZERO     : out std_logic);
73     end component;
74     signal MUX_CNTL     : MUX_OPS;
75     signal REGCNT_CNTL  : REGCNT_OPS;
76     signal STACK_CNTL   : STACK_OPS;
77     signal UPC_CNTL     : UPC_OPS;
78     signal UPC_DOUT     : std_logic_vector(N-1 downto 0);
79     signal STACK_DOUT   : std_logic_vector(N-1 downto 0);
80     signal REGCNT_DOUT  : std_logic_vector(N-1 downto 0);
81     signal MUXOUT       : std_logic_vector(N-1 downto 0);
82     signal REGCNT_ZERO  : std_logic;
```

Lines 24 to 73 are component declarations. Lines 74 to 82 declare signals to be used to connect the components. Lines 84 to 129 instantiate five components.

```
83     begin
84     Y <= MUXOUT;
85     mux0 : MUX
86     port map (
87        MUX_CNTL        => MUX_CNTL,
88        D               => D,
89        REGCNT_DOUT     => REGCNT_DOUT,
90        STACK_DOUT      => STACK_DOUT,
91        UPC_DOUT        => UPC_DOUT,
92        MUXOUT          => MUXOUT);
93     upc0 : UPC
94     port map (
95        CLK             => CLK,
96        CI              => CI,
```

```
97          UPC_CNTL        => UPC_CNTL,
98          MUXOUT          => MUXOUT,
99          UPC_DOUT        => UPC_DOUT);
100    cntl0 : CNTL
101    port map (
102         CCn             => CCn,
103         CCENn           => CCENn,
104         RLDn            => RLDn,
105         I               => I,
106         REGCNT_ZERO     => REGCNT_ZERO,
107         OEn             => OEn,
108         VECTn           => VECTn,
109         MAPn            => MAPn,
110         PLn             => PLn,
111         MUX_CNTL        => MUX_CNTL,
112         REGCNT_CNTL     => REGCNT_CNTL,
113         STACK_CNTL      => STACK_CNTL,
114         UPC_CNTL        => UPC_CNTL,
115         OE              => OE);
116    stack0 : STACK
117    port map (
118         CLK             => CLK,
119         STACK_CNTL      => STACK_CNTL,
120         STACK_DIN       => UPC_DOUT,
121         STACK_DOUT      => STACK_DOUT,
122         FULLn           => FULLn);
123    regcnt0 : REGCNT
124    port map (
125         CLK             => CLK,
126         D               => D,
127         REGCNT_CNTL     => REGCNT_CNTL,
128         REGCNT_DOUT     => REGCNT_DOUT,
129         REGCNT_ZERO     => REGCNT_ZERO);
130    end RTL;
```

The following is the top-level VHDL entity. Note that a bus such as I (D, Y) is split into individual signals. All port signal names are changed to uppercase names to facilitate the interface to the place and route tool.

```
1       ------------------- file chip2910_e.vhd
2       library IEEE;
3       use IEEE.std_logic_1164.all;
4       use work.AMPACK.all;
5       entity CHIP2910 is
6         port (
7           CLK    : in  std_logic;
8           CCN    : in  std_logic;
9           CCENN  : in  std_logic;
10          RLDN   : in  std_logic;
11          CI     : in  std_logic;
```

```
12          D11    : in  std_logic;
13          D10    : in  std_logic;
14          D9     : in  std_logic;
15          D8     : in  std_logic;
16          D7     : in  std_logic;
17          D6     : in  std_logic;
18          D5     : in  std_logic;
19          D4     : in  std_logic;
20          D3     : in  std_logic;
21          D2     : in  std_logic;
22          D1     : in  std_logic;
23          D0     : in  std_logic;
24          I3     : in  std_logic;
25          I2     : in  std_logic;
26          I1     : in  std_logic;
27          I0     : in  std_logic;
28          OEN    : in  std_logic;
29          FULLN  : out std_logic;
30          PLN    : out std_logic;
31          MAPN   : out std_logic;
32          VECTN  : out std_logic;
33          Y11    : out std_logic;
34          Y10    : out std_logic;
35          Y9     : out std_logic;
36          Y8     : out std_logic;
37          Y7     : out std_logic;
38          Y6     : out std_logic;
39          Y5     : out std_logic;
40          Y4     : out std_logic;
41          Y3     : out std_logic;
42          Y2     : out std_logic;
43          Y1     : out std_logic;
44          Y0     : out std_logic);
45   end CHIP2910;
```

The following shows an architecture for the CHIP2910 entity. Lines 4 and 5 reference the Chip Express QYH500 library. Lines 8 to 24 declare component CORE2910. Lines 25 to 55 declare components related to pads. These components are to be instantiated without any optimization.

```
1    -------------------- file chip2910_pad.vhd
2    library IEEE;
3    use IEEE.std_logic_1164.all;
4    library qyh500;
5    use qyh500.all;
6    use work.AMPACK.all;
7    architecture PAD of CHIP2910 is
8      component CORE2910
9      port (
```

```
10          CLK     : in  std_logic;
11          CCn     : in  std_logic;
12          CCENn   : in  std_logic;
13          RLDn    : in  std_logic;
14          CI      : in  std_logic;
15          D       : in  std_logic_vector(N-1 downto 0);
16          I       : in  std_logic_vector(3   downto 0);
17          OEn     : in  std_logic;
18          FULLn   : out std_logic;
19          PLn     : out std_logic;
20          MAPn    : out std_logic;
21          VECTn   : out std_logic;
22          Y       : out std_logic_vector(N-1 downto 0);
23          OE      : out std_logic_vector(N-1 downto 0));
24        end component;
25        component cdqy2    -- buffer for clock
26          port (
27            A     : in    std_logic;
28            Z     : out   std_logic);
29        end component;
30        component ipn        -- Input pad.
31          port (
32            Z     : in    std_logic;
33            INb   : out   std_logic);
34        end component;
35        component ostt04    -- TTL 4ma output pad
36          port (
37            D     : in    std_logic;
38            Z     : out   std_logic);
39        end component;
40        component osst04     -- TTL 4ma tristate output pad /slew
41          port (
42            OE    : in    std_logic;
43            D     : in    std_logic;
44            Z     : out   std_logic);
45        end component;
46        component ipsbc1     -- TTL level shifter 1x drive
47          port (
48            A     : in    std_logic;
49            Z     : out   std_logic);
50        end component;
51        component ipsbc2     -- TTL level shifter 2x drive
52          port (
53            A     : in    std_logic;
54            Z     : out   std_logic);
55        end component;
56        signal L_CLK, B_CLK, C_CLK  : std_logic;
57        signal L_CCn, C_CCn         : std_logic;
58        signal L_CI, C_CI           : std_logic;
59        signal L_CCENn, C_CCENn     : std_logic;
```

```
60    signal L_RLDn, C_RLDn      : std_logic;
61    signal L_D, C_D  : std_logic_vector(N-1 downto 0);
62    signal L_I, C_I  : std_logic_vector(3 downto 0);
63    signal C_FULLn  : std_logic;
64    signal C_PLn    : std_logic;
65    signal C_MAPn   : std_logic;
66    signal C_VECTn  : std_logic;
67    signal C_Y      : std_logic_vector(N-1 downto 0);
68    signal C_OE     : std_logic_vector(N-1 downto 0);
```

Lines 56 to 68 declare signals to be used to connect components related to pads and the core CORE2910 component. The input pad has a level shifter. For example, in line 75, output signal RLDn is connected to an internal signal L_RLDn through a pad. SIgnal L_RLDn is then connected to a level shifter in line 95. The output of the level shifter is signal C_RLDn. Signal C_RLDn is then connected to component CORE2910, using the port map in line 148.

The clock signal CLK is going through a pad to get to signal L_CLK in line 70. Signal L_CLK is connected to a level shifter to get to signal B_CLK in line 71. Line 72 instantiates a clock buffer cdqy2 with signal B_CLK as the input and signal C_CLK as the output. Signal C_CLK is then connected to component CORE2910, using the port map in line 145.

Line 115 shows a straight output pad. Signal C_FULLn is mapped to the input of the output pad and port mapped to the component CORE2910 in line 153. Lines 119 and 120 show an example of a three-state pad instantiation. A three-state pad has two inputs and one output. Both inputs are from the component CORE2910. One is the data output and the other is the three-state enable signal.

Note also that signals to be connected to the core component CORE2910 are changed to the bus format. This makes the VHDL coding easier in the internal blocks.

```
69    begin
70      clkpad   : ipn    port map (Z => CLK,     INb => L_CLK);
71      clklevel : ipsbc2 port map (A => L_CLK,   Z   => B_CLK);
72      clkbuf   : cdqy2  port map (A => B_CLK,   Z   => C_CLK);
73      ccnpad   : ipn    port map (Z => CCn,     INb => L_CCn);
74      ccennpad : ipn    port map (Z => CCENn,   INb => L_CCENn);
75      rldnpad  : ipn    port map (Z => RLDn,    INb => L_RLDn);
76      cipad    : ipn    port map (Z => CI,      INb => L_CI);
77      i3pad    : ipn    port map (Z => I3,      INb => L_I(3));
78      i2pad    : ipn    port map (Z => I2,      INb => L_I(2));
79      i1pad    : ipn    port map (Z => I1,      INb => L_I(1));
80      i0pad    : ipn    port map (Z => I0,      INb => L_I(0));
81      d11pad   : ipn    port map (Z => D11,     INb => L_D(11));
82      d10pad   : ipn    port map (Z => D10,     INb => L_D(10));
83      d9pad    : ipn    port map (Z => D9,      INb => L_D(9));
84      d8pad    : ipn    port map (Z => D8,      INb => L_D(8));
85      d7pad    : ipn    port map (Z => D7,      INb => L_D(7));
86      d6pad    : ipn    port map (Z => D6,      INb => L_D(6));
87      d5pad    : ipn    port map (Z => D5,      INb => L_D(5));
```

```
88    d4pad     : ipn    port map (Z => D4,      INb => L_D(4));
89    d3pad     : ipn    port map (Z => D3,      INb => L_D(3));
90    d2pad     : ipn    port map (Z => D2,      INb => L_D(2));
91    d1pad     : ipn    port map (Z => D1,      INb => L_D(1));
92    d0pad     : ipn    port map (Z => D0,      INb => L_D(0));
93    ------------------------------------------------------------
94    ccenlevel: ipsbc1 port map (A => L_CCENn, Z   => C_CCENn);
95    rldnlevel: ipsbc1 port map (A => L_RLDn,  Z   => C_RLDn);
96    ccnlevel : ipsbc1 port map (A => L_CCn,   Z   => C_CCn);
97    cilevel  : ipsbc1 port map (A => L_CI,    Z   => C_CI);
98    i3level  : ipsbc1 port map (A => L_I(3),  Z   => C_I(3));
99    i2level  : ipsbc1 port map (A => L_I(2),  Z   => C_I(2));
100   i1level  : ipsbc1 port map (A => L_I(1),  Z   => C_I(1));
101   i0level  : ipsbc1 port map (A => L_I(0),  Z   => C_I(0));
102   d11level : ipsbc1 port map (A => L_D(11), Z   => C_D(11));
103   d10level : ipsbc1 port map (A => L_D(10), Z   => C_D(10));
104   d9level  : ipsbc1 port map (A => L_D(9),  Z   => C_D(9));
105   d8level  : ipsbc1 port map (A => L_D(8),  Z   => C_D(8));
106   d7level  : ipsbc1 port map (A => L_D(7),  Z   => C_D(7));
107   d6level  : ipsbc1 port map (A => L_D(6),  Z   => C_D(6));
108   d5level  : ipsbc1 port map (A => L_D(5),  Z   => C_D(5));
109   d4level  : ipsbc1 port map (A => L_D(4),  Z   => C_D(4));
110   d3level  : ipsbc1 port map (A => L_D(3),  Z   => C_D(3));
111   d2level  : ipsbc1 port map (A => L_D(2),  Z   => C_D(2));
112   d1level  : ipsbc1 port map (A => L_D(1),  Z   => C_D(1));
113   d0level  : ipsbc1 port map (A => L_D(0),  Z   => C_D(0));
114   -- output pads
115   fullnpad : ostt04 port map (D => C_FULLn, Z   => FULLn);
116   vectnpad : ostt04 port map (D => C_VECTn, Z   => VECTn);
117   plnpad   : ostt04 port map (D => C_PLn,   Z   => PLn);
118   mapnpad  : ostt04 port map (D => C_MAPn,  Z   => MAPn);
119   y11pad   : osst04
120     port map (OE => C_OE(11), D => C_Y(11), Z   => Y11);
121   y10pad   : osst04
122     port map (OE => C_OE(10), D => C_Y(10), Z   => Y10);
123   y9pad    : osst04
124     port map (OE => C_OE(9), D => C_Y(9),  Z   => Y9);
125   y8pad    : osst04
126     port map (OE => C_OE(8), D => C_Y(8),  Z   => Y8);
127   y7pad    : osst04
128     port map (OE => C_OE(7), D => C_Y(7),  Z   => Y7);
129   y6pad    : osst04
130     port map (OE => C_OE(6), D => C_Y(6),  Z   => Y6);
131   y5pad    : osst04
132     port map (OE => C_OE(5), D => C_Y(5),  Z   => Y5);
133   y4pad    : osst04
134     port map (OE => C_OE(4), D => C_Y(4),  Z   => Y4);
135   y3pad    : osst04
136     port map (OE => C_OE(3), D => C_Y(3),  Z   => Y3);
137   y2pad    : osst04
```

```
138        port map (OE => C_OE(2), D => C_Y(2),   Z   => Y2);
139     y1pad    : osst04
140        port map (OE => C_OE(1), D => C_Y(1),   Z   => Y1);
141     y0pad    : osst04
142        port map (OE => C_OE(0), D => C_Y(0),   Z   => Y0);
143     core29100 : CORE2910
144     port map (
145        CLK     => C_CLK,
146        CCn     => C_CCn,
147        CCENn   => C_CCENn,
148        RLDn    => C_RLDn,
149        CI      => C_CI,
150        D       => C_D,
151        I       => C_I,
152        OEn     => OEn,
153        FULLn   => C_FULLn,
154        PLn     => C_PLn,
155        MAPn    => C_MAPn,
156        VECTn   => C_VECTn,
157        Y       => C_Y,
158        OE      => C_OE);
159  end PAD;
```

We have described VHDL code from the lower-level blocks to the top. This may not be the necessary sequence to write the VHDL code. The top level and the core level can be designed first. If a technology library has been selected, the three-state enable signals for the three-state output pads and bidirectional pads should be determined to be high or low active. This allows the correct polarity of the three-state enable signals to be generated in the lower-level blocks. The exact pads to be used should be studied to determine their drive strength. For example, most of the pads used in this example are 4 mA pads. If the technology library has not been selected or the pads are not in the synthesis and the VHDL simulation library, another VHDL architecture is recommended (as shown in the last chapter) to model the pads. This allows the complete design to be simulated with the test bench sooner. The architecture can then be modified slightly to replace the actual pad instantiation for synthesis and simulation. In both cases, the entity should remain the same so that the test bench can be configured to select any one of the architectures. After the entity of the core component CORE2910 is completed, the lower-level blocks can be designed. These lower-level blocks are then combined in the architecture portion of the core component CORE2910.

11.3 DESIGN VERIFICATION

To verify the CHIP2910 design, a test bench can be used. When AM2910 was first designed, it had to have had its purpose. The Y output is used to address a microprogram read-only memory (ROM). The output of the ROM can be used with an arithmetic logic unit (ALU) such as AM2901. For simplicity, our test bench wraps back to the D input of CHIP2910. Figure 11-5 shows the test bench block diagram. Note that there is a pipeline register. This is due to the combination paths from inputs (such as I, D) to output Y.

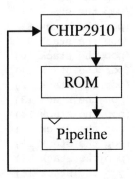

Figure 11-5 Test bench block diagram.

The ROM contains a microprogram. The following illustration shows a simple microprogram designed to verify all instructions. Each word of the ROM has 20 bits. The leftmost 4 bits are RLDn, CI, CCENn, CCn. The next 4 bits are for the instruction input I. The next 12 bits are for the D input. For example, in line 1, RLDn CI, CCENn, CCn are '1', '1', '0', and '0', respectively. The instruction is hexadecimal e which is the CONT continue instruction. The D input will be "001" in hexadecimal. Lines 2 to 19 check the stack push instruction. There are 18 pushes so that the behavior of the full stack can be tested. Lines 20 to 37 test the stack pop operations. More details of the program and instructions will be explained with simulation waveforms below.

```
1     1100 e 001    -- 0, continue
2     1100 4 002    -- 1, PUSH 1
3     0100 4 003    -- 2, PUSH 2
4     1100 4 004    -- 3, PUSH 3
5     1101 4 005    -- 4, PUSH 4
6     1100 4 006    -- 5, PUSH 5
7     1110 4 007    -- 6, PUSH 6
8     0100 4 008    -- 7, PUSH 7
9     1100 4 009    -- 8, PUSH 8
10    1111 4 00a    -- 9, PUSH 9
11    1100 4 00b    -- a, PUSH 10
12    1100 4 00c    -- b, PUSH 11
13    1100 4 00d    -- c, PUSH 12
14    1100 4 00e    -- d, PUSH 13
15    1100 4 00f    -- e, PUSH 14
16    1100 4 010    -- f, PUSH 15
```

```
17      1100 4 011    -- 10, PUSH 16
18      1100 4 012    -- 11, PUSH 17
19      1100 4 013    -- 12, PUSH 18
20      1110 d 014    -- 13, LOOP (pop the stack 1)
21      1100 d 015    -- 14, LOOP (pop the stack 2)
22      1100 d 016    -- 15, LOOP (pop the stack 3)
23      1100 d 017    -- 16, LOOP (pop the stack 4)
24      1110 d 018    -- 17, LOOP (pop the stack 5)
25      1100 d 019    -- 18, LOOP (pop the stack 6)
26      1110 d 01a    -- 19, LOOP (pop the stack 7)
27      1100 d 01b    -- 1a, LOOP (pop the stack 8)
28      1111 d 01c    -- 1b, LOOP (pop the stack 9)
29      1100 d 01d    -- 1c, LOOP (pop the stack 10)
30      0100 d 01e    -- 1d, LOOP (pop the stack 11)
31      1111 d 01f    -- 1e, LOOP (pop the stack 12)
32      1110 d 020    -- 1f, LOOP (pop the stack 13)
33      1100 d 021    -- 20, LOOP (pop the stack 14)
34      0111 d 022    -- 21, LOOP (pop the stack 15)
35      1110 d 023    -- 22, LOOP (pop the stack 16)
36      1100 d 024    -- 23, LOOP (pop the stack 17)
37      1100 d 025    -- 24, LOOP (pop the stack 18)
38      1100 e 026    -- 25, CONT, continue
39      1100 e 027    -- 26, CONT, continue
40      1100 1 035    -- 27, CJS to address 35
41      1100 0 000    -- 28
42      1100 0 000    -- 29
43      1100 0 000    -- 2a
44      1100 0 000    -- 2b
45      1100 0 000    -- 2c
46      1100 0 000    -- 2d
47      1100 0 000    -- 2e
48      1100 0 000    -- 2f
49      1100 0 000    -- 30
50      1100 0 000    -- 31
51      1100 0 000    -- 32
52      1100 0 000    -- 33
53      1100 0 000    -- 34
54      1100 2 038    -- 35, JMAP to address 38
55      1000 e bad    -- 36, should not be here, OOPS
56      1100 0 000    -- 37, JZ, jump back to 0
57      1101 3 0ff    -- 38, CJP, no jump
58      1100 3 03c    -- 39, CJP, jump to 3b
59      1100 e bad    -- 3a, OOPS!
60      1100 0 000    -- 3b, JZ
61      1100 1 040    -- 3c, CJS to 40
62      1100 1 045    -- 3d, CJS to 45
63      1100 0 000    -- 3e,
64      1100 0 000    -- 3f
65      1100 4 003    -- 40, PUSH, load counter 3
66      1100 e 040    -- 41, continue
```

```
67    1100 8 041    -- 42, RFCT, repeat
68    1101 a 043    -- 43, Condition return, no return
69    1100 a 044    -- 44, return
70    1100 4 fff    -- 45, load counter with FFF
71    1100 d 046    -- 46, pop
72    1100 c 002    -- 47, LDCT, load counter and continue
73    1100 9 048    -- 48, RPCT, repeat this instruction
74    1101 6 000    -- 49, CJV, conditional jump, no jump
75    1100 6 04c    -- 4a, CJV, jump to 4c
76    1100 0 bad    -- 4b, OOPS!
77    1100 c 04f    -- 4c, LDCT, load counter 4f
78    1101 7 000    -- 4d, JRP, jump to counter address 4f
79    1100 0 bad    -- 4e, OOPS
80    1100 7 051    -- 4f, JRP, jump to D
81    1100 0 bad    -- 50, OOPS
82    1100 c 054    -- 51, LDCT 54
83    1101 5 000    -- 52, JSRP, jump to counter
84    1100 0 bad    -- 53, OOPS
85    1100 5 056    -- 54, JSRP, jump to D
86    1100 0 bad    -- 55, OOPS
87    1100 4 003    -- 56, PUSH and load counter with 60
88    1100 e aaa    -- 57, continue
89    1101 f 05a    -- 58, TWB, three way branch to 5a when done
90    1100 e bad    -- 59
91    1100 e d0e    -- 5a
92    1100 e d0e    -- 5b
93    1100 e d0e    -- 5c
94    ...
95                  -- (RLDn CI, CCENn, CCn), I, D, -- rom address
```

The following shows a simple VHDL code for the ROM. The idea is to read the above text file into the ROM VHDL code. Lines 3 to 5 reference packages. Lines 18 to 47 constitute a **process statement**. Lines 30 to 39 read the text file and save the information in variable MEM. There is a variable READDONE to ensure that the text file is read only once. The **loop statement** starting in line 31 reads exactly WORDS lines. The text file to be read should have at least that many lines. Only a portion of the above microprogram text file is shown.

```
1     library IEEE;
2     use IEEE.std_logic_1164.all;
3     use IEEE.std_logic_arith.all;
4     use IEEE.std_logic_textio.all;
5     use STD.TEXTIO.all;
6     entity ROM is
7       generic (
8         SIZE       : integer := 20;
9         WORDS      : integer := 96;
10        ADSIZE     : integer := 7;
11        CODEFILE   : string  := "rom.in");
```

```
12      port (
13        A    : in  std_logic_vector(ADSIZE-1 downto 0);
14        DOUT : out std_logic_vector(SIZE-1 downto 0));
15      end ROM;
16      architecture BEH of ROM is
17      begin
18        check : process(A )
19          file MEMIN  : text is in CODEFILE;
20          subtype ARRAYWORD is std_logic_vector((SIZE - 1) downto
0);
21          type ARRAYTYPE is array(Natural range <>) of ARRAYWORD;
22          variable address  : integer;
23          variable INLINE   : line;
24          variable MEM      : ARRAYTYPE (0 to (WORDS - 1));
25          variable bits4    : std_logic_vector(3 downto 0);
26          variable OPCODE   : std_logic_vector(3 downto 0);
27          variable AD       : std_logic_vector(11 downto 0);
28          variable READDONE : boolean := FALSE;
29        begin
30          if (not READDONE) then
31            for i in MEM'range loop
32              readline(MEMIN, INLINE);
33              read( INLINE, bits4);
34              hread( INLINE, OPCODE);
35              hread( INLINE, AD);
36              MEM(i) := bits4 & OPCODE & AD;
37            end loop;
38            READDONE := TRUE;
39          end if;
40          if (IS_X(A)) then
41            DOUT <= (others => 'X');
42            assert FALSE report "address not valid" severity NOTE;
43          else
44            address := CONV_INTEGER(unsigned(A));
45            DOUT    <= MEM(address);
46          end if;
47        end process;
48      end BEH;
```

The test bench VHDL code is shown below. Lines 7 and 8 declare the generics PERIOD and DELAY to specify the clock period and the input delay. It was set to 40 ns and 15 ns, respectively, for the RTL and structural simulation. It is then set up to 1000 ns and 200 ns to prepare test vectors that are to be discussed below.

```
1      library IEEE;
2      use IEEE.std_logic_1164.all;
3      use IEEE.std_logic_arith.all;
4      use work.AMPACK.all;
5      entity TBCHIP2910 is
6        generic (
```

```
 7              PERIOD : time := 1000 ns;
 8              DELAY  : time := 200 ns);
 9    end TBCHIP2910;
10    architecture BEH of TBCHIP2910 is
11      component CHIP2910
12      port (
13        CLK    : in  std_logic;
14        CCN    : in  std_logic;
15        CCENN  : in  std_logic;
16        RLDN   : in  std_logic;
17        CI     : in  std_logic;
18        D11    : in  std_logic;
19        D10    : in  std_logic;
20        D9     : in  std_logic;
21        D8     : in  std_logic;
22        D7     : in  std_logic;
23        D6     : in  std_logic;
24        D5     : in  std_logic;
25        D4     : in  std_logic;
26        D3     : in  std_logic;
27        D2     : in  std_logic;
28        D1     : in  std_logic;
29        D0     : in  std_logic;
30        I3     : in  std_logic;
31        I2     : in  std_logic;
32        I1     : in  std_logic;
33        I0     : in  std_logic;
34        OEN    : in  std_logic;
35        FULLN  : out std_logic;
36        PLN    : out std_logic;
37        MAPN   : out std_logic;
38        VECTN  : out std_logic;
39        Y11    : out std_logic;
40        Y10    : out std_logic;
41        Y9     : out std_logic;
42        Y8     : out std_logic;
43        Y7     : out std_logic;
44        Y6     : out std_logic;
45        Y5     : out std_logic;
46        Y4     : out std_logic;
47        Y3     : out std_logic;
48        Y2     : out std_logic;
49        Y1     : out std_logic;
50        Y0     : out std_logic);
51      end component;
52      component ROM
53      generic (
54        SIZE        : integer := 20;
55        WORDS       : integer := 96;
56        ADSIZE      : integer := 7;
```

```
57         CODEFILE    : string  := "rom.in");
58     port (
59       A    : in  std_logic_vector(ADSIZE-1 downto 0);
60       DOUT : out std_logic_vector(SIZE-1 downto 0));
61     end component;
62     signal CLK, CCN, CCENN, RLDN, CI    : std_logic := '1';
63     signal OEN, FULLN, PLN, MAPN, VECTN : std_logic;
64     signal D, Y, FILTY : std_logic_vector(N-1 downto 0);
65     signal ROMD       : std_logic_vector(19  downto 0);
66     signal I          : std_logic_vector( 3 downto 0);
67     signal OP         : INST_TYPE;
68     signal RSTn       : std_logic;
```

Lines 70 to 74 generate the clock CLK waveform. Lines 75 to 83 generate a reset signal RSTn. There is no global reset signal in the AM2910 and CY7C910 chip. The initialization of the internal stack and microprogram counter is done through the JZ (jump to zero) instruction. In this test bench, when RSTn is asserted, the I input is forced to be 0 so that CHIP2910 can execute the JZ instruction as shown in lines 101 to 117. Lines 88 to 97 convert the invalid address from Y to a valid address so that the ROM can be addressed.

```
69     begin
70       clkgen : process
71       begin
72         CLK <= '1'; wait for PERIOD / 2;
73         CLK <= '0'; wait for PERIOD / 2;
74       end process;
75       rstngen : process
76       begin
77         RSTn <= '0';
78         for j in 1 to 5 loop
79           wait until CLK'event and CLK = '1';
80         end loop;
81         RSTn <= '1' after DELAY;
82         wait;
83       end process;
84
85       rom0 : ROM generic map ( SIZE => 20, WORDS => 128,
86         ADSIZE => 7, CODEFILE => "rom.in")
87       port map (A => FILTY(6 downto 0), DOUT => ROMD);
88       changey : process (Y)
89       begin
90         for j in FILTY'range loop
91           if (Y(j) = '0' or Y(j) = '1') then
92             FILTY(j) <= Y(j);
93           else
94             FILTY(j) <= '0';
95           end if;
96         end loop;
```

```
97      end process;
98
99      pipereg : process
100     begin
101       wait until CLK'event and CLK = '1';
102       if (RSTn = '0') then
103         D      <= "000000000000" after DELAY;
104         I      <= "0000" after DELAY;
105         RLDN   <= '0' after DELAY;
106         CI     <= '0' after DELAY;
107         CCENn  <= '0' after DELAY;
108         CCn    <= '0' after DELAY;
109       else
110         D      <= ROMD(11 downto 0) after DELAY;
111         I      <= ROMD(15 downto 12) after DELAY;
112         RLDn   <= ROMD(19) after DELAY;
113         CI     <= ROMD(18) after DELAY;
114         CCENn  <= ROMD(17) after DELAY;
115         CCn    <= ROMD(16) after DELAY;
116       end if;
117     end process;
118     OP     <= INST_TYPE'val(conv_integer(unsigned(I)));
119     chip29100 : CHIP2910
120     port map (
121         CLK   => CLK,    CCN    => CCn,     CCENN  => CCENn,
122         RLDN  => RLDn,   CI     => CI,      D11    => D(11),
123         D10   => D(10),  D9     => D(9),    D8     => D(8),
124         D7    => D(7),   D6     => D(6),    D5     => D(5),
125         D4    => D(4),   D3     => D(3),    D2     => D(2),
126         D1    => D(1),   D0     => D(0),    I3     => I(3),
127         I2    => I(2),   I1     => I(1),    I0     => I(0),
128         OEN   => OEn,    FULLN  => FULLn,   PLN    => PLn,
129         MAPN  => MAPn,   VECTN  => VECTn,   Y11    => Y(11),
130         Y10   => Y(10),  Y9     => Y(9),    Y8     => Y(8),
131         Y7    => Y(7),   Y6     => Y(6),    Y5     => Y(5),
132         Y4    => Y(4),   Y3     => Y(3),    Y2     => Y(2),
133         Y1    => Y(1),   Y0     => Y(0));
134   end BEH;
135   configuration CFG_TBCHIP2910 of TBCHIP2910 is
136     for BEH
137       for chip29100 : CHIP2910
138         use configuration work.CFG_CHIP2910;
139       end for;
140       for all : ROM
141         use entity work.ROM(BEH);
142       end for;
143     end for;
144   end CFG_TBCHIP2910;
```

Lines 135 to 144 configure the text bench. Line 138 references a configuration CFG_CHIP2910. The VHDL code is not shown here. This is left as an exercise.

Figure 11-6 shows a simulation waveform just before the RSTn signal is deasserted. The waveform shows many key signals in the design. Note that all control signals such as MUX_CNTL (STACK_CNTL, UPC_CNTL, and REGCNT_CNTL) are shown with their corresponding enumerated types. A signal OP is also shown as the instruction mnemonic that is implemented in line 118. Showing the enumerated type in the waveform makes the design verification easier. As shown in the waveform, the stack and microprogram counter are cleared. MUX_CNTL is set to SEL_ZERO so that Y would be zero. The register/counter is also loaded with zero. After the RSTn is deasserted, the instructions in the microprogram are executed. The first instruction is a CONT instruction. The microprogram counter is increased by 1. The MUX_CNTL is set to SEL_UPC. The next instruction PUSH is executed. The first PUSH instruction does not change the stack pointer SP. The EMPTYn flag is deasserted. After that, the stack pointer continues to be increased.

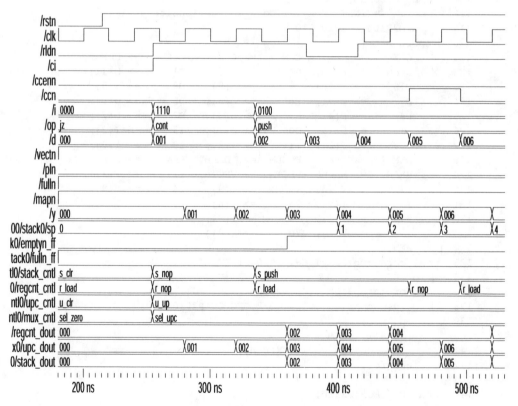

Figure 11-6 TBCHIP2910 simulation waveform — 1.

Figure 11-7 shows the next simulation waveform when the stack is full. The FULLn output is asserted after the last empty location is filled. The next PUSH instruction overwrites the value at the top of the stack 12 with 13. The stack is then popped. Note that the stack value shows 11. FULLn output is deasserted.

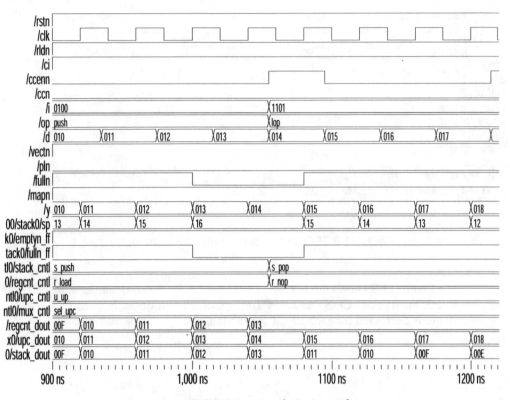

Figure 11-7 TBCHIP2910 simulation waveform - 2.

Figure 11-8 shows the next waveform where the stack is popped to empty. The EMPTYn flag is asserted. The stack pointer SP remains at 0. The stack value remains at 2 which is the first value being pushed.

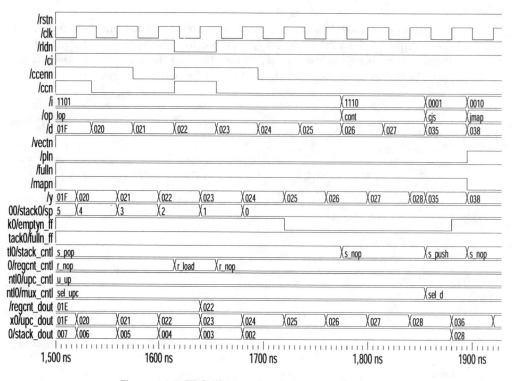

Figure 11-8 TBCHIP2910 simulation waveform — 3.

Figure 11-9 shows the next waveform where the register/counter is loaded with value 3. Instructions CONT and RFCT are executed four times. Note that the register/counter is decreased for each RFCT instruction.

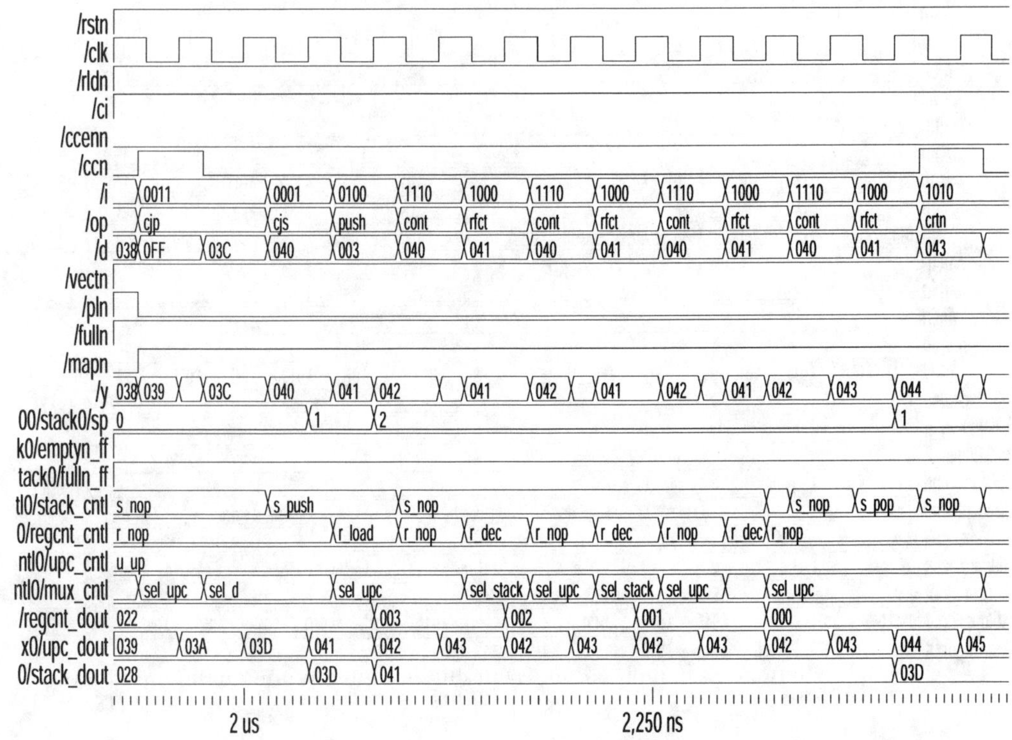

Figure 11-9 TBCHIP2910 simulation waveform — 4.

Figure 11-10 shows the next waveform where the CRTN, CJS, LDCT, RPCT, CJV, JRP, and JSRP instructions are tested. You are encouraged to verify the correct operation of each instruction.

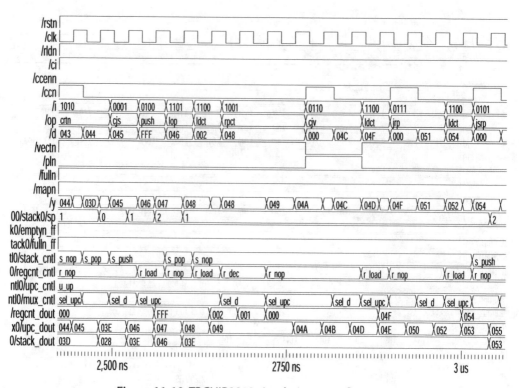

Figure 11-10 TBCHIP2910 simulation waveform — 5.

Figure 11-11 shows the last simulation waveform for the test microprogram. The three-way branch instruction TWB is tested. Finally, CI input is set to 0. Note that the microprogram stays at the same value.

There may be many functions or combinations of instructions that are not tested by this test case. This is left as an exercise. We will continue with our description of the design process.

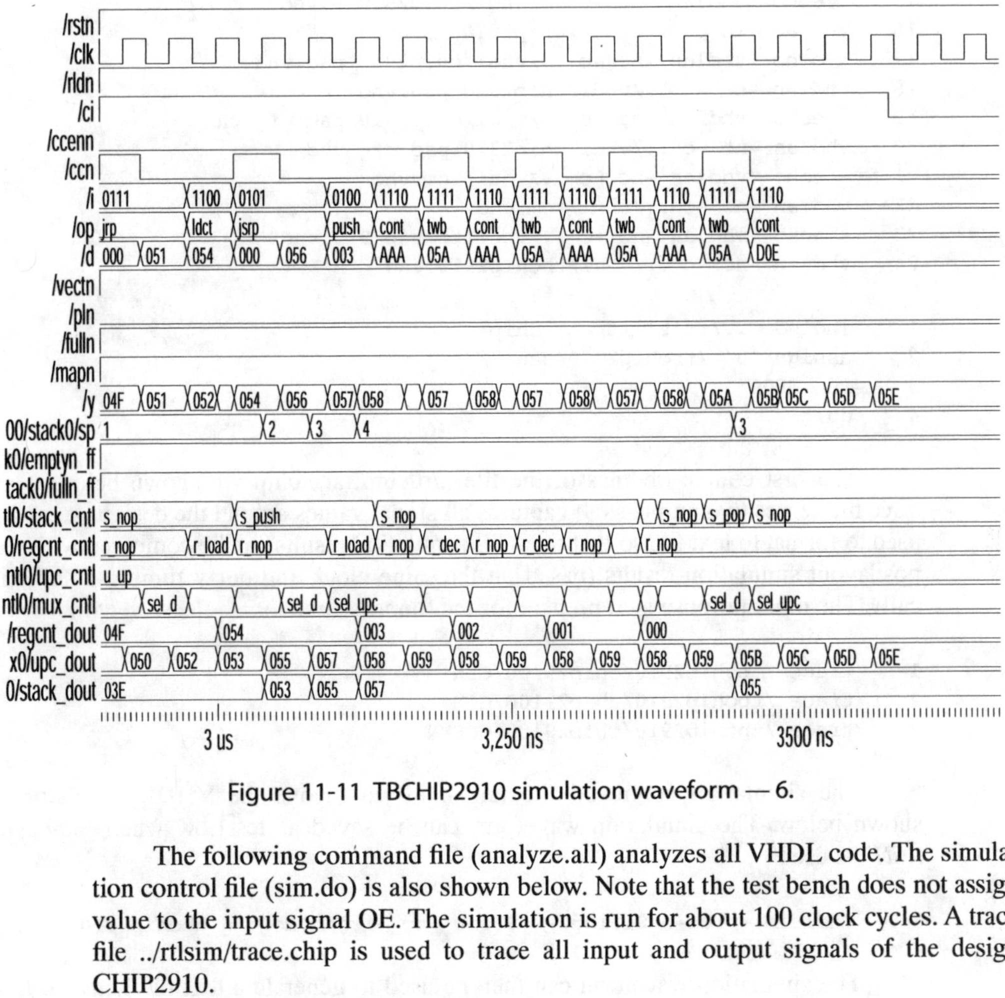

Figure 11-11 TBCHIP2910 simulation waveform — 6.

The following command file (analyze.all) analyzes all VHDL code. The simulation control file (sim.do) is also shown below. Note that the test bench does not assign value to the input signal OE. The simulation is run for about 100 clock cycles. A trace file ../rtlsim/trace.chip is used to trace all input and output signals of the design CHIP2910.

```
1       echo "vhdlan -nc -c ../vhdl/ampack.vhd"
2    vhdlan -nc -c ../vhdl/ampack.vhd
3       echo "vhdlan -nc -c ../vhdl/cntl.vhd"
4    vhdlan -nc -c ../vhdl/cntl.vhd
5       echo "vhdlan -nc -c ../vhdl/mux.vhd"
6    vhdlan -nc -c ../vhdl/mux.vhd
7       echo "vhdlan -nc -c ../vhdl/regcnt.vhd"
8    vhdlan -nc -c ../vhdl/regcnt.vhd
9       echo "vhdlan -nc -c ../vhdl/stack.vhd"
10   vhdlan -nc -c ../vhdl/stack.vhd
11      echo "vhdlan -nc -c ../vhdl/upc.vhd"
```

```
12    vhdlan -nc -c ../vhdl/upc.vhd
13      echo "vhdlan -nc -c ../vhdl/core2910.vhd"
14    vhdlan -nc -c ../vhdl/core2910.vhd
15      echo "vhdlan -nc -c ../vhdl/chip2910_e.vhd"
16    vhdlan -nc -c ../vhdl/chip2910_e.vhd
17      echo "vhdlan -nc -c ../vhdl/chip2910_pad.vhd"
18    vhdlan -nc -c ../vhdl/chip2910_pad.vhd
19      echo "vhdlan -nc -c ../vhdl/chip2910_pad_cfg.vhd"
20    vhdlan -nc -c ../vhdl/chip2910_pad_cfg.vhd
21      echo "vhdlan -nc -c ../vhdl/rom.vhd"
22    vhdlan -nc -c ../vhdl/rom.vhd
23      echo "vhdlan -nc -c ../vhdl/tbchip2910.vhd"
24    vhdlan -nc -c ../vhdl/tbchip2910.vhd

1     include ../rtlsim/trace.chip
2     assign '0' /tbchip2910/oen
3     run 3800
4     quit
```

The first couple of lines of the file ../rtlsim/trace.chip are shown below. This trace file serves two purposes. It captures all signal values around the design. It can be used to format to text file so that the RTL simulation results can be compared with the postlayout simulation results (based on the same clock and delay timing) automatically. The other purpose is to prepare for the foundry functional release vector.

```
1     trace /tbchip2910/chip29100/CLK
2     trace /tbchip2910/chip29100/CCN
3     trace /tbchip2910/chip29100/CCENN
```

The simulation command file (sim.cmd) to invoke the VHDL simulator is shown below. The simulation waveform can be saved as test1.ow when "*sim.cmd test1*" is used.

```
1     vhdlsim -pro -o $1.log -w $1.ow -nc -i sim.do cfg_tbchip2910
```

The simulation waveform can then be used to generate a text file with the following command file (wif2tab.cmd). The waveform is strobed every 40 ns, starting at time 38 ns. The wif2tab command uses a format file to specify the order of the signals to appear in the vector file.

```
1     wif2tab -f ../rtlsim/wif2tab.fmt -nc -nh -offset 38 -strobe 40
$1.ow > $1.tmp
2     sed -e 's/H/1/g' -e 's/L/0/g' -e 's/U/1/g' < $1.tmp > $1.vec
3     rm -f $1.tmp
```

The following shows the first couple lines of the wif2tab.fmt file. The wif2tab.cmd file also replaces characters 'H', 'L', 'U', to '1', '0', and '1', respectively.

```
1       /tbchip2910/chip29100/CLK;
2       /tbchip2910/chip29100/CCN;
3       /tbchip2910/chip29100/CCENN;
```

The following shows the first couple of lines of the vector text file. Note that the time stamps 38, 78, and 118 are starting at time 38 ns and every 40 ns afterward. This vector text file is then used to automatically compare with the postsynthesis and the postlayout simulation results for discrepancies.

```
1            38     0111111111111111111111101011XXXXXXXXXXXX
2            78     0000000000000000000000001011000000000000
3           118     0000000000000000000000001011000000000000
```

11.4 DESIGN SYNTHESIS

The following synthesis script file (compile.core2910.hier) is used to synthesize the CORE2910 block. It reads all lower-level blocks' VHDL codes and core2910.vhd in lines 3 to 9. It synthesizes CORE2910 block with lower-level block boundaries kept intact. Certainly, there are other ways to synthesize this design. They are left as exercises.

```
1     sh date
2     design = core2910
3     read -f vhdl VHDLSRC + "ampack.vhd"
4     read -f vhdl VHDLSRC + "upc.vhd"
5     read -f vhdl VHDLSRC + "regcnt.vhd"
6     read -f vhdl VHDLSRC + "mux.vhd"
7     read -f vhdl VHDLSRC + "stack.vhd"
8     read -f vhdl VHDLSRC + "cntl.vhd"
9     read -f vhdl VHDLSRC + design + ".vhd"
10
11    include compile.common
12
13    create_clock "CLK*" -name clk -period CLKPERIOD -waveform {0.0
HALFCLK}
14    set_clock_skew -uncertainty CLKSKEW clk
15    set_input_delay CLKPER7 -add_delay -clock clk {CI RLDn CCn
CCENn I D}
16    set_output_delay CLKPER1 -add_delay -clock clk all_outputs()
17    set_dont_touch_network {CLK }
18    set_scan_style multiplexed_flip_flop
19    group_path -name fromi -from {I }
20    group_path -name fromCCn -from {CCn }
21    group_path -name fromCCEn -from {CCENn}
22    compile -map_effort medium
23    create_schematic -all
```

```
24    write -f db -hier -o DBSRC + design + ".hier.db"
25    remove_design -design
26    remove_license VHDL-Compiler
27    sh date
```

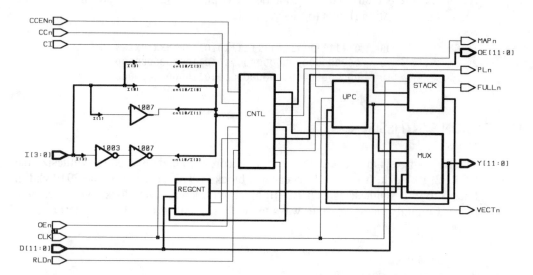

Figure 11-12 Synthesized schematic for CORE2910.

The clock period used for synthesis is 30 ns. The longest timing path between flip-flops is about 28 ns. The input signal timings are also important. The design is flattened and the following command is used to generate a timing report. The design has an area of 3,333 gates.

```
1     report_timing -from I -to *reg*/D \
2        > ../report/core.time
3     report_timing -from {CCn CCENn RLDn} -to *reg*/D \
4        >> ../report/core.time
5     report_timing -from *reg*/Q -to *reg*/D \
6        >> ../report/core.time
7     report_timing -from *reg*/QN -to *reg*/D \
8        >> ../report/core.time
9     report_timing -from I -to {Y} \
10       >> ../report/core.time
11    report_timing -from I -to {MAPn VECTn PLn} \
12       >> ../report/core.time
13    report_timing -from {RLDn CCn CCENn} -to {MAPn VECTn PLn} \
14       >> ../report/core.time
```

The following synthesis command file is used to fix the minimum timing paths. The area is increased to 3,569 gates.

```
1      /* Set operating conditions and wire load model */
2      current_design CORE2910
3      include compile.common
4      set_wire_load "estimated"
5      set_operating_conditions quick_min
6      set_dont_touch_network {CLK }
7      set_drive 0             {CLK }
8      CLKSKEW   = 1
9      create_clock "CLK_BUF*" -name clk -period CLKPERIOD -waveform
{0.0 HALFCLK}
10     set_dont_touch_network all_inputs()
11     set_dont_touch_network all_outputs()
12     set_clock_skew -ideal -uncertainty CLKSKEW clk
13     update_timing
14     set_fix_hold clk
15     compile -prioritize_min_paths -only_design_rule >
fix_hold.trans
16     write -f db -hier -o ../db/core2910_fixhold.db
```

The following synthesis script file (compile.chip2910) is used to bring in the top-level block.

```
1      sh date
2      design = chip2910
3      analyze -f vhdl VHDLSRC + design + "_e.vhd"
4      analyze -f vhdl VHDLSRC + design + "_pad.vhd"
5      elaborate CHIP2910 -arch PAD
```

Figure 11-13 shows the CHIP2910 schematic. The CORE2910 block is shown in the center. There are 2 cells for all input signals, except for the clock input that has 3 cells. Each output has one cell. The following synthesis command is used to generate a VHDL netlist with two SDF timing files, one for the worst case, the other for the best case.

```
1      ungroup -all
2      change_names -rule vhdl
3      current_design CHIP2910
4      vhdlout_top_configuration_arch_name = "structure"
5      vhdlout_write_top_configuration = "FALSE"
6      vhdlout_single_bit = "USER"
7      vhdlout_preserve_hierarchical_types = "USER"
8      vhdlout_use_packages = { \
9         IEEE.STD_LOGIC_1164.ALL \
10        QYH500.COMPONENTS.ALL \
11        }
12     write -f vhdl -hier -o ../strvhdl/chip2910.str.vhd
```

```
13    set_wire_load "estimated"
14    set_operating_conditions quick_max
15    write_timing -format sdf -output ../strvhdl/chip2910.sdf.max -
context vhdl
16    set_operating_conditions quick_min
17    write_timing -format sdf -output ../strvhdl/chip2910.sdf.min -
context vhdl
```

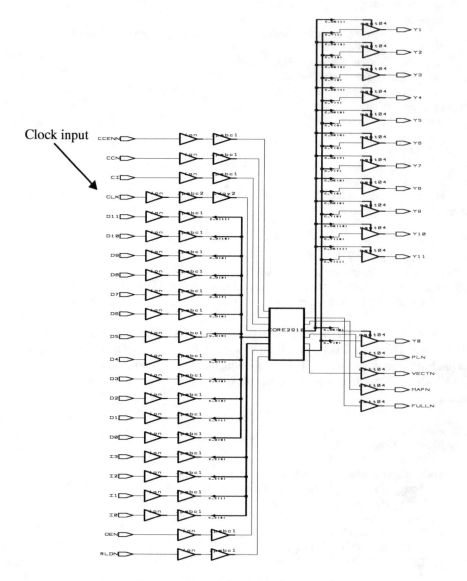

Figure 11-13 Synthesized schematic for CHIP2910.

The netlist is generated after the design is ungrouped. This is done because the design is small. In line 2, the names are changed using the rule "vhdl". This replaces special characters such as "(", "[", ")", "]" with other characters. For example, a signal core29100xMUX_CNTLx0x is used which has the original name of core29100/ MUX_CNTL[0] inside the Design Analyzer. The following shows the generated VHDL netlist for the CHIP2910 entity. The complete netlist has 5,861 lines.

```
1     library IEEE;
2     library QYH500;
3     use IEEE.STD_LOGIC_1164.ALL;
4     use QYH500.COMPONENTS.ALL;
5     entity CHIP2910 is
6        port( CLK, CCN, CCENN, RLDN, CI, D11, D10, D9, D8,
7               D7, D6, D5, D4, D3, D2,
8               D1, D0, I3, I2, I1, I0, OEN : in std_logic;
9               FULLN, PLN, MAPN, VECTN, Y11, Y10, Y9, Y8, Y7,
10              Y6, Y5, Y4, Y3, Y2, Y1, Y0 : out std_logic);
11    end CHIP2910;
```

SDF files are also generated, each has 23252 lines. The following shows a couple of lines of the SDF file. The SDF file consists of interconnect delays (line 18) and cell delays (line 30).

```
1     (DELAYFILE
2     (SDFVERSION "OVI 1.0")
3     (DESIGN "CHIP2910")
4     (DATE "Sat Mar 28 14:42:07 1998")
5     (VENDOR "qyh500m")
6     (PROGRAM "Synopsys Design Compiler cmos")
7     (VERSION "1997.01-44683")
8     (DIVIDER /)
9     (VOLTAGE 4.50:4.50:4.50)
10    (PROCESS)
11    (TEMPERATURE 125.00:125.00:125.00)
12    (TIMESCALE 1ns)
13    (CELL
14      (CELLTYPE "CHIP2910")
15      (INSTANCE)
16      (DELAY
17        (ABSOLUTE
18        (INTERCONNECT core29100xstack0xSP_regx4x/QN
core29100xU241/A (0.003:0.003:0.003))
19          (INTERCONNECT core29100xU241/Z core29100xU240/A
(0.002:0.002:0.002))
20          (INTERCONNECT core29100xstack0xSP_regx3x/QN
core29100xU239/A (0.003:0.003:0.003))
21          (INTERCONNECT core29100xU239/Z core29100xU238/A
(0.002:0.002:0.002))
```

```
22          (INTERCONNECT core29100xstack0xSP_regx2x/QN
core29100xU237/A (0.002:0.002:0.002))
23          (INTERCONNECT core29100xU237/Z core29100xU236/A
(0.002:0.002:0.002))
24     ... skipped many many lines ...
25     (CELL
26       (CELLTYPE "ipsbc1")
27       (INSTANCE d1level)
28       (DELAY
29         (ABSOLUTE
30         (IOPATH A Z (1.256:1.256:1.256) (3.504:3.504:3.504))
31         )
32       )
33     )
```

11.5 POSTSYNTHESIS TIMING VERIFICATION

The above two SDF files have their timing based on the synthesis results. They can be used for the postsynthesis, prelayout simulation. This verifies whether the netlist is generated correctly. The following command file (analyze.all) in the gatesim directory is used to analyze all VHDL code.

```
1       echo "vhdlan -nc -c ../vhdl/ampack.vhd"
2    vhdlan -nc -c ../vhdl/ampack.vhd
3       echo "vhdlan -nc -c ../strvhdl/chip2910.str.vhd"
4    vhdlan -nc -c ../strvhdl/chip2910.str.vhd
5       echo "vhdlan -nc -c ../vhdl/chip2910_str_cfg.vhd"
6    vhdlan -nc -c ../vhdl/chip2910_str_cfg.vhd
7       echo "vhdlan -nc -c ../vhdl/rom.vhd"
8    vhdlan -nc -c ../vhdl/rom.vhd
9       echo "vhdlan -nc -c ../vhdl/tbchip2910.vhd"
10   vhdlan -nc -c ../vhdl/tbchip2910.vhd
```

The following command file (cim.cmd) is used to run VHDL simulation two times: one for the worst-case timing and the other for the best-case timing. Two simulation waveform files, testmax.ow and testmin.ow, are generated.

```
1    vhdlsim -pro -o testmax.log -w testmax.ow \
2       -sdf_af -sdf_max -sdf_top /tbchip2910/chip29100 \
3       -sdf ../strvhdl/chip2910.sdf.max \
4       -nc -i ../rtlsim/sim.do cfg_tbchip2910
5    vhdlsim -pro -o testmin.log -w testmin.ow \
6       -sdf_af -sdf_min -sdf_top /tbchip2910/chip29100 \
7       -sdf ../strvhdl/chip2910.sdf.min \
8       -nc -i ../rtlsim/sim.do cfg_tbchip2910
```

The same wif2tab.cmd file in the ../rtlsim directory can be used to generate vector text files. They are then compared with the RTL simulation vector text file. They are identical, which implies that the functionality of the RTL VHDL code is the same as the postsynthesis results with the SDF timing overlaid.

11.6 PREPARING RELEASE FUNCTIONAL VECTORS

The design netlist released to the foundry requires test vectors so that the fabricated chip can be tested to ensure quality. Many foundries require various test vector formats. There are many tools and design processes to generate test vectors. Here, we will illustrate a simple test vector preparation process. Hopefully, you may use a similar process in preparing your test vectors.

We have verified that the simulation results between the RTL and postsynthesis simulations. Since RTL can be run faster, we will use the RTL simulation results to generate the test vector. Chip Express requires the test vector in 1,000 ns of clock period. We changed the PERIOD and DELAY to 1,000 ns and 200 ns in the TBCHIP2910 test bench. Run the RTL simulation and save the simulation waveform. The following wif2tab command file can be used. The wif2tab command file is similar to the one above, except for the signal values which are strobed every 500 ns starting at 490 ns. The purpose of this is to strobe 10 ns before the falling and rising of the clock.

```
1    wif2tab -f ../rtlsim/wif2tab.fmt -nc -nh -strobe 500 -offset
490 $1.ow > $1.tmp
2    sed -e 's/H/1/g' -e 's/L/0/g' -e 's/U/1/g' < $1.tmp > $1.vec
3    rm -f $1.tmp
```

A text vector file (test1.vec) is generated. The following shows the first five lines.

```
1        490    11111111111111111111101011XXXXXXXXXXXX
2        990    01111111111111111111101011XXXXXXXXXXXX
3       1490    10000000000000000000001011000000000000
4       1990    00000000000000000000001011000000000000
5       2490    10000000000000000000001011000000000000
```

Chip Express supports a tool called **ytv2ctv**. The following shows the command. It takes the **test1.vec** file with a format file **pinformat.txt** to generate the **test1.ctv** file. The pinformat.txt file is shown below. Basically, all signals are defined as "INPUT" or "OUTPUT". For other designs, there may be bidirectional signals. The tester needs to know the three-state enable signals so it can determine when it should drive the bidirectional signals, and when the bidirectional signals should be compared.

```
1    ytv2ctv -no_opt -c pinformat.txt -i test1.vec -o test1.ctv
```

```
 1    BEGIN_SIGNALS
 2    CLK       INPUT
 3    CCN       INPUT
 4    CCENN     INPUT
 5    RLDN      INPUT
 6    CI        INPUT
 7    D11       INPUT
 8    D10       INPUT
 9    D9        INPUT
10    D8        INPUT
11    D7        INPUT
12    D6        INPUT
13    D5        INPUT
14    D4        INPUT
15    D3        INPUT
16    D2        INPUT
17    D1        INPUT
18    D0        INPUT
19    I3        INPUT
20    I2        INPUT
21    I1        INPUT
22    I0        INPUT
23    OEN       INPUT
24    FULLN     OUTPUT
25    PLN       OUTPUT
26    MAPN      OUTPUT
27    VECTN     OUTPUT
28    Y11       OUTPUT
29    Y10       OUTPUT
30    Y9        OUTPUT
31    Y8        OUTPUT
32    Y7        OUTPUT
33    Y6        OUTPUT
34    Y5        OUTPUT
35    Y4        OUTPUT
36    Y3        OUTPUT
37    Y2        OUTPUT
38    Y1        OUTPUT
39    Y0        OUTPUT
40    END_SIGNALS
```

The top of the test1.ctv is shown below. Line 2 defines the clock cycle to be 1000 ns. Line 3 defines the delay time for the signal CLK is 0 ns. All other input signal have a delay of 200 ns as shown in lines 5 to 8. All output signals are strobed at 800 ns as shown in lines 9 to 11. There are no bidirectional signals. Five vectors are shown here. Note that input and output signals are separated.

```
1     @TIMING
2        CYCLE 1000
3        INTIM DT = 0
4        'CLK'
5        INTIM DT = 200
6        'CCN' 'CCENN' 'RLDN' 'CI' 'D11' 'D10' 'D9' 'D8' 'D7'
7        'D6' 'D5' 'D4' 'D3' 'D2' 'D1' 'D0' 'I3' 'I2' 'I1'
8        'I0' 'OEN'
9        OUTTIM STB = 800
10       'FULLN' 'PLN' 'MAPN' 'VECTN' 'Y11' 'Y10' 'Y9' 'Y8' 'Y7' 'Y6'
11       'Y5' 'Y4' 'Y3' 'Y2' 'Y1' 'Y0'
12       BUSTIM DT = 200 STB = 0
13    @ENDTIM
14    @SIGNAL
15    #VECTOR/CYCLE
16    ORDER
17    +  'CLK' 'CCN' 'CCENN' 'RLDN' 'CI' 'D11' 'D10' 'D9' 'D8' 'D7'
18    +  'D6' 'D5' 'D4' 'D3' 'D2' 'D1' 'D0' 'I3' 'I2' 'I1'
19    +  'I0' 'OEN'
20    +
21    +  'FULLN' 'PLN' 'MAPN' 'VECTN' 'Y11' 'Y10' 'Y9' 'Y8' 'Y7' 'Y6'
22    +  'Y5' 'Y4' 'Y3' 'Y2' 'Y1' 'Y0'
23
24    1111111111111111111110    1011XXXXXXXXXXXX
25    0111111111111111111110    1011XXXXXXXXXXXX
26    100000000000000000000     1011000000000000
27    000000000000000000000     1011000000000000
28    100000000000000000000     1011000000000000
29    000000000000000000000     1011000000000000
```

The above test1.ctv file will be sent to the foundry along with the design netlist. How do we know that the vectors are generated correctly? Chip Express has another tool called **testvhd**. The following shows the testvhd.cmd file. It takes the netlist and the test1.ctv file and generates VHDL test bench **testbench.vhd**, input vector file **test1vec.in**, and output vector file **test1vec.out**.

```
1     testvhd -tv test1.ctv -netlist \
2     ../strvhdl/chip2910.str.vhd -family qyh500 -test.vhd \
3     testbench.vhd -vhd.in test1vec.in -vhd.out \
4     test1vec.out -top_module CHIP2910
```

A file run_quick_vss is also generated as shown below. It analyzes the netlist and the new test bench VHDL code. The test bench basically reads in the **test1vec.in** file and then compares the design output with the expected **test1vec.out** file.

```
1     #!/bin/csh
2
3     # TOOL        : QUICK testvhd program
```

```
4    # TIMESTAMP    : Mar 25 16:51:42 1998
5
6    vhdlan ../strvhdl/chip2910.str.vhd testbench.vhd
7    vhdlsim -nc -i run.inc  -tres 0.01 t_test_CHIP2910
```

The above command generates an error report file vhd.log as shown below. The first two vectors are not compared. This is okay since the design has not been initialized in the first two vectors.

```
1                                    FPMVYYYYYYYYYYYY
2                                    ULAE119876543210
3                                    LNPC10
4                                    L NT
5                                    N  N
6    ERROR expected output word = 1011XXXXXXXXXXXX   $1
7    ERROR output word          = X                  $1
8    ERROR expected output word = 1011XXXXXXXXXXXX   $2
9    ERROR output word          = X                  $2
10   *****Simulation completed with   2 errors.*****
```

The package information must also be specified for the release. The following shows the package file **package.txt**. It specifies package LDCC44. Each signal is assigned a unique pin number. POWER and GROUND pins are also specified.

```
1    PACKAGE = LDCC44
2    # pin number        signal name
3    # ----------        -----------
4    1                   Y4
5    2                   D4
6    3                   Y5
7    4                   D5
8    5                   VECTN
9    6                   GROUND pad
10   7                   PLN
11   8                   MAPN
12   9                   I3
13   10                  I2
14   11                  POWER pad
15   12                  I1
16   13                  I0
17   14                  CCENN
18   15                  CCN
19   16                  RLDN
20   17                  GROUND pad
21   18                  FULLN
22   19                  D6
23   20                  Y6
24   21                  D7
25   22                  Y7
```

26	23	D8
27	24	Y8
28	25	D9
29	26	Y9
30	27	D10
31	28	Y10
32	29	D11
33	30	Y11
34	31	OEN
35	32	POWER pad
36	33	GROUND pad
37	34	CLK
38	35	CI
39	37	Y0
40	36	POWER pad
41	38	D0
42	39	Y1
43	40	D1
44	41	Y2
45	42	D2
46	43	Y3
47	44	D3

The following command **tar.cmd** prepares all the design data for release. It has the package information, pin format, test vectors, and the netlist. Ship it.

```
1   tar -cvf changbook.tar package.txt pinformat.txt test1.ctv ../
strvhdl/chip2910.str.vhd
```

11.7 POSTLAYOUT VERIFICATION

The author would like to acknowledge Ramagopal Madamala at Chip Express for providing support in running through the place and route tools. The post place and route design database is transferred electronically through file transfer protocol (FTP). The design database consists of a VHDL structural netlist, an SDF timing file, and many summary reports (such as timing setup and hold time violation, test vectors, pin table, etc.). The report files are examined and no errors are found. The VHDL structural netlist is analyzed with the same test bench. The SDF timing file is overlaid for running the same test bench. The results are captured and compared with the RTL and prelayout VHDL codes. The design is ready to be signed off for fabrication.

11.8 DESIGN MANAGEMENT

To effectively manage the design, the design directory has the following subdirectories:

db:

This stores Synopsys synthesized design database.

gatesim, gatesim.work:

Directory gatesim is used to run the structural VHDL simulation with gatesim.work as the VHDL working library.

release:

This directory stores all design data for shipping to the fabrication foundry such as the netlist, test vectors, and pin description.

strvhdl:

This contains structural VHDL codes.

fpga:

This directory is used to map the design into FPGA.

qsim, qsim.work:

Directory qsim is used to run Mentor QuickVHDL VHDL simulation with qsim.work as the VHDL working library.

report:

This contains Synopsys synthesis reports.

syn, syn.work:

Directory syn is used to run the Synopsys synthesis. It contains synthesis scripts. Directory syn.work is the corresponding VHDL working library directory.

fromchipx:

This directory contains postlayout design database from the foundry.

rtlsim, rtlsim.work:

Directory rtlsim is used to run RTL VHDL simulation with Synopsys VSS VHDL simulator with rtlsim.work as the VHDL working library.

vhdl:

This directory contains all RTL VHDL source codes.

The directory structure shown here is an example. As we have discussed in Chapter 3, there are many ways to organize the directory structure. The goal is to present design information in an organized way. There are certainly many other ways to achieve this goal.

11.9 EXERCISES

11.1 Refer to the microprogram and simulation waveforms. Verify that the microprogram is executed correctly.

11.2 What additional test programs can you suggest to test the design more thoroughly? How would you organize all your test programs effectively? How does that change the simulation commands and setups when many microprograms are to be verified?

11.3 Refer to the microprogram text file. It is written manually. It is error prone and time consuming to write such microprograms. What do you suggest to solve this problem?

11.4 Refer to the compile.core2910.hier synthesis command file. Since the design is small, it can be flattened without too much trouble. Modify the command file to flatten the design before compile. Compare the synthesis results based on these two approaches.

11.5 Another way to synthesize the design is a bottom-up approach. Develop a synthesis script file to synthesize the design. Compare the synthesis results with the approaches suggested in the last exercise.

11.6 Refer to the process by which minimum paths are fixed with the synthesis script file fixhold.scr. In design CHIP2910, where may minimum paths exist to cause the setup and hold time violation?

11.7 AM2910 has a companion, ALU chip AM2901. Obtain the AM2901 data book. Develop VHDL code. Verify and synthesize the design. Develop a test bench to combine AM2910 and Am2901 designs to verify both designs.

11.8 Target the design to a FPGA if you have access to FPGA synthesis and FPGA place and route tools.

Chapter 12

Error Detection And Correction

This chapter will cover the concepts of error detection and correction (EDAC) for binary code. An actual EDAC integrated circuit design is described and designed with VHDL. The VHDL code is synthesized to an FPGA.

12.1 ERROR DETECTION AND CORRECTION CODE

The information stored in digital computers uses codes, which are simply sets of symbols so that meanings or values are attached. When the information is transmitted from one place to another, errors may occur. Even the information stored in a memory is subject to error. For example, a memory cell may be stuck at a certain value. In the radiation environment, the storage element may be upset by the radiation particles. Therefore, the ability to detect and to correct the error in the codes is important for high reliability digital systems such as flight critical systems in airplanes, space shuttles, and missile electronic systems.

Hamming first published his paper "Error Detecting and Error Correction Codes" in *Bell System Technical Journal*, Vol. 29, No. 2, April 1950, pp. 147–160. In order to detect and correct errors, Hamming used the term *systematic* code. Each code has exactly n bits, where m bits are associated with the information while the other $k = n - m$ bits are used for error detection and correction. The efficiency of the code is measured by the *redundancy ratio* $R = n/m$. It is the ratio of the number of bits used (m bits) necessary to convey the same information (m bits).

12.2 SINGLE ERROR DETECTING CODES

The purpose of the *single error detecting code* is to be able to detect whether there exists one error in the code. The first n-1 bits can be used for the information. The nth bit can be placed as '0' or '1' so that the total number of the n bits have an even number of '1's. This is also referred to as the *even parity*. Whenever an error occurs in a single bit, the total number of '1's in the code will be an odd number. The redundancy ratio $R = n/(n-1) = 1 + 1/(n-1)$.

To get a low redundancy ratio from the above formula, *n* can be increased. However, increasing *n* will increase the probability of at least one error in the code. This results in a double-bit error that would pass the parity check and the error would be undetected.

The single error detection is often called *parity check*. The *even parity* is to maintain the total number of '1's as an even number. The *odd parity* is to maintain the total number of '1's as an odd number.

12.3   SINGLE ERROR CORRECTING CODES

Assume that we have *m*-bit of information. We want to have another *k* parity bits. In order to correct any single bit in the information, the *k* parity bits should be able to uniquely give a *check number* to indicate the position of the bit which has error. Therefore, the check number must describe $m + k + 1$ different things so that $2**k >= m + k + 1$. Intuitively, *k* parity bits can have $2**k$ different check numbers. The different check number must be able to cover the code $(m + k)$ bits with 1 more to indicate that no error is detected.

Since $n = m + k$, $2**k >= m + k + 1$ can be written as $2**(n - m) >= n + 1$. $2**m <= 2**n/(n + 1)$. Figure 12-1 shows the table with *n*, *m*, and *k* values which gives the maximum *m* for a given *n* or the minimum *n* for a given *m*.

n	1	2	3	4	5	6	7	8	9	10	11	12	13	14	15	16
m	0	0	1	1	2	3	4	4	5	6	7	8	9	10	11	11
k	1	2	2	3	3	3	3	4	4	4	4	4	4	4	4	5

Figure 12-1 Single error correcting table.

It remains to decide where these *k* parity bits are placed in the code. Since they are used to indicate a position check number, a natural choice would be at positions 1, 2, 4, 8, . . . for check bit positions. The remaining bits are for the information bits. The check bit at position 1 will check all the position numbers which are represented as a binary number having '1' at bit position 1. For example, all the odd-numbered positions would have '1' in the position 1. Parity bit at position 4 would check positions 4, 5, 6, 7, 13, 14, 15. Figure 12-2 shows the parity bit position and the positions that are checked by the parity. For example, suppose that we have 4 bits of information to represent numbers 0 to 15. From Figure 12-1, we need 3 parity bits. They will be at position 1, 2, and 4. The remaining positions are for the information bits. Figure 12-3 shows the 7-4 code. Note that even parity is used. For example, the information code 12, parity bit at position 4 would cover bit positions 4, 5, 6, 7 with even parity. Bits 4, 5, 6, and 7 would be '1', '1', '0', and '0', respectively. Parity bit at position 2 would cover bit positions 2, 3, 6, and 7. Their values would be '1', '1', '0', and '0', respectively. Parity bit at position 1 would cover bit positions 1, 3, 5, and 7. They would be '0', '1', '1', and '0', respectively.

Parity bit	Parity bit position	Position checked
1	1	1,3,5,7,9,11,13,15
2	2	2,3,6,7,10,11,14,15
3	4	4,5,6,7,12,13,14,15
4	8	8,9,10,11,12,13,14,15

Figure 12-2 Single error correcting parity bit positions.

Position							Decimal value
1	2	3	4	5	6	7	
0	0	0	0	0	0	0	0
1	1	0	1	0	0	1	1
0	1	0	1	0	1	0	2
1	0	0	0	0	1	1	3
0	1	0	0	1	0	0	4
1	0	0	1	1	0	1	5
0	0	0	1	1	1	0	6
1	1	0	0	1	1	1	7
0	0	1	1	0	0	0	8
1	1	1	0	0	0	1	9
0	1	1	0	0	1	0	10
1	0	1	1	0	1	1	11
0	1	1	1	1	0	0	12
1	0	1	0	1	0	1	13
0	0	1	0	1	1	0	14
1	1	1	1	1	1	1	15

information bits

parity bits

Figure 12-3 7-4 Single error correcting code.

Now, let's see how the error can be corrected. Suppose that information 12 with code "0111100" is sent and bit 3 is reversed. The receiving end receives the code "0101100". The same algorithm can be used to generate the parity. The parity bit at position 4 would be '1' which is the same. The parity bit at position 1 would check bit positions 1, 3, 5, and 7 which would get '1'. The parity bit position 2 would check bit positions 2, 3, 6, and 7 which would get '1'. The parity bit position 4 would check bit positions 4, 5, 6, and 7 which would get '0'. By putting the generated parity bits '0', '1', and '1' together, "011" indicates that the bit at position 3 is wrong, and it can be reversed to correct the error.

12.4 SINGLE ERROR CORRECTING AND DOUBLE ERROR DETECTING CODES

To construct a single error correcting with double error detecting code, we begin with a single error correcting code. One more position is added to check all previous positions, using an even parity check. Three cases are examined below:

- *No errors.* All parity checks, including the last, are satisfied.

- *Single error.* The last parity check fails in all cases no matter where the error occurs in the information, original check positions, or the last check position. The original checking number gives the position of the error, where now the zero value indicates the last check position.

- *Two errors.* The last parity check is satisfied, and the checking number indicates some kind of error, but there is no way to correct the errors.

Position								Decimal value
0	1	2	3	4	5	6	7	
0	0	0	0	0	0	0	0	0
0	1	1	0	1	0	0	1	1
1	0	1	0	1	0	1	0	2
1	1	0	0	0	0	1	1	3
1	1	0	0	1	1	0	0	4
1	0	1	0	0	1	0	1	5
0	1	1	0	0	1	1	0	6
0	0	0	0	1	1	1	1	7
1	1	1	1	0	0	0	0	8
0	0	0	1	0	0	0	1	9
0	1	0	1	1	0	1	0	10
0	0	1	1	0	0	1	1	11
0	0	1	1	1	1	0	0	12
0	1	0	1	0	1	0	1	13
1	0	0	1	0	1	1	0	14
1	1	1	1	1	1	1	1	15

information bits (shaded)

parity bits

Figure 12-4 Single error correcting and double error detecting code.

Figure 12-4 shows the same table as in Figure 12-3, except that parity bit position 0 is added to ensure that whole 8-bit code is even parity. The single-bit error can be corrected as above and the parity at bit position would be reversed. For double-bit errors, parity at bit position would be the same while the parity bits at position 1, 2, and 4 would indicate a non-zero value. For example, decimal code 12 is "00111100". Suppose that both bits at position 5 and 7 are wrong, the code becomes "00111001". Parity at bit position would be '0' which is the same. Parity at bit position 1 would get '0'. Parity at bit position 2 would get '1'. Parity at bit position 4 would get '0'. "010" is a nonzero value and the parity at bit position 0 is fine. This detects a double bit error. It is possible that other combinations of double-bit errors would result in the same bit pattern of "010".

12.5 ERROR DETECTING AND CORRECTING CODE DESIGN EXAMPLE

Now we have studied the basic ideas of the error detecting and correcting code. Let's see how the idea can be applied. TI has designed SN54ALS616 IC. It is a 16-bit parallel error detection and correction circuit. The data word is 16-bit. There are 6 parity bits for the single bit error correction and double-bit error detection. Figure 12-5 shows the block diagram.

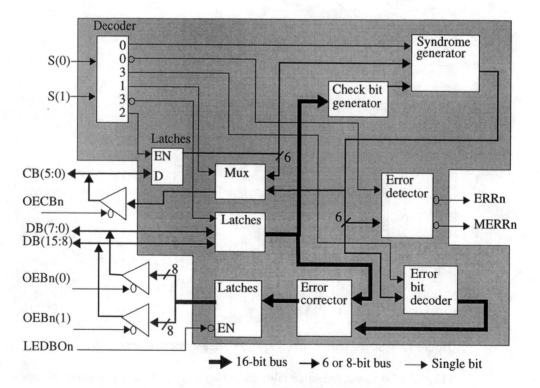

Figure 12-5 TI SN54ALS616 block diagram.

Error detection is accomplished as the input 6-bit check word and the 16-bit data word is applied to the internal parity generator and checkers. If all 6 bits of the generated parity match the input 6-bit check, it is assumed that no error has occurred and both error flags (ERRn and MERRn) will be high.

If the parity of one or more parity bits do not match, an error has occurred and the proper error flag or flags will be asserted low. If a single-bit error is detected, ERRn is asserted low. Since the single-bit error can be corrected, MERRn is not asserted low. The data word would be correct. A 2-bit error is not correctable, both ERRn and MERRn would be asserted low. Three or more simultaneous bit errors can cause this circuit to believe that there is no error.

The following shows the VHDL entity in file **edac616_e.vhd**. Input signal S (in line 8) has 2 bits to control the mode of the circuit operation. CB (in line 9) is a bi-directional signal 6-bit check bits signal. Its direction is controlled by input signal OECBn. DB (in line 11) is a bi-directional 16-bit data signal. Its direction is controlled by input signal OEBn (in line 12). Input signal LEDBOn (in line 13) controls whether the data is latched after it has been corrected. When 1 error bit is detected, output signal ERRn (in line 14) will be asserted low. MERRn (in line 15) remains high since the single bit error can be corrected. When 2 error bits are detected, both ERRn and MERRn are asserted low.

```
1       --------------------------------------------------------
2       -- TI SN54ALS616, 617 16-bit parallel EDAC circuit
3       --------------------------------------------------------
4       library IEEE;
5       use IEEE.std_logic_1164.all;
6       entity EDAC616 is
7         port(
8           S      : in    std_logic_vector( 1 downto 0);
9           CB     : inout std_logic_vector( 5 downto 0);
10          OECBn  : in    std_logic;
11          DB     : inout std_logic_vector(15 downto 0);
12          OEBn   : in    std_logic_vector( 1 downto 0);
13          LEDBOn : in    std_logic;
14          ERRn   : out   std_logic;
15          MERRn  : out   std_logic);
16      end EDAC616;
```

The following VHDL code shows the architecture code in file **edac616_rtl.vhd**. The core logic of the design is implemented in component CORE616 which is declared in lines 4 to 14. The component is instantiated in lines 26 to 30. Straight input signals and output signals are renamed in lines 31 to 38. Lines 39 to 49 describe the behavior of the bidirectional pads.

```
1       library IEEE;
2       use IEEE.std_logic_1164.all;
3       architecture RTL of EDAC616 is
4         component CORE616
5           port(
```

```
 6        S        : in     std_logic_vector( 1 downto 0);
 7        CBi      : in     std_logic_vector( 5 downto 0);
 8        CBo      : out    std_logic_vector( 5 downto 0);
 9        DBi      : in     std_logic_vector(15 downto 0);
10        DBo      : out    std_logic_vector(15 downto 0);
11        LEDBOn : in       std_logic;
12        ERRn     : out    std_logic;
13        MERRn    : out    std_logic);
14      end component;
15      signal  S_C      : std_logic_vector( 1 downto 0);
16      signal  CBi_C    : std_logic_vector( 5 downto 0);
17      signal  CBo_C    : std_logic_vector( 5 downto 0);
18      signal  OECBn_C  : std_logic;
19      signal  DBi_C    : std_logic_vector(15 downto 0);
20      signal  DBo_C    : std_logic_vector(15 downto 0);
21      signal  OEBn_C   : std_logic_vector( 1 downto 0);
22      signal  LEDBOn_C : std_logic;
23      signal  ERRn_C   : std_logic;
24      signal  MERRn_C  : std_logic;
25    begin
26      core6160 : CORE616 port map (
27        S      => S_C,        CBi     => CBi_C,
28        CBo    => CBo_C,      DBi     => DBi_C,
29        DBo    => DBo_C,      LEDBOn => LEDBOn_C,
30        ERRn   => ERRn_C,     MERRn   => MERRn_C);
31        --------------------------- input signals
32      LEDBOn_C <= LEDBOn;
33      OEBn_C   <= OEBn;
34      OECBn_C  <= OECBn;
35      S_C      <= S;
36        --------------------------- output signals
37      MERRn <= MERRn_C;
38      ERRn  <= ERRn_C;
39        --------------------------- bi-directional signals
40      DB(7 downto 0) <=
41        DBo_C(7 downto 0) when OEBn_C(0) = '0' else "ZZZZZZZZ";
42      DBi_C(7 downto 0) <= DB(7 downto 0);
43
44      DB(15 downto 8) <=
45        DBo_C(15 downto 8) when OEBn_C(1) = '0' else "ZZZZZZZZ";
46      DBi_C(15 downto 8) <= DB(15 downto 8);
47
48      CB    <= CBo_C when OECBn_C = '0' else "ZZZZZZ";
49      CBi_C <= CB;
50    end RTL;
```

The following shows the CORE616 VHDL code. Line 6 references package PACK in the working library. Lines 20 to 36 implement the parity bits according to the table as shown in lines 37 to 48. Lines 49 to 60 implement the syndrome with a procedure that is used to determine various conditions. The syndrome is enabled by the CHECK signal so that the syndrome can be generated with the exclusive NOR of the parity bits generated and the input parity bits to be compared.

```
1       ---------------------------------------------------------
2       -- TI SN54ALS616, 617 16-bit parallel EDAC circuit
3       ---------------------------------------------------------
4       library IEEE;
5       use IEEE.std_logic_1164.all;
6       use work.pack.all;
7       entity CORE616 is
8         port(
9           S       : in    std_logic_vector( 1 downto 0);
10          CBi     : in    std_logic_vector( 5 downto 0);
11          CBo     : out   std_logic_vector( 5 downto 0);
12          DBi     : in    std_logic_vector(15 downto 0);
13          DBo     : out   std_logic_vector(15 downto 0);
14          LEDBOn  : in    std_logic;
15          ERRn    : out   std_logic;
16          MERRn   : out   std_logic);
17      end CORE616;
18      architecture RTL of CORE616 is
19          -- generate check bits for 16 bits data word
20        procedure CBGEN16X6 (
21          signal D  : in  std_logic_vector(15 downto 0);
22          signal CB : out std_logic_vector(5 downto 0)) is
23        begin
24          CB(0) <= D(13) xor D(11) xor D(10) xor D(9) xor
25                   D(6)  xor D(3)  xor D(1)  xor D(0);
26          CB(1) <= D(14) xor D(12) xor D(10) xor D(9) xor
27                   D(8)  xor D(4)  xor D(2)  xor D(0);
28          CB(2) <= D(15) xor D(12) xor D(11) xor D(8) xor
29                   D(7)  xor D(5)  xor D(2)  xor D(1);
30          CB(3) <= D(15) xor D(14) xor D(13) xor D(9) xor
31                   D(8)  xor D(5)  xor D(4)  xor D(3);
32          CB(4) <= D(15) xor D(14) xor D(13) xor D(12) xor
33                   D(11) xor D(10) xor D(7)  xor D(6);
34          CB(5) <= D(7)  xor D(6)  xor D(5)  xor D(4) xor
35                   D(3)  xor D(2)  xor D(1)  xor D(0);
36        end CBGEN16X6;
37      ---------------------------------------------------------
38      -- The next table is from TI Databook SN54ALS616, 617
39      ---------------------------------------------------------
40      --                  16-bit Data Word
41      -- BIT  15 14 13 12 11 10  9  8  7  6  5  4  3  2  1  0
42      -- CB0         x     x  x  x        x        x     x  x
43      -- CB1      x     x     x  x  x           x     x     x
44      -- CB2   x        x  x        x  x     x        x  x
45      -- CB3   x  x  x           x  x        x  x  x
46      -- CB4   x  x  x  x  x  x        x  x
47      -- CB5                     x  x  x  x  x  x  x  x
```

```
48      --------------------------------------------------------------
49      procedure SYNDROME(
50        signal CHECK   : in  std_logic;
51        signal CBIN    : in  std_logic_vector(5 downto 0);
52        signal CBGEN   : in  std_logic_vector(5 downto 0);
53        signal SYNSIG  : out std_logic_vector(5 downto 0)) is
54      begin
55        if (CHECK = '1') then
56           SYNSIG <= not (CBGEN xor CBIN);
57        else
58           SYNSIG <= CBGEN;
59        end if;
60      end SYNDROME;
61      --------------------------------------------------------------
```

Lines 62 to 102 constitute a procedure, based on the syndrome value, to generate the error output signals ERRn and MERRn. It also generates a signal FLIP to indicate the bit position that is to be corrected when the single bit error is detected. Note that the single-bit error that occurs in the parity bits is not to be corrected in this design. Therefore, there is no corresponding FLIP bit in lines 90 to 92. The FLIP bit generation is enabled by signal FLIPen which is implemented in lines 96 to 98. The error generation can also be disabled when ERRen = '0' as shown in lines 99 to 101.

```
62      procedure ERRBIT16X6(
63        signal SYNSIG : in  std_logic_vector(5 downto 0);
64        signal ERRen  : in  std_logic;
65        signal FLIPen : in  std_logic;
66        signal ERRn   : out std_logic;
67        signal MERRn  : out std_logic;
68        signal FLIP   : out std_logic_vector(15 downto 0) ) is
69      begin
70        FLIP  <= (FLIP'range => '0');
71        ERRn  <= '0'; MERRn <= '1';
72        case SYNSIG is
73          -- corrected single data bit error
74          when "011100" => FLIP(0)  <= '1';
75          when "011010" => FLIP(1)  <= '1';
76          when "011001" => FLIP(2)  <= '1';
77          when "010110" => FLIP(3)  <= '1';
78          when "010101" => FLIP(4)  <= '1';
79          when "010011" => FLIP(5)  <= '1';
80          when "001110" => FLIP(6)  <= '1';
81          when "001011" => FLIP(7)  <= '1';
82          when "110001" => FLIP(8)  <= '1';
83          when "110100" => FLIP(9)  <= '1';
84          when "101100" => FLIP(10) <= '1';
85          when "101010" => FLIP(11) <= '1';
86          when "101001" => FLIP(12) <= '1';
87          when "100110" => FLIP(13) <= '1';
```

```
88            when "100101" => FLIP(14) <= '1';
89            when "100011" => FLIP(15) <= '1';
90          when "011111" | "101111" | "110111" |
91               "111011" | "111101" | "111110" =>
92            MERRn <= '1'; -- single check bit error
93       ·   when "111111" => ERRn     <= '1'; -- no errors
94           when others   => MERRn    <= '0'; -- fatal error
95         end case;
96         if (FLIPen = '0') then
97           FLIP <= (FLIP'range => '0');      ·
98         end if;
99         if (ERRen = '0')  then
100          ERRn <= '1'; MERRn <= '1';
101        end if;
102      end ERRBIT16X6;
103      -----------------------------------------------------------
104      signal SDEC0n, SDEC3n, SDEC2      :std_logic;
105      signal SDEC                       :std_logic_vector(3 downto 0);
106      signal CBLATCH,CBSIG,SYNSIG:std_logic_vector(5 downto 0);
107      signal DBLATCH,CORRECT,FLIP:std_logic_vector(15 downto 0);
108    begin
109      DECODER24(S, SDEC); SDEC2 <= SDEC(2);
110      SDEC0n <= not SDEC(0); SDEC3n <= not SDEC(3);
111      LATCH(CBi, SDEC2,  CBLATCH);
112      LATCH(DBi, SDEC3n, DBLATCH);
113      CBGEN16X6(DBLATCH, CBSIG);
114      SYNDROME(SDEC0n, CBLATCH, CBSIG, SYNSIG);
115      CBo <= SYNSIG when SDEC(1) = '0' else CBLATCH;
116      ERRBIT16X6(SYNSIG, SDEC0n, SDEC(3), ERRn, MERRn, FLIP);
117      CORRECT <= DBLATCH xor FLIP;
118      LATCHn(CORRECT, LEDBOn, DBo);
119    end RTL;
```

Lines 109 to 118 implement all circuits as shown in Figure 12-5 except for the pads which are implemented at the EDAC616 level. Line 109 implements the decoder with a procedure call. Line 110 assigns signals SDEC0n and SDEC3n. Lines 111 and 112 implement the latches for check bits and data bits. Line 113 uses a procedure call for the parity generation circuit. Line 114 generates the syndrome with another procedure call. Line 115 implements the multiplexer. Line 116 generates the FLIP, ERRn, and MERRn signals. Line 117 corrects the data using exclusive OR. Line 118 implements the latches where the latch enable signal LEDBOn is low active. The procedures, such as LATCH and LATCHn, are described in the package PACK referenced in line 6.

12.6 DESIGN VERIFICATION

To verify this design, the following test bench can be used. Lines 6 to 16 declare component EDAC616. Lines 17 to 24 declare signals that connect to the EDAC616 component. Lines 25 and 26 declare the clock signal CLK and its half-period.

```
1    library IEEE;
2    use IEEE.std_logic_1164.all;
3    entity TBEDAC616 is
4    end TBEDAC616;
5    architecture BEH of TBEDAC616 is
6      component EDAC616
7      port(
8        S     : in    std_logic_vector( 1 downto 0);
9        CB    : inout std_logic_vector( 5 downto 0);
10       OECBn : in    std_logic;
11       DB    : inout std_logic_vector(15 downto 0);
12       OEBn  : in    std_logic_vector( 1 downto 0);
13       LEDBOn : in   std_logic;
14       ERRn  : out   std_logic;
15       MERRn : out   std_logic);
16     end component;
17     signal  S      : std_logic_vector( 1 downto 0);
18     signal  CB     : std_logic_vector( 5 downto 0);
19     signal  OECBn  : std_logic;
20     signal  DB     : std_logic_vector(15 downto 0);
21     signal  OEBn   : std_logic_vector( 1 downto 0);
22     signal  LEDBOn : std_logic;
23     signal  ERRn   : std_logic;
24     signal  MERRn  : std_logic;
25     constant PERIOD2 : time := 25 ns;
26     signal   CLK     : std_logic;
```

Lines 28 to 32 generate the clock waveform. The clock signal is not used in the EDAC616 design. It is used to generate signals in relative order with the wait statements as shown in lines 44, 50, and other lines. A **process statement**, starting in line 33, is used to generate input signals for the circuit. The first **loop statement**, starting in line 52, is used to test the single bit error. The circuit operation is controlled by setting the S input signal. It is initially set to "00" in line 45. By controlling signal S to be "01", "10", and "11" respectively, the correct data is latched internally. The data input DB is then purposely reversed 1 bit and the correct data and the check bits become output. This is shown in Figure 12-6. The error bit position is rotating from the least significant bit to the most significant bit.

```
27   begin
28     clkgen : process
29     begin
30       CLK <= '0'; wait for PERIOD2;
```

```
31          CLK <= '1'; wait for PERIOD2;
32        end process;
33        dgen : process
34          variable RAND : std_logic_vector(15 downto 0) :=
35            "0000000000000000";
36          variable FLIPD : std_logic_vector(22 downto 0);
37          variable DATA  : std_logic_vector(22 downto 0);
38          variable FLIPC : std_logic_vector(5 downto 0);
39          variable CBT   : std_logic_vector(5 downto 0);
40          variable C_DAT : std_logic_vector(15 downto 0);
41        begin
42          FLIPD := "10000000000000000000000";
43          CB     <= "ZZZZZZ";
44          wait until CLK'event and CLK = '1';
45          S      <= "00";
46          DB     <= RAND;
47          C_DAT  := RAND;
48          OEBn   <= "11"; OECBn  <= '0' ;
49          LEDBOn <= '0' ;
50          wait until CLK'event and CLK = '1';
51          CBT    := CB; -- remember the generated check bits
52          for j in 0 to 22 loop
53            wait until CLK'event and CLK = '1';
54            S      <= "10";
55            DATA   := '0' & CBT & RAND;
56            DATA   := DATA xor FLIPD; -- single bit error
57            FLIPD  := FLIPD(21 downto 0) & FLIPD(22);
58            DB     <= DATA(15 downto 0);
59            CB     <= DATA(21 downto 16);
60            OEBn   <= "11"; OECBn  <= '1' ;
61            LEDBOn <= '0' ;
62            wait until CLK'event and CLK = '1';
63            S      <= "11";
64            wait until CLK'event and CLK = '1';
65            DB     <= (DB'range => 'Z');
66            CB     <= (CB'range => 'Z');
67            OEBn   <= "00"; OECBn  <= '0' ;
68            wait until CLK'event and CLK = '1';
69            assert (C_DAT = DB) report "error" severity NOTE;
70            wait until CLK'event and CLK = '1';
71            S      <= "01"; LEDBOn <= '1';
72          end loop;
```

Starting at line 73, the **loop statement** is used to test 2-bit error cases. It goes through the similar test sequence as before. Lines 78 and 79 are used to reverse the bits indexed by the loop index j and k. Figure 12-7 shows the simulation waveform.

```
73          for j in 0 to 21 loop -- pick 2 bits to reverse
74            for k in j+1 to 22 loop
75              wait until CLK'event and CLK = '1';
```

```
76                S       <= "10";
77                DATA    := '0' & CBT & RAND;
78                DATA(j) := not DATA(j);
79                DATA(k) := not DATA(k);
80                DB      <= DATA(15 downto 0);
81                CB      <= DATA(21 downto 16);
82                OEBn    <= "11"; OECBn <= '1' ;
83                LEDBOn <= '0' ;
84                wait until CLK'event and CLK = '1';
85                S       <= "11";
86                wait until CLK'event and CLK = '1';
87                DB      <= (DB'range => 'Z');
88                CB      <= (CB'range => 'Z');
89                OEBn    <= "00"; OECBn <= '0' ;
90                wait until CLK'event and CLK = '1';
91                assert (C_DAT = DB) report "error" severity NOTE;
92                wait until CLK'event and CLK = '1';
93                S       <= "01"; LEDBOn <= '1';
94             end loop;
95           end loop;
96           RAND := RAND(14 downto 0) & (RAND(15) xor not RAND(0));
97        end process;
98     -------------------------------------------------------------------
99        edac6160 : EDAC616 port map (
100          S       => S,     CB      => CB,     OECBn => OECBn,
101          DB      => DB,    OEBn    => OEBn,   LEDBOn => LEDBOn,
102          ERRn    => ERRn,  MERRn   => MERRn);
103       end BEH;
```

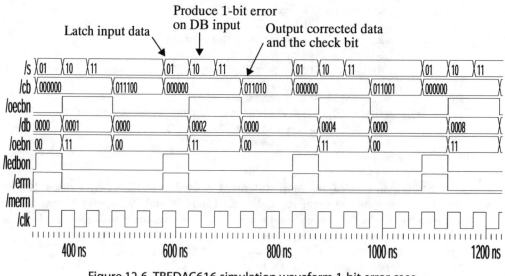

Figure 12-6 TBEDAC616 simulation waveform 1-bit error case.

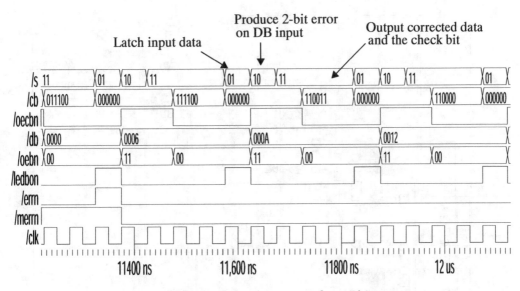

Figure 12-7 TBEDAC616 simulation waveform 2-bit error case.

Certainly, there are more simulations that can be run to fully verify the circuit. Readers are referred to the Texas Instruments (TI) databook for a more detailed description of the circuit. This is left as Exercise 12.1.

12.7 DESIGN SYNTHESIS

The CORE616 VHDL code can be synthesized with Synopsys Design Compiler; the schematic is shown in Figure 12-8. The design approaches as described in the previous two chapters can be used to generate the design netlist for ASIC place and route. Here, we will go through the FPGA design process. In an ASIC design process, pads are usually instantiated. In FPGA, pads are programmed with the programmable Input Output cells inside the FPGA. There are usually no options (such as drive capability) to select different types of pads. Each Input Output pad logic cell can be programmed to become an input, output, three-state output, or bi-directional pad. VHDL code files **core616.vhd**, **edac616_e.vhd**, and **edac616_rtl.vhd** can be read into a FPGA synthesis tool. The synthesis process is similar to the Synopsys Design Compiler. Figure 12-9 shows one schematic of the CORE616 block. Figure 12-10 shows the top-level schematic with Input Output pads and the CORE616 block.

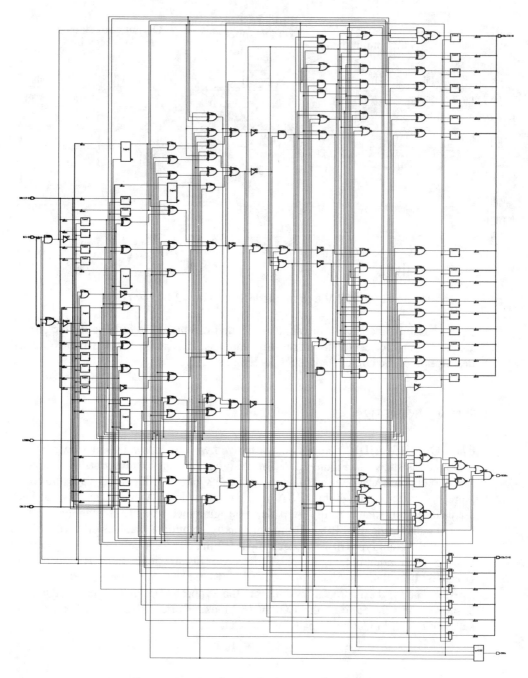

Figure 12-8 Synthesized schematic for CORE616.

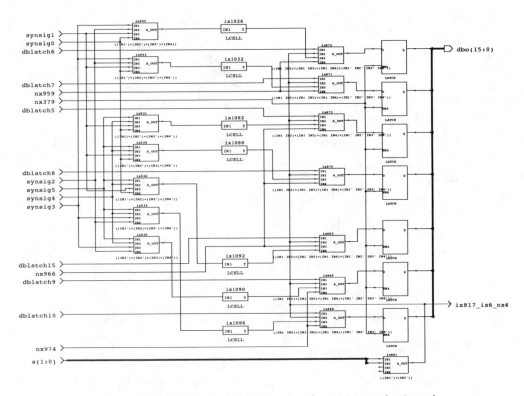

Figure 12-9 Schematic for CORE616 with a FPGA synthesis tool.

The following shows the area report. Note that the latches are programmed with 38 register cells as shown in line 11. Sixteen cells are for data input latch, 16 for data output latch, and 6 for the check bit latch.

```
1    *******************************************************
2    Cell: edac616    View: rtl    Library: work
3    *******************************************************
4    Number of ports :                          30
5    Number of nets :                           85
6    Number of instances :                      56
7    Number of references to this view :         0
8
9    Total accumulated area :
10   Number of TRIs :                           22
11   Number of DFFs :                           38
12   Number of LCs :                           110
13   Number of CASCADEs :                        4
```

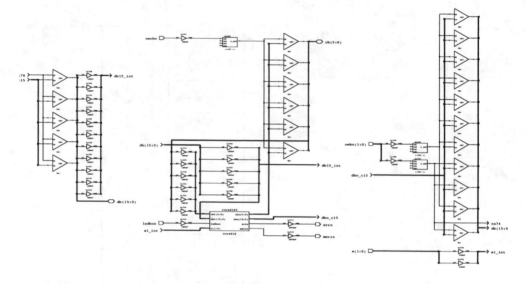

Figure 12-10 Schematic for EDAC616 with a FPGA synthesis tool.

The delay report is shown below.

	NAME	GATE	ARRIVAL	LOAD
1				
2	---			
3	clock information not specified			
4	delay thru clock network		0.00 (ideal)	
5				
6	core6160_lat_dblatch(10)/Q	LATCH	0.30 up	0.00
7	core6160_modgen_54_mi_ax_sub2r(1)/O	F2_LUT	4.40 up	0.00
8	core6160_modgen_55_10/Y	XOR2	4.40 up	0.00
9	core6160_modgen_55_12/Y	XOR2	4.40 up	0.00
10	core6160_synsig(1)/O	F4_LUT	8.50 up	0.00
11	core6160_ix818_ix132_nx12/O	F4_LUT	12.60 up	0.00
12	core6160_ix818_ix108_nx20/O	F4_LUT	16.70 up	0.00
13	core6160_nx976/O	F4_CAS	21.90 up	0.00
14	core6160_nx957/O	CASCADE2	23.00 up	0.00
15	core6160_nx977/O	F4_CAS	28.20 up	0.00
16	core6160_nx950/O	CASCADE2	29.30 up	0.00
17	merrn_dup0/O	F1_LUT	29.30 up	0.00
18	ix312/OUT	OUTBUF	33.10 up	0.00
19	merrn/		33.10 up	0.00
20	data arrival time		33.10	
21				
22	data required time (max delay specified)		30.00	

```
23      ---------------------------------------------------------------
24      data required time                              30.00
25      data arrival time                               33.10
26                                                      ----------
27      slack                                           -3.10
```

12.8 NETLIST GENERATION AND FPGA PLACE AND ROUTE

After the FPGA synthesis process is done, a netlist can be generated. Most FPGA place and route tools can take the netlist in EDIF format. The EDIF netlist is used in Altera MaxPlus2 for the place and route. It concludes with the following messages:

```
1       ***** Logic for device 'edac616' compiled without errors.
2       Device: EPF10K10LC84-3
3       FLEX 10K Configuration Scheme: Passive Serial
```

The tool automatically selects the device EPF10K10LC84-3, which is an 84-pin package. It is also possible to specify the device for the tool. This is left as Exercise 12.3.

12.9 EXERCISES

12.1 Refer to the TI databook for this circuit. How can you increase verification of the circuit to ensure the correctness of the design? How do you change the test bench to accomplish a more thorough verification?

12.2 The EDAC circuit can be extended to 32-bit. Refer to TI databook for SN54ALS632A. Develop VHDL code. Verify the design. Synthesize with a FPGA synthesis tool. Complete the FPGA place and route for the design. Generate back-annotated timing and perform the timing simulation with the same test bench.

12.3 Synthesize the VHDL code with a FPGA synthesis tool such as Exemplar, Simplicity, or Synopsys FPGA Express. Generate a netlist (it may be in EDIF format). Perform the place and route with a FPGA place and route tool such as Altera, Xilinx, or Lucent. Analyze the timing (IO timing and clock speed) of the design.

12.4 Refer to Figure 12-1. Complete the table for $n = 17$ to 32.

Chapter 13

Fixed-Point Multiplication

A multiplication circuit is usually referred to as a *multiplier*. A multiplier takes two binary operands to generate a product. A binary number can represent either a fixed-point or a floating-point number. In this chapter, we will discuss only the fixed-point multiplication algorithms implemented in hardware. An integer can be represented with an unsigned binary number or as a 2's complement number. We will use VHDL to describe the multiplier in both the unsigned and the 2's complement binary representations.

13.1 MULTIPLICATION CONCEPT

The multiplication algorithm for binary numbers is basically similar to the traditional pencil and paper method. Figure 13-1 shows an example of multiplying decimal numbers 11 and 13. The pencil and paper method shows that the first partial product of 33 formed by multiplying the least significant digit of the multiplier, which is 3 with the multiplicand 11. The second partial product 11 is formed by multiplying the second least significant digital of the multiplier, which is 1 with the multiplicand 11. The second partial product is shifted left one digit position. These two partial products are added together to form the product 143. In digital designs, binary numbers are usually used. The unsigned binary representations of the multiplicand and the multiplier are 1011 and 1101, respectively. There are 4 bits in the multiplier. Four partial products can be formed. The multiplication of two binary bits is simply a logical AND function. There is no carry necessary in getting the binary partial product (not like the decimal multiplication partial product). This simplifies the hardware. Note that the partial products are either the same as the multiplicand or zero, except that they shift left to the corresponding bit position of the multiplier. These four partial products are then added together to form the product 10001111, which is equivalent to 143 in decimal.

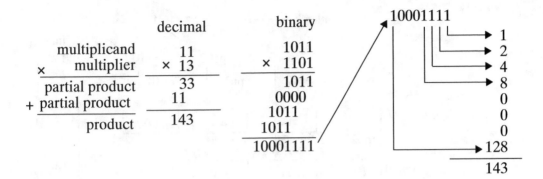

Figure 13-1 Multiplication of 1011 and 1101.

13.2 UNSIGNED BINARY MULTIPLIER

We know that partial products can be formed by an array of AND gates. Let's see how these partial products can be added together. A straightforward approach is to use three adders to add the four partial products after the partial products are shifted and filled with zeroes. For example, partial product D0 is "00001011", D1 is "00000000", D2 is "00101100", and D3 is "01011000". The product is equal to D0 + D1 + D2 + D3. However, we will use a different approach using the concept of carry save adder arrays. The concept of the *carry save adder* is to save the carry for the next stage. Figure 13-2 illustrates the concept. The partial products are indicated by PPxy where x is the partial product number formed by multiplier bit x, y is the bit position in the partial product. The adder has two outputs. One is the sum bit. The other is the carry bit. We will use PSxy to indicate the partial sum formed by the full adders in row x. PCxy is used to indicate the partial carry generated by the full adders in row x. The row number is indicated by the index j. The column number is indicated by the index k. Note that the top three rows of full adders are similar to add the partial products. The bottom row of the full adders are different. They add the PS and PC outputs of the previous row with the carry ripple through the bottom row. For the first row of full adders, they add the first two partial products. Bit PP00 is the same as the least significant bit of the product. The most significant bit of each partial product is added in the next row of full adders, while the rest of partial products are added in the current row of adders.

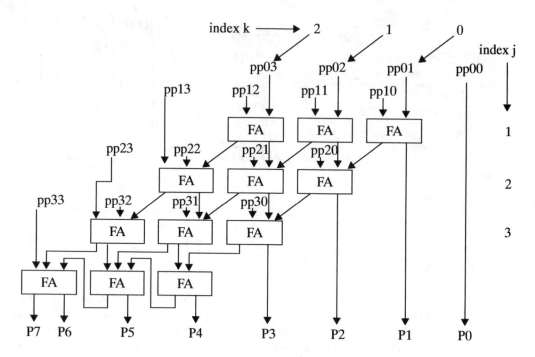

Figure 13-2 Multiplication with carry save adders.

To implement the multiplier with the carry save adders, the following VHDL code can be used. Line 5 declares multiplicand A and multiplier B with 4 bits each. The product PROD has 8 bits as shown in line 6. Line 9 declares constant N with value 4. The idea is to use this constant as a reference for the rest of the VHDL code. This constant can be changed to another value (along with the width of the ports in lines 5 and 6), and the VHDL code should work. Of course, another way to do this is to declare N as a generic. Lines 10 to 12 declare subtype and type to be used in line 12 where partial product PP, partial sum PS, and partial carry PC are declared. Lines 13 to 22 declare components full adder FA and an AND gate component ANDG. They are simple and their codes are not shown here.

```
1    library IEEE;
2    use IEEE.std_logic_1164.all;
3    entity MULT is
4      port (
5        A, B : in  std_logic_vector(3 downto 0);
6        PROD : out std_logic_vector(7 downto 0));
7    end MULT;
8    architecture RTL1 of MULT is
9      constant N : integer := 4;
```

```
10        subtype plary is std_logic_vector(N-1 downto 0);
11        type pary is array(0 to N) of plary;
12        signal PP, PC, PS : pary;
13        component FA
14        port (
15          A, B, CI : in  std_logic;
16          S, COUT  : out std_logic);
17        end component;
18        component ANDG
19        port (
20          A, B : in  std_logic;
21          C    : out std_logic);
22        end component;
```

Lines 24 to 30 use generate statements to instantiate the ANDG component. Line 29 takes advantage of the outer-level generate statement to initialize the first partial carry array. Line 31 sets the first partial sum PS(0) as the first partial product PP(0). This way, three rows of carry save adders can be generated in the same way. Line 32 sets the least significant bit of the product to be the least significant bit of the first partial product. Lines 33 to 41 use two levels of generate statements to instantiate full adder components. Note the full adder adds the current index of the partial product PP(j)(k), the partial sum of the last row (index j-1) with the column index shifted left (index k+1) PS(j-1)(k+1), and the partial carry from the last row PC(j-1)(k). The full adder outputs the partial sum PS(j)(k) and partial carry PC(j)(k). You are encouraged to convince yourself of the correctness of the index values. For each row of the first three rows of full adders, a product bit is produced. This is done in line 39. Line 40 connects the most significant bit of the partial product PP(j)(N-1) to the partial sum so that it is added in the next row.

```
23    begin
24      pgen : for j in 0 to N-1 generate
25        pgen1 : for k in 0 to N-1 generate
26          and0 : ANDG port map (
27            A => A(k), B => B(j), C => PP(j)(k));
28        end generate;
29        PC(0)(j) <= '0';
30      end generate;
31      PS(0) <= PP(0);
32      PROD(0) <= PP(0)(0);
33      addr : for j in 1 to N-1 generate
34        addc : for k in 0 to N-2 generate
35          fa0 : FA port map (
36            A => PP(j)(k), B => PS(j-1)(k+1), CI => PC(j-1)(k),
37            S => PS(j)(k), COUT => PC(j)(k));
38        end generate;
39        PROD(j)    <= PS(j)(0);
40        PS(j)(N-1) <= PP(j)(N-1);
41      end generate;
42      PC(N)(0) <= '0';
```

```
43       addlast : for k in 1 to N-1 generate
44         fa1 : FA port map (
45            A => PS(N-1)(k), B => PC(N-1)(k-1), CI => PC(N)(k-1),
46            S => PS(N)(k), COUT => PC(N)(k));
47         end generate;
48         PROD(2*N-1) <= PC(N)(N-1);
49         PROD(2*N-2 downto N) <= PS(N)(N-1 downto 1);
50     end RTL1;
```

Line 40 to 49 implement the last row of full adders. Line 42 initializes the least significant bit of the partial carry that is to be rippled through to the left. Lines 43 to 46 instantiate full adders. Note that each full adder adds the previous row partial sum PS(N-1)(k), the shifted partial carry of the previous row PC(N-1)(k-1), and the ripple carry from the last stage in the current row PC(N)(k-1). Line 49 forms the product output from the partial sum outputs. Line 48 forms the most significant bit of the product from the leftmost full adder's carry output.

The following architecture RTL2 is similar to architecture RTL1 except that the AND function to form the partial product (line 59) and the full adder function are implemented with signal assignment statements directly (lines 67 to 70, and lines 77 to 80).

```
51     architecture RTL2 of MULT is
52       constant N : integer := 4;
53       subtype plary is std_logic_vector(N-1 downto 0);
54       type pary is array(0 to N) of plary;
55       signal PP, PC, PS : pary;
56     begin
57       pgen : for j in 0 to N-1 generate
58         pgen1 : for k in 0 to N-1 generate
59           PP(j)(k) <= A(k) and B(j);
60         end generate;
61         PC(0)(j) <= '0';
62       end generate;
63       PS(0) <= PP(0);
64       PROD(0) <= PP(0)(0);
65       addr : for j in 1 to N-1 generate
66         addc : for k in 0 to N-2 generate
67           PS(j)(k) <= PP(j)(k) xor PC(j-1)(k) xor PS(j-1)(k+1);
68           PC(j)(k) <= (PP(j)(k)   and PC(j-1)(k)) or
69                       (PP(j)(k)   and PS(j-1)(k+1)) or
70                       (PC(j-1)(k) and PS(j-1)(k+1));
71         end generate;
72         PROD(j)    <= PS(j)(0);
73         PS(j)(N-1) <= PP(j)(N-1);
74       end generate;
75       PC(N)(0) <= '0';
76       addlast : for k in 1 to N-1 generate
77         PS(N)(k) <= PC(N)(k-1) xor PC(N-1)(k-1) xor PS(N-1)(k);
78         PC(N)(k) <= (PC(N)(k-1)   and PC(N-1)(k-1)) or
```

```
79                    (PC(N)(k-1)    and PS(N-1)(k)) or
80                    (PC(N-1)(k-1) and PS(N-1)(k)));
81      end generate;
82      PROD(2*N-1) <= PC(N)(N-1);
83      PROD(2*N-2 downto N) <= PS(N)(N-1 downto 1);
84    end RTL2;
```

To verify the above VHDL code, which indeed implement the multiplication function, the following test bench can be used. The test bench constructs are similar as before. Inputs A and B are enumerated with all possible values. Line 30 forms the bench mark value to be compared with the MULT component outputs. These two MULT components are instantiated in lines 46 to 49.

```
1     library IEEE;
2     use IEEE.std_logic_1164.all;
3     use IEEE.std_logic_arith.all;
4     entity TBMULT is
5     end TBMULT;
6     architecture BEH of TBMULT is
7       component MULT
8       port (
9         A, B : in  std_logic_vector(3 downto 0);
10        PROD : out std_logic_vector(7 downto 0));
11      end component;
12      constant W   : integer := 4; -- signal width
13      signal  A, B : std_logic_vector(W-1 downto 0);
14      signal PROD, PROD1, PROD2 :
15        std_logic_vector(2*W-1 downto 0);
16      constant PERIOD : time := 50 ns;
17      constant STROBE : time := PERIOD - 5 ns;
18    begin
19      p0 : process
20      begin
21        for j in 0 to 2**W-1 loop
22          A  <= conv_std_logic_vector(j,W);
23          for k in 0 to 2**W-1 loop
24            B  <= conv_std_logic_vector(k,W);
25            wait for PERIOD;
26          end loop;
27        end loop;
28        wait;
29      end process;
30      PROD <= unsigned(A) * unsigned(B);
31      check : process
32        variable err_cnt : integer := 0;
33      begin
34        wait for STROBE;
35        for j in 1 to 2**(W*2) loop
36          if (PROD /= PROD1) or (PROD /= PROD2) then
```

```
37              assert FALSE report "not compared" severity WARNING;
38                err_cnt := err_cnt + 1;
39             end if;
40            wait for PERIOD;
41          end loop;
42          assert (err_cnt = 0) report "test failed" severity ERROR;
43          assert (err_cnt /= 0) report "test passed" severity NOTE;
44          wait;
45        end process;
46        mult1 : MULT
47          port map ( A => A, B => B, PROD => PROD1);
48        mult2 : MULT
49          port map ( A => A, B => B, PROD => PROD2);
50      end BEH;
```

The following shows the configuration. Line 54 specifies the entity MULT with architecture RTL1. Architecture RTL1 (lines 55 to 76) has lower-level components ANDG and FA. These components are inside generate statements. The generate statement labels are used in the configuration, such as in lines 56 and 57. They correspond to the MULT VHDL code lines 24 and 25. Architecture RTL2 does not have lower-level components. They are configured with lines 77 to 79. Figure 13-3 shows part of the simulation waveform. The values are shown in decimal numbers.

```
51      configuration CFG_TBMULT of TBMULT is
52        for BEH
53          for mult1 : MULT
54            use entity work.MULT(RTL1);
55            for RTL1
56              for pgen
57                for pgen1
58                  for all : ANDG
59                    use entity work.ANDG(RTL);
60                  end for;
61                end for;
62              end for;
63              for addr
64                for addc
65                  for all : FA
66                    use entity work.FA(RTL);
67                  end for;
68                end for;
69              end for;
70              for addlast
71                for all : FA
72                  use entity work.FA(RTL);
73                end for;
74              end for;
75            end for;
76          end for;
```

```
77        for mult2 : MULT
78          use entity work.MULT(RTL2);
79        end for;
80      end for;
81    end CFG_TBMULT;
```

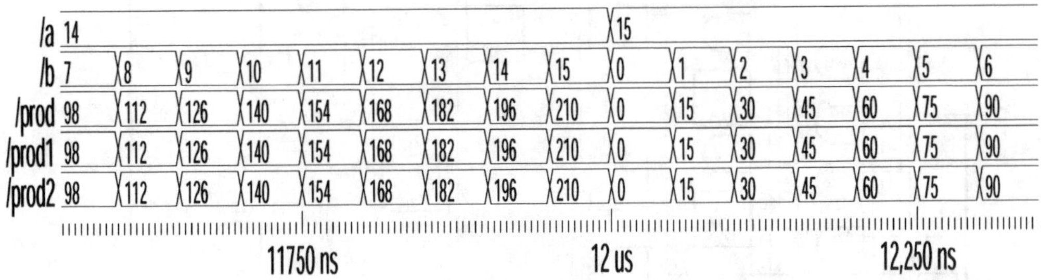

Figure 13-3 Simulation waveform for TBMULT.

MULT VHDL code can be synthesized. The following shows a synthesis script file to synthesize MULT architecture RTL1. Lines 1 and 2 read VHDL code for the AND gate and the full adder. Lines 3 to 4 compile the full adder. Lines 5 to 6 compile the AND gate. Lines 7 to 9 analyze and elaborate MULT architecture RTL1. Line 10 ungroups all the AND gates. Figure 13-4 shows the schematic result. Note that there is no "compile" command to synthesize the MULT since architecture RTL1 has only component instantiations and signal connections. There are no timing constraints set. The purpose here is to show the relationship between Figure 13-2 and Figure 13-4. Note that there is a "logic_0" connected to some of the full adders. This is due to some full adders requiring only two inputs such as the full adders in the first row. Of course, this can be further optimized to remove the "logic_0" connections.

```
1    read -f vhdl VHDLSRC + "ch13/fa.vhd"
2    read -f vhdl VHDLSRC + "ch13/andg.vhd"
3    current_design = FA
4    compile
5    current_design = ANDG
6    compile
7    design = mult
8    analyze -f vhdl VHDLSRC + "ch13/" + design + ".vhd"
9    elaborate MULT -architecture RTL1
10   ungroup and*
```

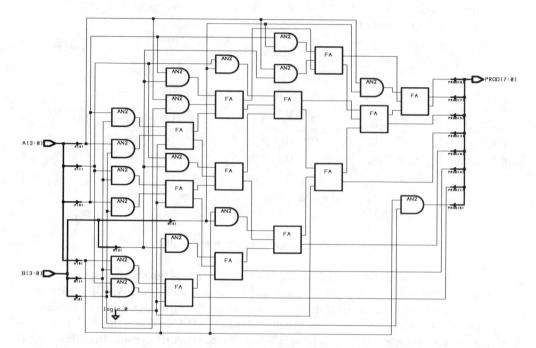

Figure 13-4 Synthesized schematic of MULT RTL1.

Another way to synthesize the MULT architecture RTL1 is to read in all VHDL code, to ungroup the lower-level components, and then synthesize the whole. The following synthesis script shows such an example. Lines 1 to 5 read in VHDL code. Line 6 ungroups (or flatten) the lower-level blocks. Lines 8 to 11 specify timing constraints. Line 12 synthesizes the whole design. Line 13 generates design statistics. Figure 13-5 shows the synthesized schematic. Note that there are no "logic_0" connections.

```
1     read −f vhdl VHDLSRC + "ch13/fa.vhd"
2     read −f vhdl VHDLSRC + "ch13/andg.vhd"
3     design = mult
4     analyze −f vhdl VHDLSRC + "ch13/" + design + ".vhd"
5     elaborate MULT −architecture RTL1
6     ungroup −all
7     include compile.common
8     create_clock −name clk −period CLKPERIOD −waveform {0.0
HALFCLK}
9     set_clock_skew −uncertainty CLKSKEW clk
10    set_input_delay  INPDELAY −add_delay −clock clk all_inputs()
11    set_output_delay CLKPER8  −add_delay −clock clk all_outputs()
12    compile −map_effort medium
```

```
13     include compile.report
```

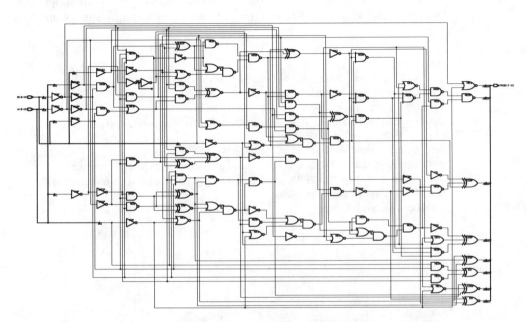

Figure 13-5 Synthesized schematic of MULT RTL1, ungrouped.

Figure 13-5 has the following area and timing statistics.

```
1     Number of ports:              16
2     Number of nets:              101
3     Number of cells:              93
4     Number of references:         28
5
6     Combinational area:       214.000000
7     Net Interconnect area:    619.990540
8
9     Total cell area:          214.000000
10    Total area:               833.990540
11
12    Operating Conditions: WCCOM    Library: lca300k
13    MULT                  B9X9                  lca300k
14      clock clk (rise edge)          30.00       30.00
15      clock network delay (ideal)     0.00       30.00
16      clock uncertainty              -1.00       29.00
17      output external delay         -24.00        5.00
18      data required time                          5.00
19    -----------------------------------------------------
20      data required time                          5.00
21      data arrival time                          -8.33
```

```
22      --------------------------------------------------------
23      slack (VIOLATED)                                 -3.33
```

Architecture RTL2 can be synthesized with the following synthesis script. Figure 13-6 shows the synthesized schematic. The area and timing statistics are also shown.

```
1       design = mult
2       analyze -f vhdl VHDLSRC + "ch13/" + design + ".vhd"
3       elaborate MULT -architecture RTL2
4       include compile.common
5       create_clock -name clk -period CLKPERIOD -waveform {0.0
HALFCLK}
6       set_clock_skew -uncertainty CLKSKEW clk
7       set_input_delay  INPDELAY -add_delay -clock clk all_inputs()
8       set_output_delay CLKPER8  -add_delay -clock clk all_outputs()
9       compile -map_effort medium
10      include compile.report
```

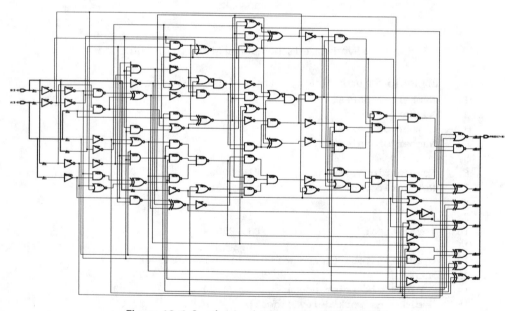

Figure 13-6 Synthesized schematic of MULT RTL2.

```
1       Number of ports:              16
2       Number of nets:               96
3       Number of cells:              88
4       Number of references:         23
```

```
 5
 6    Combinational area:          184.000000
 7    Net Interconnect area:       596.905762
 8
 9    Total cell area:             184.000000
10    Total area:                  780.905762
11
12    Operating Conditions: WCCOM   Library: lca300k
13    MULT                    B9X9              lca300k
14
15       clock clk (rise edge)           30.00      30.00
16       clock network delay (ideal)      0.00      30.00
17       clock uncertainty               -1.00      29.00
18       output external delay          -24.00       5.00
19       data required time                          5.00
20       ------------------------------------------------
21       data required time                          5.00
22       data arrival time                          -8.72
23       ------------------------------------------------
24       slack (VIOLATED)                           -3.72
```

MULT VHDL code can be easily extended to multiply data with more bits. This is left as Exercise 13.2. The carry save adder multiplier can be analyzed as follows. To multiply N-bits by M-bits, we need $(N - 1)*M$ full adders (some of them can be half adders). The longest timing delay is coming from the first partial product (1 AND gate delay), going down the full adder row ($M - 1$ rows), and then the ripple carry through the last row ($N - 1$ stages). Suppose each full adder takes about two gate delays, the longest timing delay is about $2*(M - 1 + N - 1) + 1$ gate delays. Can we improve on this?

13.3 2'S COMPLEMENT MULTIPLICATION

The above MULT VHDL code works for binary numbers with the unsigned format. The leftmost bit is not the sign bit. Popular digital systems and computers have been using the 2's complement format to represent numbers. The leftmost bit is considered as the sign bit. Also, the number of partial products are the same as the number of bits in multiplier B. In 1951, Booth [Booth] suggested a recoding scheme to reduce the number of partial products for the 2's complement numbers. As an example, let's multiply 100101 (decimal −27) with 011001 (decimal 25). The multiplier 011001 is appended with another bit 0 as the rightmost bit to become 0110010. Now we form a group of 3 bits from right to left with 1 bit overlapped as shown in Figure 13-7. These 3-bit patterns are called the *recoding bits*. Each 3-bit recoding pattern corresponds to what the partial product should be when it is formed as shown in Figure 13-7. The original multiplier is appended a '0' at the right end to form three sets of recoding patterns from right to left as 010, 100, 011. The corresponding partial products are multiplied by 1, −2, and 2, respectively. Since the recoding patterns are shifted 2 bits to the left. The corresponding weighted values of the partial products should be multiplied by 1, 4, and 16, respectively: $1 \times 1 + (-2)*4 + 2*16 = 25$, which is the same as the multiplier value 011001 in decimal.

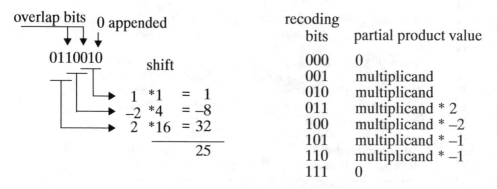

Figure 13-7 Booth recoding table.

Figure 13-8 shows how $-27*25$ can be performed with the Booth recoding scheme. The first partial product is the same as the multiplicand 100101. It is then sign extended to make the partial product 12 bits wide. The second partial product is weighted -2. The 2's complement of the multiplicand represents the negative value of the multiplicand. The multiplication of 2 can be accomplished by appending 0 (shift left 1 bit) at the right end. The result is then sign extended. The third partial product is weighted 2. This can be achieved by shifting left 1 bit. Again, the result is sign extended. Note that the second partial product is shifted 2 bits to the left since each set of the recoding bits advance 2 bits to the left. The third partial product is shifted 4 bits to the left. The partial products are then added together to form the product.

```
sign ──────▶ 111111 1100101  ◀─────  multiplicand = 100101,
   extended   0000 0110110    ◀─────  2's complement =011011 * 2
-27*25 = -675  1 1001010      ◀─────  multiplicand = 100101 * 2
              ─────────────
              110101011101
```
2's complement = 001010100011= 512 + 128 + 32 + 2 + 1 = 675 decimal

Figure 13-8 Booth recoding multiplication example.

The Booth recoding scheme can be implemented in the following VHDL code. Here, we assume that the multiplicand has 16 bits. Since Booth recoding can be at most 2 or -2 times the multiplicand, the output of the recoder uses 17 bits. Lines 13 to

18 define a 2's complement function simply by adding 1 to the inverted input data. Lines 20 to 34 use a case statement to implement the Booth recoding table as shown in Figure 13-7.

```
1     library IEEE;
2     use IEEE.std_logic_1164.all;
3     use IEEE.std_logic_unsigned.all;
4     entity RECODER is
5       port (
6         DIN  : in  std_logic_vector(15 downto 0);
7         BIT3 : in  std_logic_vector( 2 downto 0);
8         DOUT : out std_logic_vector(16 downto 0));
9     end RECODER;
10    architecture RTL of RECODER is
11      constant N : integer := 16;
12      subtype bitn1 is std_logic_vector(N downto 0);
13      function COMP2 (D : in bitn1) return bitn1 is
14        variable Dn : bitn1;
15      begin
16        Dn := not D;
17        return(Dn+1);
18      end COMP2;
19    begin
20      process (DIN, BIT3)
21      begin
22        case BIT3 is
23          when "001" | "010" =>
24            DOUT <= DIN(N-1) & DIN;
25          when "101" | "110" =>
26            DOUT <= COMP2(DIN(N-1) & DIN);
27          when "100" =>
28            DOUT <= COMP2(DIN & '0');
29          when "011" =>
30            DOUT <= DIN & '0';
31          when others =>
32            DOUT <= (DOUT'range => '0');
33        end case;
34      end process;
35    end RTL;
```

The RECODER VHDL code can be synthesized with the following synthesis script. Figure 13-9 shows the synthesized schematic.

```
1     design = recoder
2     analyze -f vhdl VHDLSRC + "ch13/" + design + ".vhd"
3     elaborate RECODER -architecture RTL
4     include compile.common
5     create_clock -name clk -period CLKPERIOD -waveform {0.0
HALFCLK}
```

```
6    set_clock_skew -uncertainty CLKSKEW clk
7    set_input_delay  INPDELAY -add_delay -clock clk all_inputs()
8    set_output_delay CLKPER9  -add_delay -clock clk all_outputs()
9    compile -map_effort medium
10   include compile.report
```

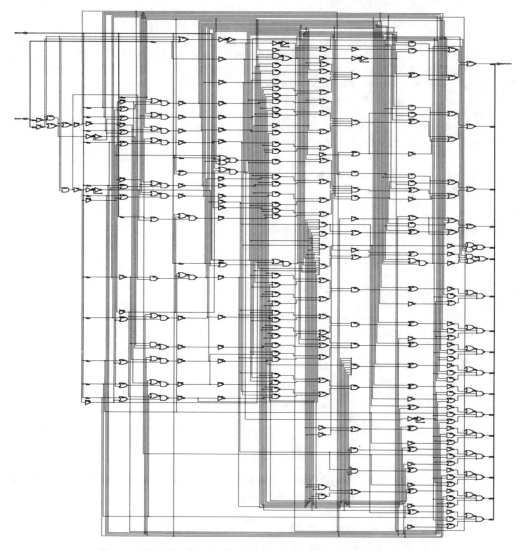

Figure 13-9 Synthesized schematic for RECODER, ungrouped.

13.4 WALLACE TREE ADDERS

The recoding techniques proposed by Booth as described in the last section reduce the number of partial products to half. This reduces the number of full adders rows with the carry save adders. We have analyzed the timing delay of the carry-save-adders-based multiplier. The speed is limited by the number of full adders that the first partial product has to go through. In 1964, Wallace [Wallace] noticed that A + B + C + D = (A + B)+(C + D). The carry save adder approach adds the partial products sequentially: the first two partial products are added before the third partial product is added. A + B + C + D requires three full adder delays. Wallace suggested that the partial products can be parallel added. Partial products A and B, C and D can be added at the same time. Their results are then added together. This requires only two full adder delays. Figure 13-10 shows the difference in the two approaches.

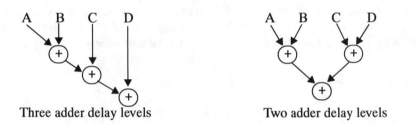

Three adder delay levels   Two adder delay levels

Figure 13-10 Parallel adder tree.

In 1989, Nagamatsu [Nagamatsu] published an improved parallel structure for a 32×32 multiplier based on the Wallace tree adders approach. Instead of using full adders, a unit adder is used. Each unit adder adds four data inputs and one carry input. It generates 1 sum bit and two carry outputs. The following VHDL code describes the unit adder. Line 5 declares X as the 4-bit data input. Line 6 declares the carry input. Line 7 declares the sum output. Line 8 declares the carry output for the next row in the next left column of the unit adders. Line 9 declares the carry output to the next unit adder directly to the left of the current column. The idea is to get the CO to the left unit adder faster than the S and C outputs. Line 16 generates CO outputs based on X only. Lines 17 and 19 calculate two temporary results temp1 and temp2, based only on X. Output S is obtained with temp1 and CI input in line 18. Line 20 generates output C.

```
1     library IEEE;
2     use IEEE.std_logic_1164.all;
3     entity UNITADD is
4       port (
5           X    : in  std_logic_vector(3 downto 0);
6           CI   : in  std_logic;
7           S    : out std_logic;
```

```
8           C    : out std_logic;
9           CO   : out std_Logic);
10      end UNITADD;
11      architecture RTL of UNITADD is
12      begin
13        process (X, CI)
14          variable temp1, temp2 : std_logic;
15        begin
16          CO    <= (X(0) or X(1)) and (X(2) or X(3));
17          temp1 := (X(0) xor X(1)) xor (X(2) xor X(3));
18          S     <= CI xor temp1;
19          temp2 := (X(0) and X(1)) nor (X(2) and X(3));
20          C     <= (temp1 and CI) or (temp1 nor temp2);
21        end process;
22      end RTL;
```

Figure 13-11 shows the unit adders organization for adding eight partial products. Note that the first two rows of unit adders add partial products. The first row unit adders add partial products P4, P3, P2, and P1. The second row of unit adders add partial products P8, P7, P6, and P5. The third row of unit adders add the sum outputs from the first two rows and the carry bits from the unit adders in the right column.

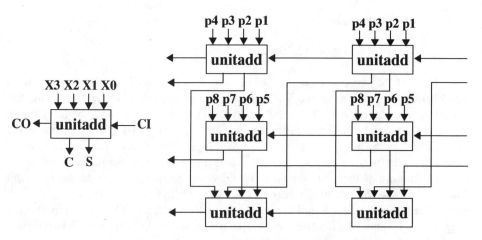

Figure 13-11 Unit adders organization for eight partial products.

The unit adder VHDL code can be synthesized with the following synthesis script. Note that the input and output delays are specified for faster CO output. Figure 13-12 shows the synthesized schematic.

```
1       design = unitadd
2       analyze –f vhdl VHDLSRC + "ch13/" + design + ".vhd"
3       elaborate UNITADD –architecture RTL
```

```
4    include compile.common
5    create_clock -name clk -period CLKPERIOD -waveform {0.0
HALFCLK}
6    set_clock_skew -uncertainty CLKSKEW clk
7    set_input_delay  CLKPER1 -add_delay -clock clk all_inputs()
8    set_input_delay  CLKPER2 -add_delay -clock clk {CI}
9    set_output_delay CLKPER8 -add_delay -clock clk {CO}
10   set_output_delay CLKPER7 -add_delay -clock clk {C}
11   compile -map_effort medium
12   include compile.report
```

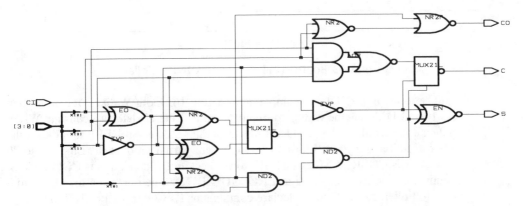

Figure 13-12 Synthesized schematic for UNITADD.

13.5 BOOTH-WALLACE TREE MULTIPLIER

Now, we can put the suggestions by Booth and Wallace together. The partial products are generated with the Booth recoder. The partial products are added with the Wallace tree adders, similar to the carry save adder approach. The last row of carry and sum outputs is added together with the carry skewed to the left by one position. Figure 13-13 shows a 16 × 16 multiplier, using Booth recoder and Wallace tree adders. The multiplicand comes from the left to go into eight Booth recoders. Each recoder takes 3 bits from the multiplier with '0' appended at the right end. The recoder has a 17-bit output. Each recoder output is shifted to its correct position, sign extended, and zero filled in the right end. There are three rows of unit adders. Each row has 32 unit adders. The carry and sum outputs of the last row are added with the carry output bits shifted left one bit position to add with the sum bits. The output of the adder forms the product.

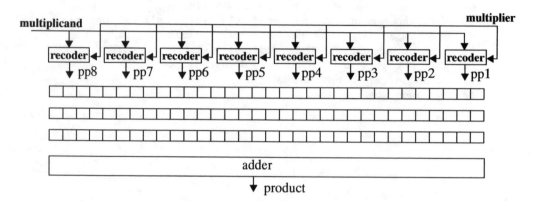

Figure 13-13 Booth - Wallace tree 16 x 16 multiplier.

The following VHDL code implements a 16 × 16 Booth-Wallace tree multiplier. Lines 10 and 11 define the widths for input A and B. Line 12 defines a constant WIDE to indicate the width of the product. Line 13 defines a constant DEEP to indicate the number of unit adder rows. Lines 14 to 16 declare partial product PP. Lines 18 to 20 declare signals COSIG, CSIG, and SSIG to connect unit adders outputs CO, C, and S, respectively. Line 21 declares signal RECODE_BITS to be one more than the width of the multiplier. Lines 22 to 35 declare components RECODER and UNITADD.

```
1     library IEEE;
2     use IEEE.std_logic_1164.all;
3     use IEEE.std_logic_unsigned.all;
4     entity MULTBW16 is
5       port (
6         A, B  : in  std_logic_vector(15 downto 0);
7         PROD  : out std_logic_vector(31 downto 0));
8     end MULTBW16;
9     architecture RTL of MULTBW16 is
10      constant M    : integer := 16;          -- A width
11      constant N    : integer := 16;          -- B    width
12      constant WIDE : integer := M+N;         -- product width
13      constant DEEP : integer := (N/2-4)/2+1; -- adder levels
14      subtype PRODTYPE is std_logic_vector(WIDE - 1 downto 0);
15      type    PPTYPE is array(M/2 downto 1) of PRODTYPE;
16      signal  PP : PPTYPE; -- partial products
17      signal  PRODUCT1 : std_logic_vector(WIDE downto 0);
18      subtype COLTYPE is std_logic_vector(WIDE downto 0);
19      type    INTTYPE is array (DEEP - 1 downto 0) of COLTYPE;
20      signal  COSIG, CSIG, SSIG : INTTYPE; -- for CO, C, S
21      signal  RECODE_BITS : std_logic_vector(N downto 0);
22      component RECODER
```

```
23      port (
24        DIN  : in  std_logic_vector(M-1 downto 0);
25        BIT3 : in  std_logic_vector(2 downto 0);
26        DOUT : out std_logic_vector(M downto 0));
27      end component;
28      component UNITADD
29      port (
30        X   : in  std_logic_vector(3 downto 0);
31        CI  : in  std_logic;
32        S   : out std_logic;
33        C   : out std_logic;
34        CO  : out std_Logic);
35      end component;
```

Line 37 appends '0' with the multiplier for the recoding bits. Lines 40 to 44 initialize the rightmost column values for COSIG, CSIG, and SSIG. Lines 48 to 52 instantiate component RECODER eight times. Note that each group of the 3 recoding bits overlap one bit with the last group of the recoder bits. The output of the recoder is shifted 2 bits each time. Line 53 performs the sign extension for each partial product. Lines 56 to 58 place '0' in the right end of the partial product to fill the empty positions when they are shifted to the left.

```
36    begin
37      RECODE_BITS <= B & '0'; -- recoding bits, append '0'
38
39        -- initialize COSIG, CSIG, and SSIG signals column 0
40      init : for i in 0 to DEEP-1 generate
41        COSIG(i)(0) <= '0';
42        CSIG(i)(0)  <= '0';
43        SSIG(i)(0)  <= '0';
44      end generate;
45
46        -- generating booth recoders for partial products
47      ppgen : for k in 1 to N / 2 generate
48        recoder0 : RECODER
49        port map (
50          DIN  => A,
51          BIT3 => RECODE_BITS(k*2 downto k*2-2),
52          DOUT => PP(k)(M+k*2-2 downto k*2-2)); --shift 2-bit
53        signext : for i in M+N-1 downto M+k*2-1 generate
54          PP(k)(i) <= pp(k)(M+k*2-2); -- sign extension to MSB
55        end generate;
56        zeroext : for i in 0 to k*2-3 generate
57          PP(k)(i) <= '0'; -- zero extention to LSB
58        end generate;
59      end generate;
```

The purpose of lines 60 to 86 is to instantiate all unit adders. The outer level of the generate statement generates unit adders by columns from right to left. The inner level of the generate statement generate unit adders by rows from top to bottom. Lines 78 to 83 instantiate the unit adder. To properly assign the values to the unit adder input X (in line 79), lines 66 to 77 are used. Basically, the first two rows of unit adders take the X inputs from the partial products. The last row of the unit adder takes the X input from the previous two rows of unit adders sum (S) outputs and the previous column unit adders carry (C) outputs. Lines 66 to 83 are enclosed by a block statement in lines 62, 63, 64, and 84.

```
60      columns: for col in 1 to WIDE    generate -- LSB to MSB
61        rows: for row in 0 to DEEP-1 generate -- top to bottom
62          oneunit : block -- one unit adder
63            signal tempx : std_logic_vector(3 downto 0);
64          begin
65              -- setup inputs for the unit adder
66            genx1 : if (row < 2) generate
67                tempx(3) <= PP(row*4 + 4)(col-1);
68                tempx(2) <= PP(row*4 + 3)(col-1);
69                tempx(1) <= PP(row*4 + 2)(col-1);
70                tempx(0) <= PP(row*4 + 1)(col-1);
71            end generate;
72            genx2 : if (row = 2) generate
73                tempx(3) <= SSIG(row-2)(col);
74                tempx(2) <= CSIG(row-2)(col-1);
75                tempx(1) <= SSIG(row-1)(col);
76                tempx(0) <= CSIG(row-1)(col-1);
77            end generate;
78            unitadd1 : UNITADD port map (
79                X   => tempx,
80                CI  => COSIG(row)(col-1),
81                S   => SSIG(row)(col),
82                C   => CSIG(row)(col),
83                CO  => COSIG(row)(col));
84          end block;
85        end generate;
86      end generate;
87      add0 : block
88      begin
89        PRODUCT1 <= (SSIG(DEEP-1)(WIDE) &
90                     SSIG(DEEP-1)(WIDE downto 1)) +
91                    (CSIG(DEEP-1)(WIDE-1) &
92                     CSIG(DEEP-1)(WIDE-1 downto 0));
93      end block;
94      PROD <= PRODUCT1(WIDE-1 downto 0);
95    end RTL;
```

Lines 88 to 91 add the carry and sum outputs from the last row of the unit adders. Note that the carry bits are shifted one bit position to be added with the corresponding sum bits output. Line 92 forms the product output. Lines 96 to 102 is another architecture DW that simply uses the multiply operator "*" with inputs A and B converted into signed type.

```
96    library IEEE;
97    use IEEE.std_logic_1164.all;
98    use IEEE.std_logic_arith.all;
99    architecture DW of MULTBW16 is
100   begin
101     PROD <= signed(A) * signed(B);
102   end DW;
```

13.6 BOOTH-WALLACE TREE MULTIPLIER VERIFICATION

The following test bench can be used to verify the Booth-Wallace tree multiplier. The test bench constructs are similar as before. Line 13 declares a constant STEPS to be used in line 22 to 24. The purpose is to step through A and B with a controlled number. If STEPS is set to 1, the test bench will run through all combinations of A and B, which may take a long time. The test bench uses architecture DW as the bench mark to compare with architecture RTL. The MULTBW16 components are instantiated in lines 48 to 51. They are configured in lines 53 to 78. Note that the configuration uses generate statement labels as shown in lines 58, 63, and 64. Figure 13-14 shows a portion of the simulation waveform.

```
1     library IEEE;
2     use IEEE.std_logic_1164.all;
3     use IEEE.std_logic_arith.all;
4     entity TBMULTBW16 is
5     end TBMULTBW16;
6     architecture BEH of TBMULTBW16 is
7       component MULTBW16
8       port (
9         A, B : in  std_logic_vector(15 downto 0);
10        PROD : out std_logic_vector(31 downto 0));
11      end component;
12      constant W    : integer := 16; -- signal width
13      constant STEPS : integer := 191;
14      signal  A, B : std_logic_vector(W-1 downto 0);
15      signal PROD, PROD1 :
16        std_logic_vector(2*W-1 downto 0);
17      constant PERIOD : time := 50 ns;
18      constant STROBE : time := PERIOD - 5 ns;
19    begin
20      p0 : process
21      begin
22        for j in 0 to (2**W-1)/STEPS loop
23          A  <= conv_std_logic_vector(j*STEPS,W);
24          for k in 0 to (2**W-1)/STEPS loop
```

```
25              B  <= conv_std_logic_vector(k*STEPS,W);
26                wait for PERIOD;
27              end loop;
28            end loop;
29          wait;
30        end process;
31        check : process
32          variable err_cnt : integer := 0;
33        begin
34          wait for STROBE;
35          for j in 1 to 2**(W) loop
36            for k in 1 to 2**(W) loop
37            if (PROD /= PROD1) then
38                assert FALSE report "not compared" severity WARNING;
39                err_cnt := err_cnt + 1;
40              end if;
41              wait for PERIOD;
42            end loop;
43          end loop;
44          assert (err_cnt = 0) report "test failed" severity ERROR;
45          assert (err_cnt /= 0) report "test passed" severity NOTE;
46          wait;
47        end process;
48        mult1 : MULTBW16
49          port map ( A => A, B => B, PROD => PROD1);
50        mult2 : MULTBW16
51          port map ( A => A, B => B, PROD => PROD);
52      end BEH;
53      configuration CFG_TBMULTBW16 of TBMULTBW16 is
54        for BEH
55          for mult1 : MULTBW16
56            use entity work.MULTBW16(RTL);
57            for RTL
58              for ppgen
59                for all : RECODER
60                  use entity work.RECODER(RTL);
61                end for;
62              end for;
63              for columns
64                for rows
65                  for oneunit
66                    for all : UNITADD
67                      use entity work.UNITADD(RTL);
68                    end for;
69                  end for;
70                end for;
71              end for;
72            end for;
73          end for;
74          for mult2 : MULTBW16
```

```
75          use entity work.MULTBW16(DW);
76        end for;
77      end for;
78    end CFG_TBMULTBW16;
```

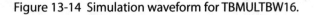

/a	11E8							
/b	7C99	7D58	7E17	7ED6	7F95	8054	8113	81D2
/prod	08B713A8	08C46FC0	08D1CBD8	08DF27F0	08EC8408	F711E020	F71F3C38	F72C9850
/prod1	08B713A8	08C46FC0	08D1CBD8	08DF27F0	08EC8408	F711E020	F71F3C38	F72C9850

421,200 ns   421,300 ns   421,400 ns   421500 ns

Figure 13-14 Simulation waveform for TBMULTBW16.

13.7 BOOTH-WALLACE TREE MULTIPLIER SYNTHESIS

The following synthesis script can be used to synthesize the MULTBW16 architecture DW. Figure 13-15 shows the synthesized schematic. As expected, a Design-
Ware multiplier is used. Figure 13-16 shows a portion of the schematic after Figure
13-15 is flattened. The area and timing statistics are shown below.

```
1     design = multbw16
2     analyze -f vhdl VHDLSRC + "ch13/" + design + ".vhd"
3     elaborate MULTBW16 -architecture DW
4     include compile.common
5     create_clock -name clk -period CLKPERIOD -waveform {0.0
HALFCLK}
6     set_clock_skew -uncertainty CLKSKEW clk
7     set_input_delay  INPDELAY -add_delay -clock clk all_inputs()
8     set_output_delay CLKPER4  -add_delay -clock clk all_outputs()
9     compile -map_effort medium
10    include compile.report
```

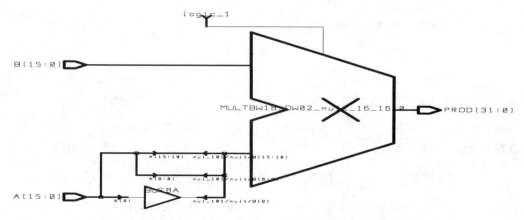

Figure 13-15 Synthesized schematic for TBMULTBW16, DW.

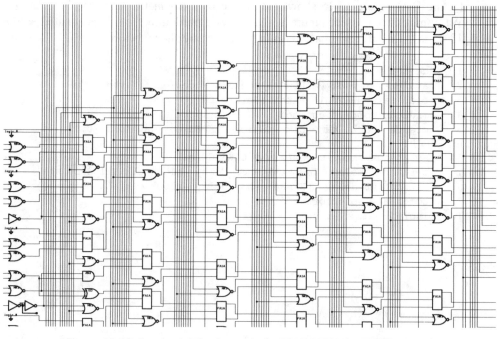

Figure 13-16 Synthesized schematic for TBMULTBW16, DW, flattened.

```
1    Number of ports:              64
2    Number of nets:             1106
3    Number of cells:             792
4    Number of references:         45
5
6    Combinational area:       2720.000000
7    Net Interconnect area:    5899.803711
8
9    Total cell area:          2720.000000
10   Total area:               8619.803711
11
12   Operating Conditions: WCCOM   Library: lca300k
13   MULTBW16               B9X9              lca300k
14
15     clock clk (rise edge)           30.00       30.00
16     clock network delay (ideal)      0.00       30.00
17     clock uncertainty               -1.00       29.00
18     output external delay          -12.00       17.00
19     data required time                          17.00
20     -----------------------------------------------
21     data required time                          17.00
22     data arrival time                          -39.22
23     -----------------------------------------------
24     slack (VIOLATED)                           -22.22
```

Architecture RTL can be synthesized with the following synthesis script. Lines 1 and 2 read in VHDL code for UNITADD and RECODER. Line 11 flattens lower-level components. The area and timing statistics follow. Note that the area is larger and the speed is faster than the architecture DW synthesis results.

```
1    read -f vhdl VHDLSRC + "ch13/" + "unitadd.vhd"
2    read -f vhdl VHDLSRC + "ch13/" + "recoder.vhd"
3    design = multbw16
4    analyze -f vhdl VHDLSRC + "ch13/" + design + ".vhd"
5    elaborate MULTBW16 -architecture RTL
6    include compile.common
7    create_clock -name clk -period CLKPERIOD -waveform {0.0
HALFCLK}
8    set_clock_skew -uncertainty CLKSKEW clk
9    set_input_delay  INPDELAY -add_delay -clock clk all_inputs()
10   set_output_delay CLKPER4  -add_delay -clock clk all_outputs()
11   ungroup -all
12   compile -map_effort medium
13   include compile.report
```

```
1    Number of ports:              64
2    Number of nets:             1324
3    Number of cells:            1274
4    Number of references:         63
```

```
5
6      Combinational area:          3503.000000
7      Net Interconnect area:       11093.873047
8
9      Total cell area:             3503.000000
10     Total area:                  14596.873047
11
12     Operating Conditions: WCCOM   Library: lca300k
13     MULTBW16                    B9X9                    lca300k
14       Point                                  Incr       Path
15       --------------------------------------------------------
16       clock clk (rise edge)                  0.00       0.00
17       clock network delay (ideal)            0.00       0.00
18       input external delay                   2.50       2.50 f
19       A[0] (in)                              0.00       2.50 f
20       U398/Z (IVP)                           0.69       3.19 r
21       U403/Z (BUF8A)                         0.44       3.64 r
22       U980/U19/Z (ND3P)                      0.62       4.26 f
23       U980/U22/Z (NR2P)                      0.54       4.81 r
24       U980/U41/Z (ND2P)                      0.44       5.25 f
25       U980/U42/Z (ENP)                       0.64       5.89 f
26       U612/Z (AN2P)                          0.46       6.36 f
27       U613/Z (OR2P)                          0.64       7.00 f
28       U367/Z (EN3P)                          1.58       8.58 f
29       U625/Z (MUX21H)                        0.87       9.46 f
30       U507/Z (EO)                            0.91      10.37 f
31       add_90/plus/plus/U79/Z (IVP)           0.34      10.71 r
32       add_90/plus/plus/U88/Z (ND2)           0.59      11.30 f
33       add_90/plus/plus/U127/Z (ND2)          0.36      11.66 r
34       add_90/plus/plus/U128/Z (AO7P)         0.56      12.23 f
35       add_90/plus/plus/U130/Z (AO2P)         0.68      12.91 r
36       add_90/plus/plus/U142/Z (AO3P)         0.77      13.68 f
37       add_90/plus/plus/U145/Z (IVP)          0.20      13.89 r
38       add_90/plus/plus/U148/Z (AO7P)         0.53      14.42 f
39       add_90/plus/plus/U149/Z (ND3P)         0.39      14.81 r
40       add_90/plus/plus/U222/Z (AN3)          1.02      15.83 r
41       add_90/plus/plus/U224/Z (OR2)          0.51      16.34 r
42       add_90/plus/plus/U226/Z (AN2P)         0.60      16.94 r
43       add_90/plus/plus/U71/Z (AO7P)          0.58      17.53 f
44       add_90/plus/plus/U227/Z (NR2)          0.73      18.25 r
45       add_90/plus/plus/U11/Z (MUX21L)        0.57      18.83 f
46       add_90/plus/plus/U230/Z (ND2P)         0.40      19.23 r
47       PROD[22] (out)                         0.00      19.23 r
48       data arrival time                                19.23
49
50       clock clk (rise edge)                 30.00      30.00
51       clock network delay (ideal)            0.00      30.00
52       clock uncertainty                     -1.00      29.00
53       output external delay                -12.00      17.00
54       data required time                               17.00
```

```
55        -------------------------------------------------------
56        data required time                       17.00
57        data arrival time                       -19.23
58        -------------------------------------------------------
59        slack (VIOLATED)                         -2.23
```

Another way to synthesize the architecture RTL is to use the following synthesis script. Lines 1 and 2 read the synthesized database for components UNITADD and RECODER. Line 11 tells the synthesis tools not to touch components UNITADD and RECODER. Figure 13-17 shows the synthesized schematic. It has the following area report. Note that the area is much larger than the previous two approaches. One of the reasons is that some unit adders are not necessary since their inputs are connected to "logic_0". Another reason is that the synthesis tools cannot optimize across the lower-level blocks' boundary when they are told not to touch it.

```
1        read ../db/recoder.db
2        read ../db/unitadd.db
3        design = multbw16
4        analyze -f vhdl VHDLSRC + "ch13/" + design + ".vhd"
5        elaborate MULTBW16 -architecture RTL
6        include compile.common
7        create_clock -name clk -period CLKPERIOD -waveform {0.0
HALFCLK}
8        set_clock_skew -uncertainty CLKSKEW clk
9        set_input_delay  INPDELAY -add_delay -clock clk all_inputs()
10       set_output_delay CLKPER4  -add_delay -clock clk all_outputs()
11       set_dont_touch {UNITADD RECODER}
12       compile -map_effort medium
13       include compile.report
```

```
1        Number of ports:          64
2        Number of nets:          512
3        Number of cells:         133
4        Number of references:      6
5
6        Combinational area:      7680.000000
7        Net Interconnect area:  22210.068359
8
9        Total cell area:         7680.000000
10       Total area:             29890.068359
```

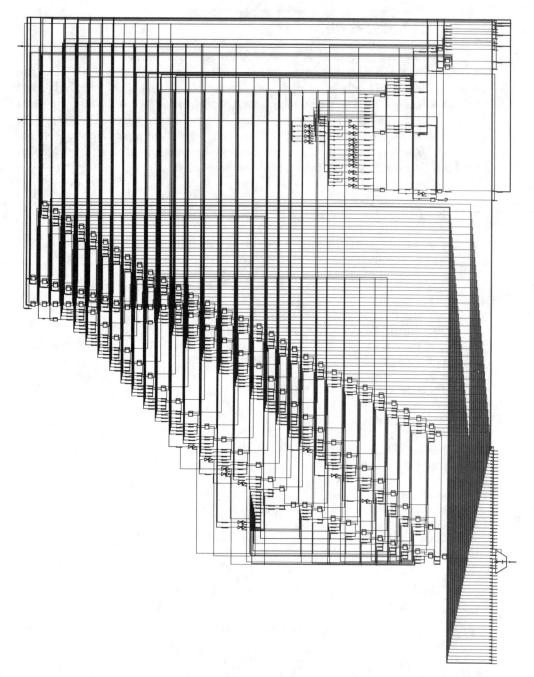

Figure 13-17 Synthesized schematic for MULTBW16, RTL, dont_touch.

Figure 13-17 can be flattened and recompiled with the following synthesis script.

```
1    remove_attribute {UNITADD RECODER } dont_touch
2    ungroup -all -flatten
3    compile
```

```
1    Number of ports:              64
2    Number of nets:             2346
3    Number of cells:            2271
4    Number of references:         61
5
6    Combinational area:       4945.000000
7    Net Interconnect area:   16142.838867
8
9    Total cell area:          4945.000000
10   Total area:              21087.839844
11
12   Operating Conditions: WCCOM    Library: lca300k
13   -----------------------------------------------------
14   MULTBW16                B9X9                lca300k
15     clock clk (rise edge)           30.00    30.00
16     clock network delay (ideal)      0.00    30.00
17     clock uncertainty               -1.00    29.00
18     output external delay          -12.00    17.00
19     data required time                       17.00
20   -----------------------------------------------------
21     data required time                       17.00
22     data arrival time                       -23.23
23   -----------------------------------------------------
24     slack (VIOLATED)                         -6.23
```

13.8 MULTIPLICATION WITH SHIFT AND ADD

Another way to perform the multiplication is to use shift and add operations. The following VHDL code shows such an example. A and B are multiplied with 4 bits each. RSTn and CLK are the signals for the flip-flops asynchronous reset and clock input. ST is the signal to tell the MULTSA circuit to start the multiplication.

```
1    library IEEE;
2    use IEEE.std_logic_1164.all;
3    use IEEE.std_logic_arith.all;
4    use IEEE.std_logic_unsigned.all;
5    use work.pack.all;
6    entity MULTSA is
7      port (
8        RSTn, CLK, ST : in  std_logic;
```

```
9          A, B          : in  std_logic_vector(3 downto 0);
10         PROD          : out std_logic_vector(7 downto 0));
11    end MULTSA;
```

The idea is to shift A bit-by-bit, and to add to a register if the corresponding bit in the multiplier is '1'. A counter is used to know when to shift and add, and when to stop. The signal CNT is used to implement the counter. It has 2 bits. The counter is enabled by the signal EN to increase the counter value by 1 as shown in lines 25 to 27. The signal SA is used to store the shifted value of A. When the start input signal ST is asserted, SA is loaded with A by shifting 1 bit as shown in line 30. The rest of the bits are loaded with '0' as shown in line 29. During the multiplication cycles as indicated by the EN signal asserted, SA is shifted left by one bit position as shown in line 32.

```
12    architecture RTL of MULTSA is
13      constant N : integer := 4;
14      signal CNT : std_logic_vector(1 downto 0);
15      signal FF, SA : std_logic_vector(2*N-1 downto 0);
16      signal EN, B1 : std_logic;
17    begin
18      ff0 : process (RSTn, CLK)
19      begin
20        if (RSTn = '0') then
21          FF  <= (FF'range => '0');
22          SA  <= (SA'range => '0');
23          CNT <= (CNT'range => '0');
24        elsif (CLK'event and CLK = '1') then
25          if (EN = '1') then
26            CNT <= CNT + 1;
27          end if;
28          if (ST = '1') then
29            SA <= (SA'range => '0');
30            SA(N downto 1) <= A;
31          elsif (EN = '1') then
32            SA <= SA(SA'length-2 downto 0) & '0';
33          end if;
34          if (ST = '1') and (B1 = '1') then
35            FF <= (FF'range => '0');
36            FF(N-1 downto 0) <= A;
37          elsif (EN = '1') and (B1 = '1') then
38            FF <= FF + SA;
39          elsif (ST = '1') then
40            FF <= (FF'range => '0');
41          end if;
42        end if;
43      end process;
44      B1 <= B(conv_integer(CNT));
45      EN <= '1' when (ST = '1') or
46                     (REDUCE_OR(CNT) = '1') else '0';
47      PROD <= FF;
48    end RTL;
```

The register FF is used to store the incremental shift-and-add accumulated value. When it is started (ST = '1'), the counter CNT should stay at "00". The multiplier bit indexed by the counter CNT is indicated by B1 as shown in line 44. Lines 34 to 36 load the FF value with the multiplicand value A when the least significant bit of the multiplier B is '1' and ST is '1'. Lines 37 and 38 add the value in SA to the FF value when EN = '1' and the current bit position of the multiplier B1 is '1'. Line 39 clears FF to all zeroes when ST is asserted to '1' but the least significant bit of the multiplier is '0' as line 34 is evaluated FALSE, otherwise, FF stays its value. Lines 45 and 46 specify the EN signal. It is asserted when ST is '1' or the counter CNT is not equal to "00". The following test bench can be used to verify the MULTSA VHDL code. The constructs are similar as before. Lines 30 to 33 generates all combinations of A and B. Lines 34 to 38 assert signal ST to be '1' for a clock cycle wide. Line 39 waits for the multipier to shift and add to complete before the next combinations of A and B are supplied. Note that line 49 waits for the same amount of time before the results are compared.

```
1    library IEEE;
2    use IEEE.std_logic_1164.all;
3    use IEEE.std_logic_arith.all;
4    entity TBMULTSA is
5    end TBMULTSA;
6    architecture BEH of TBMULTSA is
7      component MULTSA
8      port (
9        RSTn, CLk, ST : in  std_logic;
10       A, B : in  std_logic_vector(3 downto 0);
11       PROD : out std_logic_vector(7 downto 0));
12     end component;
13     constant PERIOD2 : time := 50 ns;
14     constant PERIOD  : time := PERIOD2 * 2;
15     constant STROBE  : time := PERIOD - 5 ns;
16     constant DELAY   : time := 40 ns;
17     constant W   : integer := 4; -- signal width
18     signal  A, B : std_logic_vector(W-1 downto 0);
19     signal PROD, PROD1 : std_logic_vector(2*W-1 downto 0);
20     signal RSTn, CLK, ST : std_logic;
21   begin
22     RSTn <= '0', '1' after 10 * PERIOD2;
23     clkgen : process
24     begin
25       CLK <= '0'; wait for PERIOD2;
26       CLK <= '1'; wait for PERIOD2;
27     end process;
28     p0 : process
29     begin
30       for j in 0 to 2**W-1 loop
31         A <= conv_std_logic_vector(j,W) after DELAY;
32         for k in 0 to 2**W-1 loop
```

```
33              B <= conv_std_logic_vector(k,W) after DELAY;
34              ST <= '0' after DELAY;
35              wait until CLK'event and CLK = '1';
36              ST <= '1' after DELAY;
37              wait until CLK'event and CLK = '1';
38              ST <= '0' after DELAY;
39              wait for (W+3) * PERIOD;
40            end loop;
41         end loop;
42       end process;
43       PROD <= unsigned(A) * unsigned(B);
44       check : process
45         variable err_cnt : integer := 0;
46       begin
47         for j in 1 to 2**(W*2) loop
48           wait until ST'event and ST = '1';
49           wait for (W+3) * PERIOD;
50           if (PROD /= PROD1) then
51             assert FALSE report "not compared" severity WARNING;
52             err_cnt := err_cnt + 1;
53           end if;
54         end loop;
55         assert (err_cnt = 0) report "test failed" severity ERROR;
56         assert (err_cnt /= 0) report "test passed" severity NOTE;
57         wait;
58       end process;
59       multsa0 : MULTSA
60         port map (RSTn => RSTn, CLK => CLK, ST => ST,
61                   A => A, B => B, PROD => PROD1);
62     end BEH;
63     configuration CFG_TBMULTSA of TBMULTSA is
64       for BEH
65         for multsa0 : MULTSA
66           use entity work.MULTSA(RTL);
67         end for;
68       end for;
69     end CFG_TBMULTSA;
```

Figure 13-18 shows the simulation waveform where 1011 and 1101 are multiplied. To multiply N by N bits, N clock cycles are required. It needs a counter and two registers of size $N*2$ bits. However, only one adder is needed. Figure 13-19 shows the synthesized schematic. The area comparison between a shift-and-add multiplier with a Booth-Wallace tree multiplier is left as Exercise 13.8.

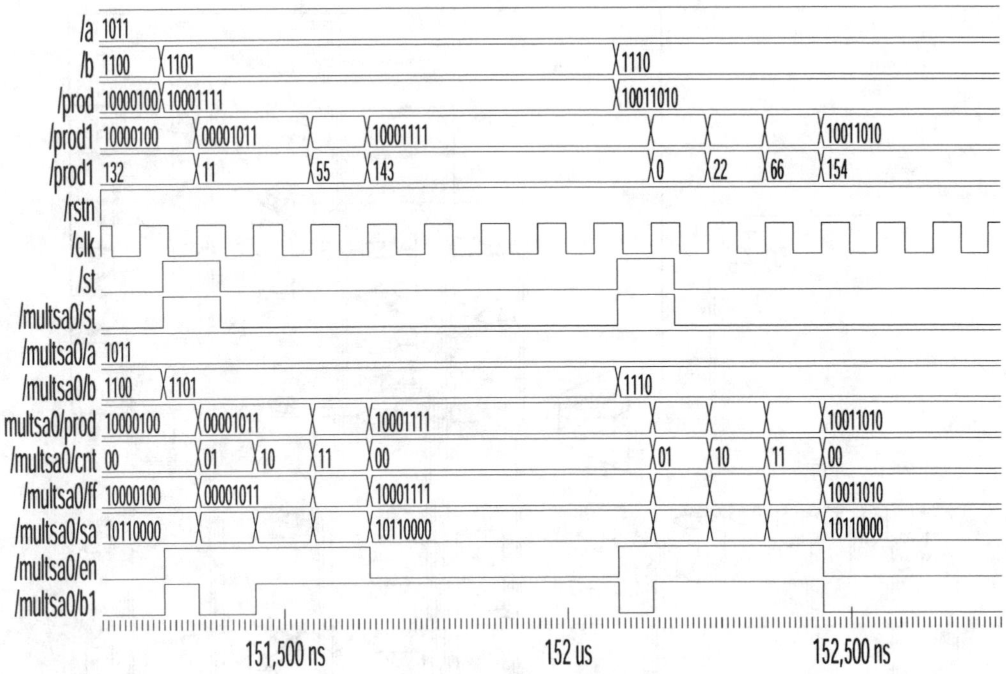

Figure 13-18 Simulation waveform for MULTSA.

There are many other multiplication algorithms. For example, Baugh and Wooley [Baugh] presented an efficient way to handle the sign extension of the partial products. Stelling, Martel, Oklobdzija, and Ravi [Stelling] suggested that optimal partial products can be added with full adders by connecting the carry and sum bits according to their timing delays at each column. In [Oklobdzija], the results showed that a carry look ahead adder as the final adder usually achieves a slower circuit. This is because the signal arrival times after the tree of adders are not the same. A combination of ripple carry adder, conditional sum adders, carry look ahead adders, and carry select adders would achieve a faster circuit for the multiplier final adder.

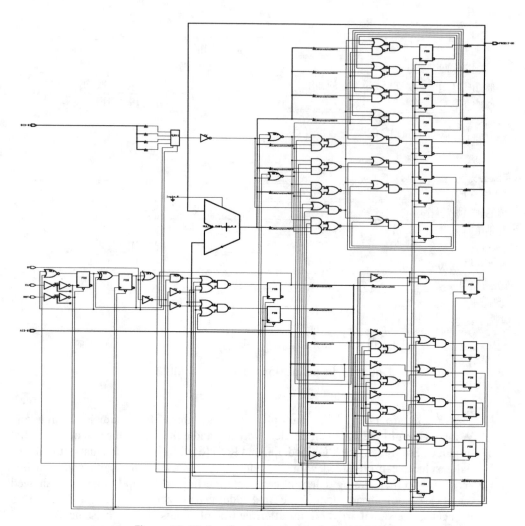

Figure 13-19 Synthesized schematic for MULTSA.

13.9 EXERCISES

13.1   Refer to MULT VHDL code. Carry save adders are used to add the partial products. As
        mentioned in the text, N - 1 adders can be used after partial products are shifted and
        filled with 0 for some bits. Write VHDL code to implement this for a 16 × 16 multiplier.
        Verify the correctness of your VHDL code. Synthesized the VHDL code and compare
        with the result of MULT RTL2 using 16-bit by 16-bit.

13.2 Refer to MULT and TBMULT VHDL code. Modify VHDL codes to multiply a 16-bit multiplier and multiplicand. Verify the functional correctness of both architectures. Develop synthesis script to synthesize both architectures and compare synthesis results. The TBMULT VHDL code enumerates all possible combinations of A and B values. How many different combinations are there? How long does it take to run all combinations? What do you suggest to change the TBMULT VHDL code?

13.3 Refer to MULTBW16 VHDL code. Develop a 32 × 32 Booth-Wallace tree multiplier. Develop a test bench to verify the VHDL code. Develop synthesis scripts to synthesize the design. Report your results in class. Synthesize the 32 × 32 multiplier with the DesignWare component. Compare the DesignWare results with your results.

13.4 Refer to MULTBW16 VHDL code. Develop VHDL code for a M × N Booth-Wallace tree multiplier where *M* and *N* are even numbers. Develop a test bench to verify your design. Develop synthesis scripts to synthesize a 24 × 12 multiplier.

13.5 Refer to MULTBW16 VHDL code line 62. Why is the block statement used to enclose lines 66 to 83?

13.6 Refer to MULTBW16 RTL synthesis results. The adder to add the last row unit adders carry and sum output takes about half of the total timing delay. Can a carry look ahead adder improve the speed? Change the VHDL code to include a carry look ahead adder. Synthesize the VHDL code and compare the synthesis results.

13.7 Refer to MULTBW16 VHDL code. Another approach to synthesize the design can be as follows: Rather than flatten the design with the ungroup command, line 11 uses the uniquify command. In this case, each RECODER and UNITADD component has its own identity. During the synthesize, these components may be synthesized differently. Try the following synthesis script (file compile.multbw.rtl.1) and compare the results with the synthesis approaches presented in the text.

```
1    read -f vhdl VHDLSRC + "ch13/" + "unitadd.vhd"
2    read -f vhdl VHDLSRC + "ch13/" + "recoder.vhd"
3    design = multbw16
4    analyze -f vhdl VHDLSRC + "ch13/" + design + ".vhd"
5    elaborate MULTBW16 -architecture RTL
6    include compile.common
7    create_clock -name clk -period CLKPERIOD -waveform {0.0
HALFCLK}
8    set_clock_skew -uncertainty CLKSKEW clk
9    set_input_delay  INPDELAY -add_delay -clock clk all_inputs()
10   set_output_delay CLKPER4  -add_delay -clock clk all_outputs()
11   uniquify
12   compile -map_effort medium
13   include compile.report
```

13.8 Refer to MULTSA VHDL code. Change it to be a 16 × 16 multiplier. Develop a test bench to verify the design. Develop synthesis scripts to synthesize the design. Compare the area with the 16 × 16 multipliers synthesis results.

13.9 Refer to MULTSA VHDL code. When the multiplier value is small, there are many leading '0' bits. When there are no more '1' bits, the multiplication is done. It is possible to complete the multiplication without going through the same number of clock cycles as the number of bits in the multiplier. This is called early termination. Modify the VHDL code to have an extra output signal RDY to indicate that the multiplication is done. Verify and synthesize the VHDL code.

13.10 A 32 × 32 multiplier may take longer than one clock cycle for your design. A pipeline multiplier is usually used. For example, flip-flops can be inserted to capture the partial products at the first stage. The second-stage flip-flops capture the results of the last row of unit adders. The third stage performs only the addition of the carry and sum outputs from the last row of the Wallace tree unit adders. Estimate the number of flip-flops required and the clock speed improvement that can be made. Develop such a VHDL code. Synthesize the VHDL code. Compare the area and timing results with a multiplier without the pipeline.

13.11 Refer to the [Baugh] paper, implement the multiplier algorithm with VHDL. Verify and synthesize the VHDL code.

13.12 Refer to the [Stelling] paper, implement the multiplier algorithm with VHDL. Verify and synthesize the VHDL code.

13.10 REFERENCES

[Baugh] Baugh, C. R. and Wooley, B. A., "A Two's Complement Parallel Array Multiplication Algorithm" in *IEEE Transactions on Computers*, C-22, 1973, pp. 1045–1047.

[Booth] Booth, A. D., "A Signed Binary Multiplication Technique" in *J. Mech. Appl. Math.*, Vol. 4, Part 2, 1951.

[Nagamatsu] Nagamatsu, M., "A 15 ns 32x32 bit CMOS Multiplier with an Improved Parallel Structure" in IEEE Custom Integrated Circuits Conference, 1989.

[Oklobdzija] Oklobdzija, V. G., "Design Strategies for the Final Adder in a Parallel Multiplier" in 29th Annual Asilomar Conference on Signals, Systems and Computers, 1995.

[Stelling] Stelling, P. F., Martel, C. U., Oklobdzija, V. G., and Ravi, R., "Optimal Circuits for Parallel Multiplier" in *IEEE Transactions on Computers*, Vol. 47, No. 3, 1998, pp. 273–285.

[Wallace] Wallace, C. S., "A Suggestion for Fast Multipliers" in *IEEE Trans. Electronic Computers*, Vol. EC-13, 1964.

Chapter 14

Fixed-Point Division

In this chapter, fixed-point binary numbers division is introduced. A division circuit in general requires more hardware than an adder or a multiplier. We will study how a divider can be optimized through better VHDL coding and synthesis.

14.1 BASIC DIVISION CONCEPT

A basic fixed-point binary number division algorithm is similar to the paper and pencil approach. Figure 14-1 shows how two 8-bit numbers 01011001 (decimal 89) and 00001001 (decimal 9) can be divided. In the division, the *dividend* A is divided by the *divisor* B. The result is called the *quotient* Q. The *remainder* R may not be zero if the divisor is not a factor of the dividend. In the division, equation A = B * Q + R holds. For this example, the quotient Q is 1001 (decimal 9) and the remainder R is 1000 (decimal 8). 89 = 9 * 9 + 8.

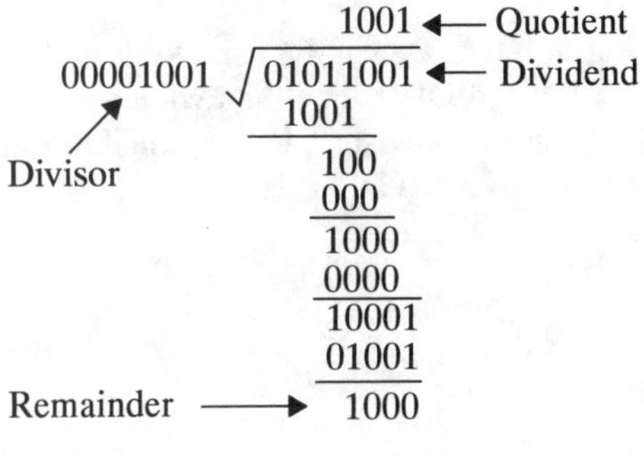

```
                           1001 ◄── Quotient
             00001001 √ 01011001 ◄── Dividend
                    ↗        1001
            Divisor          100
                             000
                             1000
                             0000
                             10001
                             01001
        Remainder ─────►     1000
```

Figure 14-1 Paper and pencil division: 01011001 / 00001001.

The above paper and pencil approach works well. However, our brains are calculating where to put the first 1 in the quotient. We may look at the dividend and the divisor. The most significant 4 bits of the divisor are zero. We figure that the first quotient bit that can be 1 is at least four bit positions from the left. It is harder for the hardware to figure this one out. The hardware is more adept at the systematic approach. One idea that we can take from the paper and pencil approach is to subtract the dividend by the divisor at the right bit positions. Since it is not easy for the hardware to figure out where the first 1 in the quotient should be, we can start from the very beginning. Figure 14-2 shows the divisor is left-shifted bit by bit as array L. The dividend is first zero extended to double the width as the partial remainder. The division starts at bit position 7 which is the leftmost position of the original divisor. The partial remainder is compared with the shifted divisor. If the partial remainder is no less than the left-shifted divisor, the quotient has '1' in the bit position, the partial remainder is subtracted from the left-shifted divisor. Otherwise, the quotient bit position has '0' and the partial remainder remains the same. This process is repeated from bit position 7 down to bit position 0. As shown in Figure 14-2, quotient "00001001" is obtained with "00001000" as the remainder. These results match the paper and pencil approach shown in Figure 14-1.

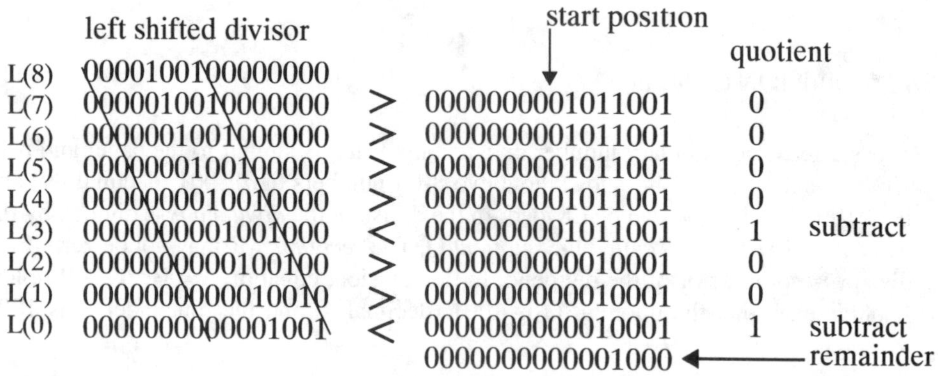

left shifted divisor   start position

				quotient	
L(8)	00001001000000000				
L(7)	00001001000000000	>	0000000001011001	0	
L(6)	00000100100000000	>	0000000001011001	0	
L(5)	00000010010000000	>	0000000001011001	0	
L(4)	00000001001000000	>	0000000001011001	0	
L(3)	00000000100100000	<	0000000001011001	1	subtract
L(2)	00000000010010000	>	0000000000010001	0	
L(1)	00000000001001000	>	0000000000010001	0	
L(0)	00000000000100100	<	0000000000010001	1	subtract
			0000000000001000		remainder

Figure 14-2 Hardware division approach: 01011001 / 00001001.

The following VHDL code can be used to describe the above approach. Entity DIV8 has inputs DIVA and DIVB of 8-bits wide as shown in lines 6 and 7. Output DIV0 is asserted '1' when the divisor DIVB is equal to zero. Output DIVQ is the quotient result. It is set to zero when DIVB is zero.

```
1    library IEEE;
2    use IEEE.std_logic_1164.all;
3    use IEEE.std_logic_unsigned.all;
4    entity DIV8 is
5      port (
6        DIVA  : in  std_logic_vector(7 downto 0);
7        DIVB  : in  std_logic_vector(7 downto 0);
8        DIV0  : out std_logic;
9        DIVQ  : out std_logic_vector(7 downto 0));
10   end DIV8;
```

Line 8 declares constant N as the width of the data. Lines 13 to 15 declare arrays R and L to hold the values of partial remainders and left-shifted divisors. Line 16 declares signal Q to store the quotient bits. Signal Z0 is used to set all '0' bits as shown in line 19. Line 17 declares signal Z to indicate whether the divisor is equal to zero. Lines 19 and 20 initialize signal L(0), and R(N). Lines 22 to 24 form the left-shifted divisors. Note that they are just connections to various divisors bits. There are no circuits required. Lines 26 to 29 compute the quotient bit by bit from bit position N-1 downto 0. Line 27 sets the quotient bit to '1' when the partial remainder is greater or equal to the corresponding left-shifted divisor. Line 28 computes the partial remainder. It may remain the same if the quotient bit is '0', otherwise, the partial remainder is subtracted from the corresponding left-shifted divisor. Line 30 detects whether the divisor is zero. Line 31 forms the output DIV0. Line 32 generates the quotient output DIVQ.

```
11    architecture RTL of DIV8 is
12      constant N : integer := 8;
13      subtype bit16 is std_logic_vector(2*N-1 downto 0);
14      type vect8   is array (N downto 0) of bit16;
15      signal R, L : vect8;
16      signal Q,Z0 : std_logic_vector(N-1 downto 0);
17      signal Z : std_logic;
18    begin
19      Z0 <= (Z0'range => '0');
20      L(0)   <= Z0 & DIVB; R(N) <= Z0 & DIVA;
21
22      sh0 : for j in 1 to N generate
23        L(j) <= L(j-1)(2*N-2 downto 0) & '0';
24      end generate;
25
26      st0 : for j in N-1 downto 0 generate
27        Q(j) <= '1' when (R(j+1) >= L(j)) else '0';
28        R(j) <= R(j+1)-L(j) when Q(j) = '1' else R(j+1);
29      end generate;
30      Z    <= '1' when DIVB = Z0 else '0';
31      DIV0 <= '1' when Z = '1'  else '0';
32      DIVQ <= (DIVQ'range => '0') when Z = '1' else Q;
33    end RTL;
```

The following test bench can be used to verify the DIV8 VHDL code. Lines 31 to 42 generate the expected results. Note that lines 38 to 40 use the type conversion functions in the std_logic_arith package. Figure 14-3 shows a portion of the simulation waveform where values "01011001" are divided by "00001001".

```
1     library IEEE;
2     use IEEE.std_logic_1164.all;
3     use IEEE.std_logic_arith.all;
4     entity TBDIV8 is
5     end TBDIV8;
```

```
6      architecture BEH of TBDIV8 is
7        component DIV8
8        port (
9          DIVA  : in  std_logic_vector(7 downto 0);
10         DIVB  : in  std_logic_vector(7 downto 0);
11         DIV0  : out std_logic;
12         DIVQ  : out std_logic_vector(7 downto 0));
13       end component;
14       constant W   : integer := 8; -- signal width
15       signal A, B, Q, Q0 : std_logic_vector(W-1 downto 0);
16       signal DIV, DIV0 : std_logic;
17       constant PERIOD : time := 50 ns;
18       constant STROBE : time := PERIOD - 5 ns;
19     begin
20       p0 : process
21       begin
22         for j in 0 to 2**W-1 loop
23           A  <= conv_std_logic_vector(j,W);
24           for k in 0 to 2**W-1 loop
25             B  <= conv_std_logic_vector(k,W);
26             wait for PERIOD;
27           end loop;
28         end loop;
29         wait;
30       end process;
31       cgen : process (A, B)
32       begin
33         if (conv_integer(unsigned(B)) = 0) then
34           DIV <= '1';
35           Q   <= (Q'range => '0');
36         else
37           DIV <= '0';
38           Q   <=conv_std_logic_vector(
39                   (conv_integer(unsigned(A)) /
40                     conv_integer(unsigned(B))), W);
41         end if;
42       end process;
43       check : process
44         variable err_cnt : integer := 0;
45       begin
46         wait for STROBE;
47         for j in 1 to 2**(W*2) loop
48           if (Q /= Q0) or (DIV /= DIV0) then
49             assert FALSE report "not compared" severity WARNING;
50             err_cnt := err_cnt + 1;
51           end if;
52           wait for PERIOD;
53         end loop;
54         assert (err_cnt = 0) report "test failed" severity ERROR;
55         assert (err_cnt /= 0) report "test passed" severity NOTE;
56         wait;
57       end process;
```

```
58      div80 : DIV8
59          port map (DIVA => A, DIVB => B, DIV0 => DIV0, DIVQ => Q0);
60      end BEH;
61      configuration CFG_TBDIV8 of TBDIV8 is
62        for BEH
63          for div80 : DIV8
64            use entity work.DIV8(RTL);
65          end for;
66        end for;
67      end CFG_TBDIV8;
```

/a	01011001			
/b	00001000	00001001	00001010	00001011
/q	00001011	00001001	00001000	
/q0	00001011	00001001	00001000	
/div				
/div0				
/div80/r				
(8)	0000000001011001			
(7)	0000000001011001			
(6)	0000000001011001			
(5)	0000000001011001			
(4)	0000000001011001			
(3)	0000000000011001	0000000000010001	0000000000001001	0000000000000001
(2)	0000000000011001	0000000000010001	0000000000001001	0000000000000001
(1)	0000000000001001	0000000000010001	0000000000001001	0000000000000001
(0)	0000000000000001	0000000000001000	0000000000001001	0000000000000001
/div80/l				
(8)	0000100000000000	0000100100000000	0000101000000000	0000101100000000
(7)	0000010000000000	0000010010000000	0000010100000000	0000010110000000
(6)	0000001000000000	0000001001000000	0000001010000000	0000001011000000
(5)	0000000100000000	0000000100100000	0000000101000000	0000000101100000
(4)	0000000010000000	0000000010010000	0000000010100000	0000000010110000
(3)	0000000001000000	0000000001001000	0000000001010000	0000000001011000
(2)	0000000000100000	0000000000100100	0000000000101000	0000000000101100
(1)	0000000000010000	0000000000010010	0000000000010100	0000000000010110
(0)	0000000000001000	0000000000001001	0000000000001010	0000000000001011
/div80/q	00001011	00001001	00001000	
/div80/z0	00000000			
/div80/z				

1,139,600 ns   1139650 ns   1139700 ns   1139750 ns

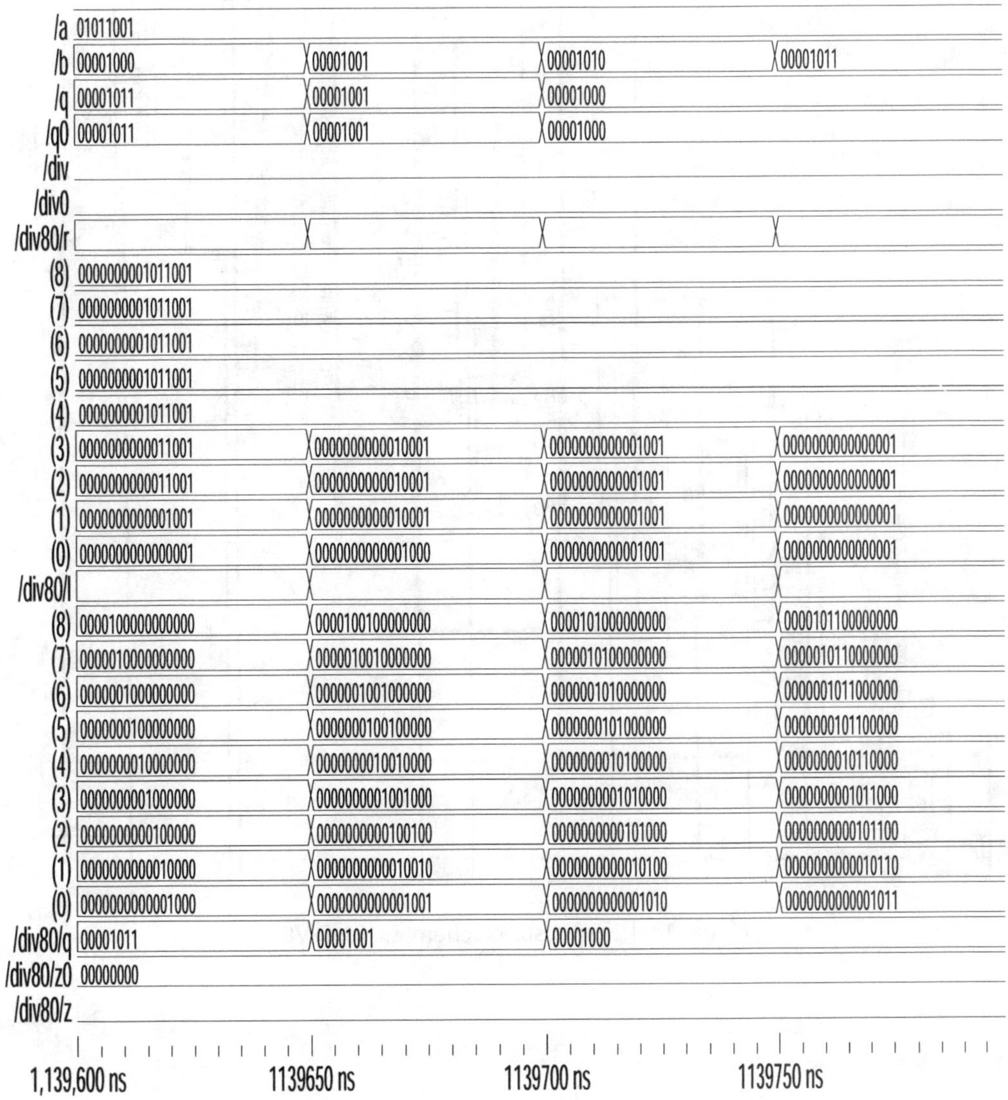

Figure 14-3 Simulation waveform for DIV8.

The DIV8 VHDL code can be synthesized. The schematic is shown in Figure 14-4. Note that there are eight comparators and only seven subtracters. Why not eight subtracters?

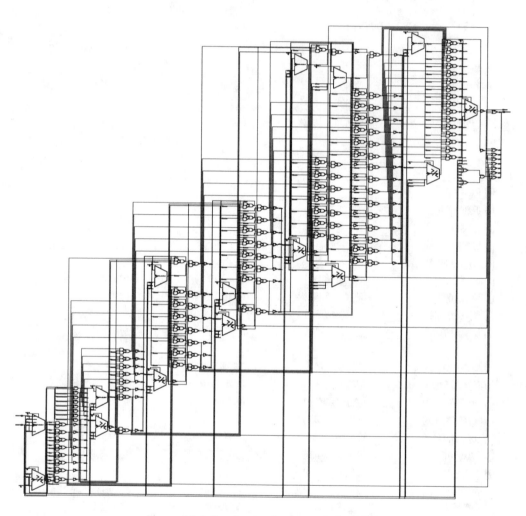

Figure 14-4 Synthesized schematic for DIV8.

14.2 32-BIT DIVIDER

The approach presented in the last section can have more bits for the dividend and divisor. Let's consider a 32-bit divider in this section. The DIV32 VHDL entity is shown below.

```
1    library IEEE;
2    use IEEE.std_logic_1164.all;
3    entity DIV32 is
4      port (
5        DIVA  : in  std_logic_vector(31 downto 0);
6        DIVB  : in  std_logic_vector(31 downto 0);
7        DIV0  : out std_logic;
8        DIVQ  : out std_logic_vector(31 downto 0));
9    end DIV32;
```

The DIV8 approach can be extended with the following architecture. However, instead of treating the dividend and divisor as unsigned numbers as in DIV8, here we treat the numbers in a 2's complement format. Line 13 defines a signal with all bits as '0'. Line 14 uses the concurrent procedure call COMP2 to generate the 2's complement numbers for the divisor and the dividend. The procedure COMP2 has been defined in package PACK which is referenced in line 3. Lines 15 and 16 select the original number or its 2's complement format based on whether the original number is negative or positive. The goal is to use the approach described for DIV8 that deals with unsigned numbers. Signals A2 and B2 will be positive with the leftmost bit as '0'. Line 17 computes the sign bit of the result. If the divisor and the dividend have different signs (in the leftmost bit), the result is negative. Lines 18 to 33 are about the same as in DIV8 except that the result is taken its 2's complement when the result is negative.

```
1    library IEEE;
2    use IEEE.std_logic_unsigned.all;
3    use work.pack.all;
4    architecture FLAT of DIV32 is
5      constant N : integer := 32;
6      subtype bit2n is std_logic_vector(2*N-1 downto 0);
7      type vectn1  is array (N downto 0) of bit2n;
8      signal R, L : vectn1;
9      signal Q,Q1,A1,B1 : std_logic_vector(N-1 downto 0);
10     signal A2,B2,Z0 : std_logic_vector(N-1 downto 0);
11     signal S, Z : std_logic;
12   begin
13     Z0 <= (Z0'range => '0');
14     COMP2(DIVA, A1); COMP2(DIVB, B1);
15     A2    <= DIVA when DIVA(N-1) = '0' else A1;
16     B2    <= DIVB when DIVB(N-1) = '0' else B1;
17     S     <= DIVA(N-1) xor DIVB(N-1);
18     L(0)  <= Z0 & B2; R(N) <= Z0 & A2;
```

```
19
20      sh0 : for j in 1 to N-1 generate
21        L(j) <= L(j-1)(2*N-2 downto 0) & '0';
22      end generate;
23
24      st0 : for j in N-1 downto 0 generate
25        Q(j) <= '1' when (R(j+1) >= L(j)) else '0';
26        R(j) <= R(j+1)-L(j) when Q(j) = '1' else R(j+1);
27      end generate;
28
29      Z    <= '1' when DIVB = Z0 else '0';
30      DIV0 <= Z;
31      COMP2(Q, Q1);
32      DIVQ <= (DIVQ'range => '0') when Z = '1' else
33                Q1 when S = '1' else Q;
34    end FLAT;
```

The above VHDL code can be synthesized. However, lines 24 to 27 are a generate statement. Lines 25 to 26 will be expanded 32 times. This is called "unwrapping" in the synthesis process. There will be thirty-two 64-bit wide comparators and thirty-two 64-bit wide subtracters after they are unwrapped. This will take a long time to complete the synthesis. The next section suggests another architecture.

14.3 DESIGN PARTITION

The following architecture HIER VHDL code is similar to the above architecture FLAT. The statements inside the architecture FLAT generate statement are placed in a subcomponent STAGE. The STAGE component is then instantiated with the same generate statement as shown in lines 30 to 33. The STAGE component can be synthesized and saved first. When architecture HIER is synthesized, STAGE can be referenced and not to be resynthesized again with the "set_dont_touch STAGE" command. This makes the synthesis process faster.

```
1     library IEEE;
2     use IEEE.std_logic_unsigned.all;
3     use work.pack.all;
4     architecture HIER of DIV32 is
5       constant N : integer := 32;
6       subtype bit2n  is std_logic_vector(2*N-1 downto 0);
7       type    vectn1 is array (N downto 0) of bit2n;
8       signal R, L : vectn1;
9       signal Q,Q1,A1,B1 : std_logic_vector(N-1 downto 0);
10      signal A2,B2,T,Z0 : std_logic_vector(N-1 downto 0);
11      signal S, Z       : std_logic;
12      component STAGE
13      port (
14        R, L : in  std_logic_vector(63 downto 0);
15        Q0   : out std_logic;
```

```
16              RO     : out std_logic_vector(63 downto 0));
17        end component;
18   begin
19        Z0     <= (Z0'range => '0');
20        COMP2(DIVA, A1); COMP2(DIVB, B1);
21        A2     <= DIVA when DIVA(N-1) = '0' else A1;
22        B2     <= DIVB when DIVB(N-1) = '0' else B1;
23        S      <= DIVA(N-1) xor DIVB(N-1);
24        L(0) <= Z0 & B2; R(N) <= Z0 & A2;
25        Z      <= '1' when DIVB = Z0 else '0';
26        DIV0 <= Z;
27        sh0 : for j in 1 to N-1 generate
28          L(j) <= L(j-1)(2*N-2 downto 0) & '0';
29        end generate;
30        st0 : for j in N-1 downto 0 generate
31          st : STAGE port map
32             (R  => R(j+1), L => L(j), Q0 => Q(j), RO => R(j));
33        end generate;
34        COMP2(Q, Q1);
35        DIVQ <= (DIVQ'range => '0') when Z = '1' else
36               Q1                  when S = '1' else Q;
37   end HIER;
```

14.4 DESIGN OPTIMIZATION

The STAGE design can be modeled with the following VHDL code. Architecture RTA (from lines 10 to 16) are similar as before. Figure 14-5 shows part of the
synthesized schematic. There are a comparator and a subtracter as expected from lines
13 and 15. Can we improve on this?

```
1    library IEEE;
2    use IEEE.std_logic_1164.all;
3    use IEEE.std_logic_unsigned.all;
4    entity STAGE is
5      port (
6        R, L  : in  std_logic_vector(63 downto 0);
7        Q0    : out std_logic;
8        RO    : out std_logic_vector(63 downto 0 ));
9    end STAGE;
10   architecture RTA of STAGE is
11     signal T : std_logic;
12   begin
13     T  <= '1' when (R >= L) else '0';
14     Q0 <= T;
15     RO <= R-L when (T = '1') else R;
16   end RTA;
```

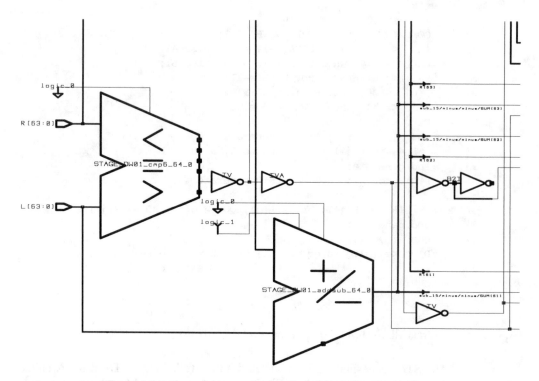

Figure 14-5 Part of the synthesized schematic for STAGE RTA.

The following architecture RTB infers a subtracter in line 21. There is no comparator inferred. Figure 14-6 shows part of the synthesized schematic.

```
17    architecture RTB of STAGE is
18      signal T : std_logic;
19      signal SUB   : std_logic_vector(63 downto 0);
20    begin
21      SUB <= R - L;
22      T   <= '1' when (SUB(63) = '0') else '0';
23      QO  <= T;
24      RO  <= SUB when (T = '1') else R;
25    end RTB;
```

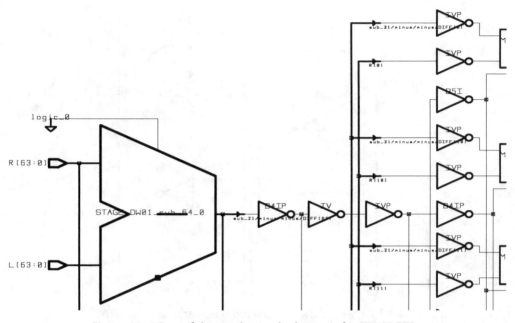

Figure 14-6 Part of the synthesized schematic for STAGE RTB.

The following architecture RTC suggests another way to implement the STAGE design. It is based on the observation that the leftmost 32 bits of the dividend are always zero. Lines 31 and 32 check whether the leftmost 32 bits of the signal L (divisor) are all zero. Signal MSB is set to '1' when the leftmost 32 bits are not all zero. The quotient bit is set to '1' when MSB is '0' and the leftmost bit of the subtraction result is also '0'. This means that the rightmost 32 bits of the dividend are greater or equal to the rightmost 32 bits of the divisor, and the leftmost 32 bits of the divisor are all zero. Line 35 generates the partial remainder to go to the next stage. Figure 14-7 shows part of the synthesized schematic. Note that the subtracter is 32 bits wide.

```
26    architecture RTC of STAGE is
27      signal T, MSB : std_logic;
28      signal SUB    : std_logic_vector(31 downto 0);
29    begin
30      SUB <= R(31 downto 0) - L(31 downto 0);
31      MSB <= '0' when L(63 downto 31) =
32           "00000000000000000000000000000000" else '1';
33      T   <= '1' when (MSB = '0') and (SUB(31) = '0') else '0';
34      QO  <= T;
35      RO  <= "00000000000000000000000000000000" & SUB
36           when (T = '1') else R;
37    end RTC;
```

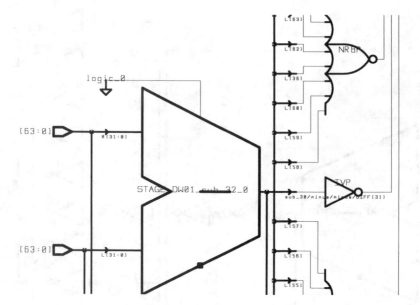

Figure 14-7 Part of the synthesized schematic for STAGE RTC.

All three architectures are synthesized. The following shows the synthesis area and timing results. The timing results are close. Architecture RTC is the fastest with the smallest area.

```
1     -- STAGE architecture RTA
2     Number of ports:              193
3     Number of nets:               977
4     Number of cells:              846
5     Number of references:          46
6
7     Combinational area:       1631.000000
8     Net Interconnect area:    6173.522949
9
10    Total cell area:          1631.000000
11    Total area:               7804.522949
12
13    Operating Conditions: WCCOM    Library: lca300k
14    STAGE                  B9X9              lca300k
15
16       clock clk (rise edge)        30.00       30.00
17       clock network delay (ideal)   0.00       30.00
18       clock uncertainty            -1.00       29.00
19       output external delay       -12.00       17.00
```

```
20     data required time                          17.00
21     ------------------------------------------------------
22     data required time                          17.00
23     data arrival time                          -18.65
24     ------------------------------------------------------
25     slack (VIOLATED)                            -1.65


1      -- STAGE architecture RTB
2      Number of ports:              193
3      Number of nets:               851
4      Number of cells:              722
5      Number of references:          43
6
7      Combinational area:        1401.000000
8      Net Interconnect area:     4976.413574
9
10     Total cell area:           1401.000000
11     Total area:                6377.413574
12
13     Operating Conditions: WCCOM    Library: lca300k
14     STAGE                      B9X9                lca300k
15       clock clk (rise edge)              30.00       30.00
16       clock network delay (ideal)         0.00       30.00
17       clock uncertainty                  -1.00       29.00
18       output external delay             -12.00       17.00
19       data required time                            17.00
20     ------------------------------------------------------
21       data required time                            17.00
22       data arrival time                            -18.25
23     ------------------------------------------------------
24       slack (VIOLATED)                              -1.25


1      -- STAGE architecture RTC
2      Number of ports:              193
3      Number of nets:               498
4      Number of cells:              366
5      Number of references:          41
6
7      Combinational area:         770.000000
8      Net Interconnect area:     2756.979248
9
10     Total cell area:            770.000000
11     Total area:                3526.979248
12
13     Operating Conditions: WCCOM    Library: lca300k
14     STAGE                      B9X9                lca300k
15
16       clock clk (rise edge)              30.00       30.00
17       clock network delay (ideal)         0.00       30.00
18       clock uncertainty                  -1.00       29.00
```

```
19      output external delay              -12.00        17.00
20      data required time                               17.00
21      -----------------------------------------------------
22      data required time                               17.00
23      data arrival time                               -16.06
24      -----------------------------------------------------
25      slack (MET)                                       0.94
```

14.5 DESIGN VERIFICATION

To verify the DIV32 with various architectures, the following shows the behavioral VHDL model for architecture BEH of DIV32. Lines 9 to 11 check whether the divisor is a valid number. Line 12 checks whether the divisor is equal to zero. Lines 13 to 14 set the outputs when the divisor is zero. Lines 17 to 19 use the operator "/" to calculate the answer with type conversions. Note that DIVA (DIVB) is changed to the signed format before it is converted to an integer. The result is then converted to std_logic_vector.

```
1      library IEEE;
2      use IEEE.std_logic_arith.all;
3      architecture BEH of DIV32 is
4      begin
5        p0 : process(DIVA, DIVB)
6          variable X : boolean;
7        begin
8          X := FALSE;
9          for j in DIVB'range loop
10           X := X or IS_X(DIVB(j));
11         end loop;
12         if (DIVB = (DIVB'range => '0')) or X then
13           DIV0 <= '1';
14           DIVQ <= (DIVQ'range => '0');
15         else
16           DIV0 <= '0';
17           DIVQ <= conv_std_logic_vector
18                      (conv_integer(signed(DIVA)) /
19                       conv_integer(signed(DIVB)), 32);
20         end if;
21       end process;
22     end BEH;
```

The following test bench can be used to verify the DIV32 design. Lines 21 to 26 declare component DIV32. Lines 16 to 20 declare signals to be used. Lines 10 to 15 defines several constants. Lines 5 and 6 reference packages for doing text file input and output. Lines 27 to 33 declare procedure SETAB to convert integers to std_logic_vector format. Lines 35 to 44 instantiate components DIV32 five times.

```vhdl
1    library IEEE;
2    use IEEE.std_logic_1164.all;
3    use IEEE.std_logic_arith.all;
4    use IEEE.std_logic_unsigned.all;
5    use IEEE.std_logic_textio.all;
6    use STD.TEXTIO.all;
7    entity TBDIV32 is
8    end TBDIV32;
9    Architecture BEH of TBDIV32 is
10     constant PERIOD   : time    := 30 ns;
11     constant BIGINT   : integer := 2**30-1+2**30;
12     constant NUMA      : integer := 177;
13     constant NUMB      : integer := 131;
14     constant STEPA     : integer := BIGINT / NUMA;
15     constant STEPB     : integer := BIGINT / NUMB;
16     signal A, B,      BEH_DIVQ : std_logic_vector(31 downto 0);
17     signal RTF_DIVQ, RTA_DIVQ : std_logic_vector(31 downto 0);
18     signal RTB_DIVQ, RTC_DIVQ : std_logic_vector(31 downto 0);
19     signal BEH_DIV0, RTF_DIV0, RTA_DIV0 : std_logic;
20     signal RTB_DIV0, RTC_DIV0           : std_logic;
21     component DIV32
22     port (
23        DIVA, DIVB : in  std_logic_vector(31 downto 0);
24        DIV0       : out std_logic;
25        DIVQ       : out std_logic_vector(31 downto 0));
26     end component;
27     procedure SETAB (
28        INTA, INTB  : in  integer;
29        signal A, B : out std_logic_vector(31 downto 0)) is
30     begin
31       A <= conv_std_logic_vector(INTA, 32);
32       B <= conv_std_logic_vector(INTB, 32);
33     end SETAB;
34   begin
35     BEH0 : DIV32 port map ( DIVA => A, DIVB => B,
36            DIV0 => BEH_DIV0, DIVQ => BEH_DIVQ);
37     FLAT : DIV32 port map ( DIVA => A, DIVB => B,
38            DIV0 => RTF_DIV0, DIVQ => RTF_DIVQ);
39     RTA0 : DIV32 port map ( DIVA => A, DIVB => B,
40            DIV0 => RTA_DIV0, DIVQ => RTA_DIVQ);
41     RTB0 : DIV32 port map ( DIVA => A, DIVB => B,
42            DIV0 => RTB_DIV0, DIVQ => RTB_DIVQ);
43     RTC0 : DIV32 port map ( DIVA => A, DIVB => B,
44            DIV0 => RTC_DIV0, DIVQ => RTC_DIVQ);
45
46     p0 : process
47       variable  ind1, ind2: integer := 0;
48       file infile : text is in "test.in";
49       variable lin : line;
50       variable P1, P2, N1, N2 : integer;
```

```
51    begin
52      while (not endfile(infile)) loop
53        readline(infile, lin);
54        read(lin, ind1);
55        read(lin, ind2);
56        SETAB(ind1, ind2, A, B);
57        wait for PERIOD;
58      end loop;
59
60      for j in 1 to NUMA loop
61        P1 := BIGINT-j*STEPA; N1 := -P1;
62        for k in NUMB downto 1 loop
63          P2 := BIGINT-k*STEPB; N2 := -P2;
64          for m in 0 to 3 loop
65            case m is
66              when 0 => SETAB(P1, P2, A, B);
67              when 1 => SETAB(N1, P2, A, B);
68              when 2 => SETAB(P1, N2, A, B);
69              when 3 => SETAB(N1, N2, A, B);
70            end case;
71            wait for PERIOD;
72          end loop;
73        end loop;
74      end loop;
75      wait;
76    end process;
77    c0 : process
78    begin
79      wait for PERIOD - 3 ns;
80      assert (BEH_DIV0 = RTB_DIV0) and (BEH_DIV0 = RTA_DIV0) and
81             (BEH_DIV0 = RTC_DIV0) and (BEH_DIV0 = RTF_DIV0) and
82             (BEH_DIVQ = RTB_DIVQ) and (BEH_DIVQ = RTA_DIVQ) and
83             (BEH_DIVQ = RTC_DIVQ) and (BEH_DIVQ = RTF_DIVQ)
84        report "Results do not match" severity error;
85    end process;
86  end BEH;
```

Lines 52 to 58 read dividend and divisor values from a text file. The idea is to use a file to verify the design before the values are automatically generated with the loop statements. The values generated by the loop statement may be harder to verify. By changing the values of NUMA, NUMB, STEPA, STEPB (lines 12 to 15), the number of dividend and divisor pairs can be adjusted. Note that in lines 64 to 72, each pair of values are repeated for the four combinations of positive and negative values. The following VHDL code configures each of the five component instantiations. One is the architecture BEH (lines 89 to 91). One is the architecture FLAT in lines 92 to 94. Lines 95 to 104 hierarchically configure the DIV32 architecture HIER with subcomponent STAGE configured with architecture RTA. Lines 105 to 125 configure DIV32 architecture HIER with subcomponent STAGE configured as architectures RTB and RTC.

```
87   configuration CFG_TBDIV32 of TBDIV32 is
88     for BEH
89       for BEH0 : DIV32
90         use entity WORK.DIV32(BEH);
91       end for;
92       for FLAT : DIV32
93         use entity WORK.DIV32(FLAT);
94       end for;
95       for RTA0 : DIV32
96         use entity WORK.DIV32(HIER);
97         for HIER
98           for st0
99             for all : STAGE
100              use entity WORK.STAGE(RTA);
101            end for;
102          end for;
103        end for;
104      end for;
105      for RTB0 : DIV32
106        use entity WORK.DIV32(HIER);
107        for HIER
108          for st0
109            for all : STAGE
110              use entity WORK.STAGE(RTB);
111            end for;
112          end for;
113        end for;
114      end for;
115      for RTC0 : DIV32
116        use entity WORK.DIV32(HIER);
117        for HIER
118          for st0
119            for all : STAGE
120              use entity WORK.STAGE(RTC);
121            end for;
122          end for;
123        end for;
124      end for;
125    end for;
126  end CFG_TBDIV32;
```

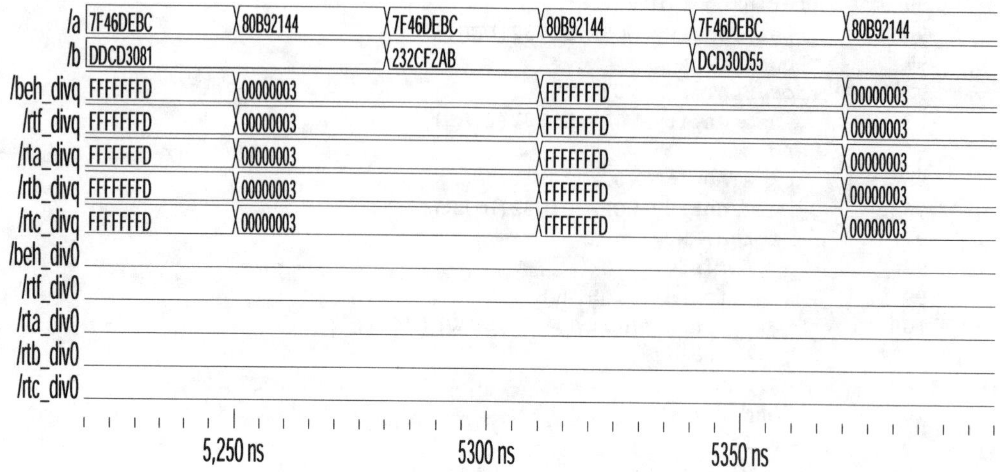

/a	7F46DEBC	80B92144	7F46DEBC	80B92144	7F46DEBC	80B92144
/b	DDCD3081		232CF2AB		DCD30D55	
/beh_divq	FFFFFFFD	00000003	FFFFFFFD		00000003	
/rtf_divq	FFFFFFFD	00000003	FFFFFFFD		00000003	
/rta_divq	FFFFFFFD	00000003	FFFFFFFD		00000003	
/rtb_divq	FFFFFFFD	00000003	FFFFFFFD		00000003	
/rtc_divq	FFFFFFFD	00000003	FFFFFFFD		00000003	
/beh_div0						
/rtf_div0						
/rta_div0						
/rtb_div0						
/rtc_div0						

5,250 ns   5300 ns   5350 ns

Figure 14-8 Simulation waveform for TBDIV32.

14.6 DESIGN SYNTHESIS

There are many ways to synthesize DIV32 architecture HIER. The first approach can be described as follows:

• Analyze STAGE VHDL code and elaborate STAGE with the selected architecture.

• Synthesize STAGE and then save the synthesis result in the database.

• Analyze DIV32 entity div32_e.vhd and architecture HIER div32_hier.vhd. Elaborate DIV32 architecture HIER.

• Read the synthesized STAGE data base. Use the "set_dont_touch {STAGE}" command.

• Set the timing constraints and then synthesize with the compile command.

The second approach can be:

• Analyze STAGE VHDL code and elaborate STAGE with the selected architecture.

• Analyze DIV32 entity div32_e.vhd and architecture HIER div32_hier.vhd. Elaborate DIV32 architecture HIER.

• Set the timing constraints. Use the command uniquify to specify that each instance of component STAGE can be compiled separately. Use the compile command to synthesize DIV32.

The synthesis of DIV32 in the above two approaches is left as Exercise 14.6. Another approach to synthesize the DIV32 is to change the DIV32 architecture HIER to be the following. It is identical to HIER architecture VHDL code, except that a **block statement** (lines 19 to 35) is used to enclose the statements that are outside the **generate statement** for the STAGE component instantiation.

```
1    library IEEE;
2    use IEEE.std_logic_unsigned.all;
3    use work.pack.all;
4    architecture HIERG of DIV32 is
5      constant N : integer := 32;
6      subtype bit2n  is std_logic_vector(2*N-1 downto 0);
7      type    vectn1 is array (N downto 0) of bit2n;
8      signal R, L : vectn1;
9      signal Q,Q1,A1,B1 : std_logic_vector(N-1 downto 0);
10     signal A2,B2,T,Z0 : std_logic_vector(N-1 downto 0);
11     signal S, Z       : std_logic;
12     component STAGE
13     port (
14       R, L  : in  std_logic_vector(63 downto 0);
15       QO    : out std_logic;
16       RO    : out std_logic_vector(63 downto 0));
17     end component;
18   begin
19     g0 : block
20     begin
21       Z0   <= (Z0'range => '0');
22       COMP2(DIVA, A1); COMP2(DIVB, B1);
23       A2   <= DIVA when DIVA(N-1) = '0' else A1;
24       B2   <= DIVB when DIVB(N-1) = '0' else B1;
25       S    <= DIVA(N-1) xor DIVB(N-1);
26       L(0) <= Z0 & B2; R(N) <= Z0 & A2;
27       Z    <= '1' when DIVB = Z0 else '0';
28       DIV0 <= Z;
29       sh0 : for j in 1 to N-1 generate
30         L(j) <= L(j-1)(2*N-2 downto 0) & '0';
31       end generate;
32       COMP2(Q, Q1);
33       DIVQ <= (DIVQ'range => '0') when Z = '1' else
34               Q1                  when S = '1' else Q;
35     end block;
36     st0 : for j in N-1 downto 0 generate
37       st : STAGE port map
38         (R => R(j+1), L => L(j), QO => Q(j), RO => R(j));
39     end generate;
40   end HIERG;
```

The following simple synthesis script shows how the block statement can be grouped as a block and then synthesize the small block. Line 6 groups the statements inside the block statement labeled g0. Line 7 sets the current design to g0. Lines 9 to 13 specify timing constraints and then synthesize block g0. After block g0 is synthesized, DIV32 contains all synthesized block since STAGE has been synthesized and read in. Line 14 sets the current design back to DIV32. Line 15 treats STAGE and g0 blocks as synthesized. Line 16 synthesizes the DIV32 design to add buffers for high-fanout nets. Figure 14-9 shows part of the synthesized schematic. Note that several buffers are added to reduce the loading on originally high fanout nets. Breaking down the design into smaller pieces reduces the synthesis computing run time.

```
1     read ../db/stage.db
2     design = div32
3     analyze -f vhdl VHDLSRC + "ch14/" + design + "_e.vhd"
4     analyze -f vhdl VHDLSRC + "ch14/" + design + "_hierg.vhd"
5     elaborate DIV32 -architecture HIERG
6     group -hdl_block g0
7     current_design g0
8     include compile.common
9     create_clock -name clk -period CLKPERIOD -waveform {0.0
HALFCLK}
10    set_clock_skew -uncertainty CLKSKEW clk
11    set_input_delay  INPDELAY -add_delay -clock clk all_inputs()
12    set_output_delay CLKPER1  -add_delay -clock clk all_outputs()
13    compile -map_effort medium
14    current_design DIV32
15    set_dont_touch {g0 STAGE}
16    compile
```

The synthesized results of the DIV32 architecture HIERG can be analyzed. The following shows the results using command report_reference. Note that lines 16 and 17 indicate that blocks STAGE and g0 have dont_touch attributes. In practice, attributes b (line 1) indicates that the design has an unknown block box. Attribute u (line 9) indicates the design contains an unmapped logic. These usually indicate something was not done properly.

```
1          b - black box (unknown)
2         bo - allows boundary optimization
3          d - dont_touch
4         mo - map_only
5          h - hierarchical
6          n - noncombinational
7          r - removable
8          s - synthetic operator
9          u - contains unmapped logic
10
11    Reference Library   Unit Area Count   Total Area Attributes
12    ------------------------------------------------------------
13    B2I     lca300k    2.000000     1    2.000000
14    B2IP    lca300k    4.000000    36  144.000000
15    IVDA    lca300k    1.000000     2    2.000000
16    STAGE            770.000000    32 24640.000000  h, d
17    g0               669.000000     1   669.000000  h, d
18    ------------------------------------------------------------
19    Total 5 references                  25457.000000
```

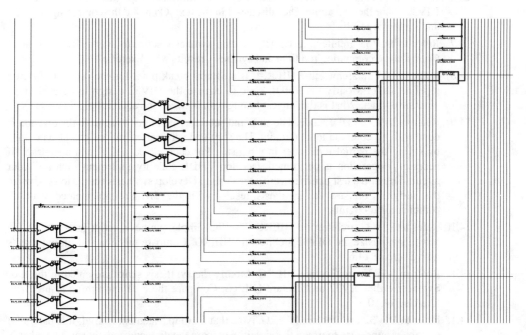

Figure 14-9 Part of synthesized schematic for DIV32 HIERG.

The idea of shift and subtract can be done in many clock cycles, the same way MULTSA is in the last chapter. This is left as Exercise 14.9.

There are many other division algorithms. Readers are referred to [Wilson] and [Swartzlander].

14.7 EXERCISES

14.1 Refer to DIV8 VHDL code and its synthesized schematic. Why are there only seven subtracters while there are eight comparators? From the VHDL code, there are the same number of operators ">=" and "-" in lines 27 and 28.

14.2 Refer to DIV8 VHDL code. The output DIVQ is set to zero when the divisor is zero. What would change if the design simply specifies that when divided by zero, the result is undefined. In other words, we do not care what the quotient output is as long as DIV0 is asserted. What hardware can be expected to be saved? Synthesize the design to verify your answers. Note that in line 32, a simpler statement would be "DIV0 <= Z;".

14.3 Refer to DIV8 VHDL code. What can be changed to also include a remainder output? Synthesize the VHDL code and compare with DIV8's results.

14.4 Refer to STAGE VHDL code. Develop synthesis script files to synthesize all three architectures. Compare the area and timing results.

14.5   Refer to TBDIV32 VHDL code. Analyze all VHDL codes. Run the TBDIV32 VHDL simulation. How long does it take to complete the simulation?

14.6   Refer to DIV32 architecture HIER. Develop synthesis script files to synthesize the DIV32 using the two approaches discussed in the text. Compare the computing run time and synthesis results.

14.7   Refer to DIV32 architecture FLAT. Develop synthesis script to synthesize the design. Compare the computing run time and synthesis results with Exercise 14.6.

14.8   Refer to DIV32 architecture HIER. It is a natural break point for each stage to insert a pipeline stage. Modify the VHDL code to change the DIV32 design to be a 32-stage pipeline divider. What is the longest timing path between the adjacent pipeline stages?

14.9   Refer to DIV32 VHDL code. The shift and subtract operations to achieve the division is done with all combinational circuits. The shift and subtract operations can also be done by saving the partial remainder in a register. For each clock, a shift and a subtract (if necessary) operation is performed. Develop VHDL code to implement such a divider circuit. Develop a test bench to verify the design. Develop synthesis scripts to synthesize the design. Compare the synthesis results with the divider implemented with all combinational circuits.

14.10 Refer to DIV32 architecture HIERG. Use the simple synthesis script presented in the text to synthesize the DIV32 architecture HIERG. Compare the area and timing results with Exercise 14.6.

14.11 Refer to DIV32 architecture HIERG. Modify the synthesis script presented in the text. Synthesize the DIV32 architecture HIERG. Compare the area and timing results with Exercise 14.10.

14.12 Refer to DIV32 architecture HIERG. Note that a group statement is used. It is natural to use two groups. One to group the circuit to prepare the dividend and divisor. The other is used for the circuit to prepare the final result. Modify the VHDL code to use two groups. Develop a synthesis script to synthesize the updated VHDL code. Compare the computing run time and synthesis results with Exercise 14.11.

14.8 REFERENCE

[Swartzlander] Swartzlander, E. E., *Computer Arithmetic,* IEEE Computer Society Press, 1990.

[Wilson] Wilson, J. B., and Ledley, R. S., "An Algorithm for Rapid Binary Division" in *IRE Transactions on Electronic Computer*, EC-10, 1961, pp. 662-670.

Chapter 15

Floating-Point Arithmetic

In this chapter, the floating-point number representation is introduced. The concepts of the addition and multiplication of floating-point numbers are discussed. VHDL code is shown to illustrate the implementations.

15.1 FLOATING-POINT NUMBER REPRESENTATION

A floating-point number y can be written in floating-point format as the following formula:

$$y = y(m) * 2^{y(e)}$$

where $y(m)$ is the mantissa portion, and $y(e)$ is the exponent part of the floating-point number y. The following shows the generic floating-point format. It has an exponent field e, a sign field, and a fraction field. The exponent field is a 2's complement number. The sign field and the fraction field can be considered as one unit and referred to as the mantissa field. The 2's complement fraction is combined with the sign bit and the implied most significant bit to create the mantissa. The mantissa represents a normalized 2's complement number. A normalized representation always has the most significant nonsign bit as the inverse of the sign bit. Therefore, the most significant bit can be implied (no need to store the implied bit) to provide additional precision. The value of a floating point number is given as:

$y = 01.f * 2^e$ if $s = 0$ (0 is the sign bit, 1 is the implied most significant nonsign bit.

$y = 10.f * 2^e$ if $s = 1$ (1 is the sign bit, 0 is the implied most significant nonsign bit.

$y = 0$ if e is the most negative 2's complement value of the specified exponent field width, and $s = 0$, and $f = 0$. For example, for a common 32-bit floating-point number,

$y = 01.f * 2^e$ if $s = 0$

$y = 10.f * 2^e$ if $s = 1$

$y = 0$ if $e = -128$, $s = 0$, and $f = 0$

This 32-bit floating point format has the following range and precision:

Least positive: $y = 1 * 2^{-127} = 5.8774717 * 10^{-39}$

Most positive: $y = (2 - 2^{-23}) * 2^{127} = 3.4028234 * 10^{38}$

Least negative: $y = (-1 - 2^{-23}) * 2^{-127} = -5.8774724 * 10^{-39}$

Most negative: $y = -2 * 2^{127} = -3.4028236 * 10^{38}$

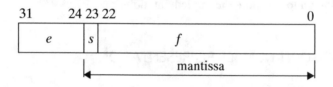

Figure 15-1 Generic floating-point format.

15.2 FLOATING-POINT ADDITION

To add two floating-point numbers, the exponents of the two floating-point numbers should be the same. This is done by shifting the mantissa of the floating-point number that has the smaller exponent. The number of bit positions to be shifted right is the same as the difference of the two exponents. The following shows an example of how to add 1.5 and 0.5 in floating-point format. The mantissa is assumed to be 32 bits wide. Number x (value 1.5) has the exponent with value 0. Number y (value 0.5) has the exponent with value -1 due to normalization. The implied bit is placed back for addition. To make the exponents the same value the y mantissa can be shifted right 1 bit position and the exponent is increased from -1 to 0. Two mantissas can be added. The result is then normalized by shifting the sum right 1 bit position. The exponent is increased from 0 to 1. The resulting mantissa is stored as 32 bits of 0 since the implied bit is not stored.

Figure 15-3 shows another example of adding 2 floating-point numbers. Note that the sum of the mantissa has '1' in each bit position. The sum requires shifting left 32 bits so that the sign bit is different from the most significant nonsign bit of the mantissa. The result is a normalized number.

implied bit

$$x = 1.5 = 01.1000000000000000000000000000000 * 2^0$$
$$y = 0.5 = 01.0000000000000000000000000000000 * 2^{-1}$$

$$01.1000000000000000000000000000000 * 2^0$$
$$+ \quad 00.1000000000000000000000000000000 * 2^0 \quad \text{shift } y \text{ right 1 bit}$$
$$\overline{010.0000000000000000000000000000000 * 2^0} \quad \text{sum result}$$
$$z = 2 = \quad 01.0000000000000000000000000000000 * 2^1 \quad \text{shift right 1 bit}$$

Figure 15-2 Floating point numbers addition example.

implied bit

$$01.111111111111111111111111111111 * 2^{127}$$
$$+ \quad 10.000000000000000000000000000000 * 2^{127}$$
$$\overline{11.111111111111111111111111111111 * 2^{127}} \quad \text{sum result}$$
$$10.000000000000000000000000000000 * 2^{95} \quad \text{shift left 32 bit}$$

Figure 15-3 Floating point numbers addition example.

Assuming that we use Figure 15-1 as the floating-point number format, the floating-point number addition algorithm can be described below:
1. Compare $x(\exp)$ and $y(\exp)$. If $x(\exp)$ is less or equal to $y(\exp)$ then $z(\exp)$ is assigned as $y(\exp)$, otherwise, $z(\exp)$ is assigned as $x(\exp)$. Without loss of generality, let's assume that $x(\exp) <= y(\exp)$.
2. Subtract exponents. Let d be $y(\exp) - x(\exp)$.
3. Align mantissas. Shift $x(\text{man})$ right d bit positions.
4. Add mantissas. Let $z(\text{man})$ be $x(\text{man}) + y(\text{man})$.
5. Test special cases of $z(\text{man})$. If $z(\text{man})$ is 0, $z(\exp)$ is set to -128. Go to step 8.
6. If $z(\text{man})$ overflows, shift $z(\text{man})$ right 1 bit position. Increase $z(\exp)$ by 1. Go to step 8.
7. Let k be the number of leading nonsignificant sign bits. $z(\text{man})$ is left-shifted k bits. $z(\exp)$ is subtracted by k.
8. If $z(\exp)$ underflows, set 0 to be the result ($z(\exp) = -128$, $z(\text{man}) = 0$). If $z(\exp)$ overflows, set the result to be the most positive value if $z(\text{man})$ is positive, set the

result to be the most negative value if z(man) is negative. If z(exp) does not overflow or underflow, no adjustment of value is required. Value z is the final result.

Let's see how we can map the above algorithm into a hardware implementation. In steps 1 and 2, the exponents need to be compared. We need to determine which exponent is bigger, which mantissa is to be shifted right, and how many bit positions to shift. To shift the mantissa in step 3, a right-shift-only barrel shifter can be used. Step 4 can be implemented with an adder. In step 5, testing 0 for the mantissa is easy. In step 6, if the mantissa overflows, the exponent is increased by 1. In step 7 for normal situations, normalization is necessary. This requires the mantissa to be shifted left so that the sign bit is not the same as the first leading mantissa. A right-shift-only barrel shifter can be used. In the last step, the result is adjusted for zero, underflow, and overflow conditions.

The exponent comparison can be implemented with the following VHDL code. Both exponents XEXP and YEXP are input ports as declared in lines 6 and 7. Line 8 declares output signal BIGX to indicate when XEXP is no less than YEXP. Line 9 declares output signal SHIFTD to indicate the number of bit positions to be shifted in the shift-right-only barrel shifter. Line 10 declares output signal TOOSMALL to indicate that the difference between the two exponents is large. The mantissa that corresponds to the smaller exponent will be insignificant and truncated after the mantissa is shifted more than 32 bits. Line 11 declares output signal BIGEXP to be the exponent used for the mantissa alignment reference. Lines 18 and 19 subtract two exponents with sign extension. Lines 20 to 26 determine which exponent is bigger by setting the BIGXsig signal. Based on the value of BIGXsig, signals BIGEXP, SHIFTD, and TOOSMALL are accordingly assigned in lines 27 to 31.

```
1    library IEEE;
2    use IEEE.std_logic_1164.all;
3    use IEEE.std_logic_unsigned.all;
4    entity EXPCOMP is
5      port (
6        XEXP     : in  std_logic_vector(7 downto 0);
7        YEXP     : in  std_logic_vector(7 downto 0);
8        BIGX     : out std_logic;
9        SHIFTD   : out std_logic_vector(4 downto 0);
10       TOOSMALL : out std_logic;
11       BIGEXP   : out std_logic_vector(7 downto 0));
12   end EXPCOMP;
13   architecture RTL of EXPCOMP is
14     signal X_Y, Y_X, SHIFTCNT : std_logic_vector(8 downto 0);
15     signal BIGXsig : std_logic;
16     signal SIGN2 : std_logic_vector(1 downto 0);
17   begin
18     X_Y   <= (Xexp(7) & Xexp) - (Yexp(7) & Yexp);
19     Y_X   <= (Yexp(7) & Yexp) - (Xexp(7) & Xexp);
20     SIGN2 <= Xexp(7) & Yexp(7);
21         -- determine which exp is bigger
22     with SIGN2 select
23       BIGXsig <= not X_Y(8) when "00",
```

```
24                    '1'        when "01",
25                    '0'        when "10",
26                    X_Y(8)     when others;
27      BIGEXP   <= Xexp when BIGXsig = '1' else Yexp;
28      SHIFTCNT <= X_Y  when BIGXsig = '1' else Y_X;
29      SHIFTD   <= SHIFTCNT(4 downto 0);
30      TOOSMALL <= SHIFTCNT(8) or SHIFTCNT(7) or
31                  SHIFTCNT(6) or SHIFTCNT(5);
32      BIGX     <= BIGXsig;
33   end RTL;
```

To prepare the mantissas, the following VHDL code is used. Based on inputs (as declared in lines 5 to 8) X, Y, BIGX, and TOOSMALL, this block generates output signal TOSHIFT as the mantissa with the smaller exponent. Signal TOSHIFT goes to the right-shift-only barrel shifter. The mantissa with the bigger mantissa is output through BIGMAN signal. Lines 18 to 19 determine the implied bit. When the floating-point number is 0 (exponent = -128, and the mantissa is 0), the implied bit should be zero, otherwise, the implied bit is the inverse of the corresponding sign bit. Lines 23 to 25 generate outputs TOSHIFT and BIGMAN based on TOOSMALL and BIGX input signals.

```
1    library IEEE;
2    use IEEE.std_logic_1164.all;
3    entity MANMUX is
4      port (
5         X        : in  std_logic_vector(31 downto 0);
6         Y        : in  std_logic_vector(31 downto 0);
7         BIGX     : in  std_logic;
8         TOOSMALL : in  std_logic;
9         TOSHIFT  : out std_logic_vector(24 downto 0);
10        BIGMAN   : out std_logic_vector(24 downto 0));
11   end MANMUX;
12   architecture RTL of MANMUX is
13     signal XMAN1, YMAN1 : std_logic_vector(24 downto 0);
14     signal X0, Y0       : std_logic;
15     constant F0 : std_logic_vector(31 downto 0) :=
16       "10000000" & "00000000" & "00000000" & "00000000";
17   begin
18     X0 <= '0' when X = F0 else not X(23);
19     Y0 <= '0' when Y = F0 else not Y(23);
20     XMAN1 <= X(23) & X0 & X(22 downto 0);
21     YMAN1 <= Y(23) & Y0 & Y(22 downto 0);
22     ------------------------------------------------------------
23     TOSHIFT <= (TOSHIFT'range => '0') when TOOSMALL = '1' else
24                YMAN1 when BIGX = '1' else XMAN1;
25     BIGMAN  <= XMAN1 when BIGX = '1' else YMAN1;
26   end RTL;
```

Barrel shifters have been described in Chapter 5. The right-shift-only barrel shifter VHDL code is shown below. The input data is 32 bits as declared in line 5. During the right shift of a 2's complement number, the sign bit is shifted right. Signal SIGNS (declared in line 16) is filled with the same value as the SIGN input. Based on the index value of i, $2**(i-1)$ sign bits are filled on the left side of the shifted value as shown in lines 24 and 25. The left-shift-only barrel shifter is similar, except that '0' is shifted in from the right end; therefore, the VHDL code is not presented here.

```
1    library IEEE;
2    use IEEE.std_logic_1164.all;
3    entity BARRELR is
4      port (
5        IN0    : in  std_logic_vector(31 downto 0);
6        SIGN   : in  std_logic;
7        S      : in  std_logic_vector( 4 downto 0);
8        Y      : out std_logic_vector(31 downto 0));
9    end BARRELR;
10   architecture RTL of BARRELR is
11     constant  N   : integer := 32;
12     constant  M   : integer := 5;
13     type arytype is array(M downto 0) of
14       std_logic_vector(N-1 downto 0);
15     signal INTSIG, SIG01, SIG10 : arytype;
16     signal SIGNS  : std_logic_vector(N-1 downto 0);
17   begin
18     sgen : for i in 0 to N-1 generate
19       SIGNS(i) <= SIGN;
20     end generate;
21     INTSIG(0) <= IN0;
22     muxgen : for i in 1 to M generate
23       SIG10(i)  <= INTSIG(i-1);
24       SIG01(i)  <= SIGNS(2**(i-1)-1 downto 0) &
25                    INTSIG(i-1)(N-1 downto 2**(i-1)) ;
26       INTSIG(i) <= SIG10(i) when S(i-1) = '0' else SIG01(i);
27     end generate;
28     Y <= INTSIG(M);
29   end RTL;
```

The following LEADSIGN VHDL code is similar to the leading zero circuit described in Chapter 5. The purpose of the LEADSIGN circuit is to find out how many leading mantissa bits are the same as the sign bit. In other words, the circuit finds the first significant mantissa bit that is not the same as the sign bit. The following VHDL code is similar to the LEADZERO VHDL in Chapter 5, except for line 16 and the input data size (25 compared to 32).

```
1    library IEEE;
2    use IEEE.std_logic_1164.all;
3    entity LEADSIGN is
```

```
4        port (
5          DIN  : in  std_logic_vector(24 downto 0);
6          CNT  : out std_logic_vector( 4 downto 0));
7       end LEADSIGN;
8       architecture RTL of LEADSIGN is
9       begin
10         p0 : process (DIN)
11           variable NUM : std_logic_vector(4 downto 0);
12           variable Cin, Cout : std_logic;
13         begin
14           NUM := (others => '0');
15           for j in 23 downto 0 loop
16             if (DIN(j) = DIN(24)) then
17               Cin := '1';
18               for k in 0 to 4 loop
19                 Cout   := NUM(k) and Cin;
20                 NUM(k) := NUM(k) xor Cin;
21                 Cin    := Cout;
22               end loop;
23             else
24               exit;
25             end if;
26           end loop;
27           CNT <= NUM;
28         end process;
29       end RTL;
```

The result adjustment can be implemented with the following VHDL code. The mantissa sum result from the adder comes in as input signal MANSUM as described in line 6. The result from the left-shift-only barrel shifter comes in as input signal MANSFT. Line 8 declares input SHIFTK to indicate how many bits the mantissa has been shifted so that the exponent can be adjusted accordingly. Lines 12 to 14 declare output signals to indicate the overflow, underflow, zero conditions. Line 15 declares the output RESULT whose implied bit has been removed.

```
1        library IEEE;
2        use IEEE.std_logic_1164.all;
3        use IEEE.std_logic_unsigned.all;
4        entity ADJUST is
5          port (
6            MANSUM  : in  std_logic_vector(25 downto 0);
7            MANSFT  : in  std_logic_vector(24 downto 0);
8            SHIFTK  : in  std_logic_vector( 4 downto 0);
9            BIGEXP  : in  std_logic_vector( 7 downto 0);
10           XSIGN   : in  std_logic;
11           YSIGN   : in  std_logic;
12           FOVF    : out std_logic; -- result overflow
13           FUNDF   : out std_logic; -- result underflow
14           FZERO   : out std_logic; -- result zero
15           RESULT  : out std_logic_vector(31 downto 0));
16       end ADJUST;
```

Lines 23 and 24 check whether the adding mantissas has an overflow. Line 25 checks whether the mantissa is zero. Lines 27 to 58 is a process statement. Line 31 sets the default values for the overflow, underflow, and zero flags. Lines 32 to 45 implement the condition when the mantissa sum has an overflow. The exponent is increased by 1 in line 33. The exponent overflow is checked in line 34. When overflow occurs, the overflow flag is set in line 35. The result is set to the most positive or the most negative value depending on whether the sign of the original mantissa as shown in lines 36 to 38 and lines 40 and 41, respectively. Line 44 sets the result value with the implied bit removed. Note that the sum of the mantissa is shifted right 1 bit. Lines 46 to 48 set the zero result and zero flag.

```
17    architecture RTL of ADJUST is
18      constant ZERO25 : std_logic_vector(24 downto 0) :=
19                        (others => '0');
20      signal MANOVF, MANZERO : std_logic;
21      signal EXPOVF, EXPUNDF : std_logic;
22    begin
23      MANOVF  <= '1' when (XSIGN = YSIGN) and
24                          (XSIGN /= MANSUM(24)) else '0';
25      MANZERO <= '1' when MANSUM(24 downto 0) = ZERO25 else '0';
26    -----------------------------------------------------------------
27      expgen : process (BIGEXP, MANOVF, MANZERO, SHIFTK,
28                        XSIGN, MANSFT, MANSUM)
29        variable EXPADJ : std_logic_vector(7 downto 0);
30      begin
31        FOVF <= '0'; FUNDF <= '0'; FZERO <= '0';
32        if (MANOVF = '1') then
33          EXPADJ := BIGEXP + 1;
34          if (BIGEXP = "01111111") then
35            FOVF <= '1';
36            if (XSIGN = '0') then -- positive
37              RESULT <= "01111111" & "01111111" &
38                        "11111111" & "11111111";
39            else
40              RESULT <= "01111111" & "10000000" &
41                        "00000000" & "00000000";
42            end if;
43          else
44            RESULT <= EXPADJ & MANSUM(25) & MANSUM(23 downto 1);
45          end if;
46        elsif (MANZERO = '1') then
47          RESULT <= "10000000" & ZERO25(23 downto 0);
48          FZERO  <= '1';
49        else
50          EXPADJ := BIGEXP - ("000" & SHIFTK);
51          if (BIGEXP(7) = '1') and (EXPADJ(7) = '0') then
52            RESULT <= "10000000" & ZERO25(23 downto 0);
53            FUNDF  <= '1'; FZERO <= '1';
```

```
54            else
55               RESULT <= EXPADJ & MANSFT(24) & MANSFT(22 downto 0);
56            end if;
57         end if;
58      end process;
59   end RTL;
```

Line 50 to 56 implement the situation when the mantissa sum is neither zero nor overflow. The mantissa is shifted by the number of nonsignificant sign bits. Line 50 adjusts the exponent. Line 51 checks for underflow. Line 52 sets the underflow value to be zero. Line 53 sets the underflow and zero flag. Otherwise, line 55 sets the result to be the output of the left-shift-only barrel shifter with the implied bit removed.

The following VHDL code instantiates the blocks described above. Note that the mantissa has 25 bits while the barrel shifters BARRELR and BARRELL take 32 bits input. The widths of these signals are matched with lines 84, 88, 93, and 96. This is left as Exercise 15.1. Figure 15-4 shows the block diagram and schematic for the FADD design.

```
1    library IEEE;
2    use IEEE.std_logic_1164.all;
3    entity FADD is
4      port (
5        X, Y      : in  std_logic_vector(31 downto 0);
6        FOVF      : out std_logic; -- result overflow
7        FUNDF     : out std_logic; -- result underflow
8        FZERO     : out std_logic; -- result zero
9        Z         : out std_logic_vector(31 downto 0));
10   end FADD;
11   architecture RTL of FADD is
12     component EXPCOMP
13     port (
14       XEXP     : in  std_logic_vector(7 downto 0);
15       YEXP     : in  std_logic_vector(7 downto 0);
16       BIGX     : out std_logic;
17       SHIFTD   : out std_logic_vector(4 downto 0);
18       TOOSMALL : out std_logic;
19       BIGEXP   : out std_logic_vector(7 downto 0));
20     end component;
21     component MANMUX
22     port (
23       X        : in  std_logic_vector(31 downto 0);
24       Y        : in  std_logic_vector(31 downto 0);
25       BIGX     : in  std_logic;
26       TOOSMALL : in  std_logic;
27       TOSHIFT  : out std_logic_vector(24 downto 0);
28       BIGMAN   : out std_logic_vector(24 downto 0));
29     end component;
30     component BARRELR
31     port (
```

```
32      IN0     : in  std_logic_vector(31 downto 0);
33      SIGN    : in  std_logic;
34      S       : in  std_logic_vector( 4 downto 0);
35      Y       : out std_logic_vector(31 downto 0));
36    end component;
37    component MANADD
38    port (
39      BIGMAN    : in  std_logic_vector(24 downto 0);
40      SMALLMAN  : in  std_logic_vector(24 downto 0);
41      MANSUM    : out std_logic_vector(25 downto 0));
42    end component;
43    component LEADSIGN
44    port (
45      DIN  : in  std_logic_vector(24 downto 0);
46      CNT  : out std_logic_vector( 4 downto 0));
47    end component;
48    component BARRELL
49    port (
50      IN0  : in  std_logic_vector(31 downto 0);
51      S    : in  std_logic_vector( 4 downto 0);
52      Y    : out std_logic_vector(31 downto 0));
53    end component;
54    component ADJUST
55    port (
56      MANSUM   : in  std_logic_vector(25 downto 0);
57      MANSFT   : in  std_logic_vector(24 downto 0);
58      SHIFTK   : in  std_logic_vector( 4 downto 0);
59      BIGEXP   : in  std_logic_vector( 7 downto 0);
60      XSIGN    : in  std_logic;
61      YSIGN    : in  std_logic;
62      FOVF     : out std_logic; -- result overflow
63      FUNDF    : out std_logic; -- result underflow
64      FZERO    : out std_logic; -- result zero
65      RESULT   : out std_logic_vector(31 downto 0));
66    end component;
67    signal BIGX, TOOSMALL     : std_logic;
68    signal SHIFTD, SHIFTK     : std_logic_vector( 4 downto 0);
69    signal BIGEXP             : std_logic_vector( 7 downto 0);
70    signal LEFTDIN, LEFTDOUT  : std_logic_vector(31 downto 0);
71    signal RIGHTDIN,RIGHTDOUT : std_logic_vector(31 downto 0);
72    signal TOSHIFT, BIGMAN    : std_logic_vector(24 downto 0);
73    signal SMALLMAN, MANSFT   : std_logic_vector(24 downto 0);
74    signal MANSUM             : std_logic_vector(25 downto 0);
75  begin
76    expcomp0 : EXPCOMP port map (
77      XEXP     => X(31 downto 24), YEXP    => Y(31 downto 24),
78      BIGX     => BIGX,            SHIFTD => SHIFTD,
79      TOOSMALL => TOOSMALL,        BIGEXP => BIGEXP);
80    manmux0 : MANMUX port map (
81      X        => X,              Y            => Y,
```

```
 82        BIGX     => BIGX,           TOOSMALL => TOOSMALL,
 83        TOSHIFT  => TOSHIFT,        BIGMAN   => BIGMAN);
 84     RIGHTDIN <= TOSHIFT & "0000000";
 85     barrelr0 : BARRELR port map (
 86        IN0    => RIGHTDIN, SIGN   => TOSHIFT(24),
 87        S      => SHIFTD,   Y      => RIGHTDOUT);
 88     SMALLMAN <= RIGHTDOUT(31 downto 7);
 89     manadd0 : MANADD port map (
 90        BIGMAN=>BIGMAN, SMALLMAN=>SMALLMAN, MANSUM => MANSUM);
 91     leadsign0 : LEADSIGN port map (
 92        DIN  => MANSUM(24 downto 0), CNT  => SHIFTK);
 93     LEFTDIN <= "0000000" & MANSUM(24 downto 0);
 94     barrell0 : BARRELL port map (
 95        IN0    => LEFTDIN, S => SHIFTK, Y   => LEFTDOUT);
 96     MANSFT   <= LEFTDOUT(24 downto 0);
 97     adjust0 : ADJUST port map (
 98        MANSUM   => MANSUM, MANSFT  => MANSFT,
 99        SHIFTK   => SHIFTK, BIGEXP  => BIGEXP,
100        XSIGN    => X(23),  YSIGN   => Y(23),
101        FOVF     => FOVF,   FUNDF   => FUNDF,
102        FZERO    => FZERO,  RESULT  => Z);
103     end RTL;
```

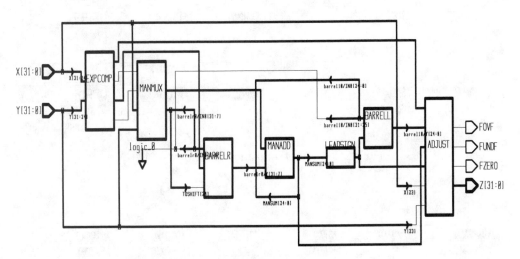

Figure 15-4 FADD block diagram and schematic.

Figure 15-5 shows several test cases of adding two floating-point numbers. The first test case is to add hexadecimal values 00400000 and FF000000. 00400000 has the exponent value 0, and the mantissa value is 01.100 followed by 20 bits of '0' in binary. Therefore, 00400000 represents 1.5 in decimal value. Similarly, FF000000 represents exponent value of –1, and the mantissa is 01.000 followed by 20 bits of '0' in binary. Therefore, FF000000 represents 0.5 in decimal value. The result is 2.0. The representation of 2.0 in floating-point format in hexadecimal after the implied bit removed is 01000000. The second test case is to add 7F7FFFFF and 7F8000000. Both exponents are 127 in decimal. The addition of mantissas is shown below:

```
  01.111_1111_1111_1111_1111_1111
+ 10.000_0000_0000_0000_0000_0000
= 11.111_1111_1111_1111_1111_1111

  10.000_0000_0000_0000_0000_0000  :shift left 24 bits
```

The mantissa result should be normalized by shifting left 24 bit positions. The exponent result is 103 (127 – 24), which is 67 in hexadecimal value.

The third, fourth, and fifth test cases are adding floating-point number 0 (represented as 80000000 in hexadecimal) with another floating-point number. The result should be equal to the other floating-point number. The zero flag is asserted for the fifth test case since the result is 0. The sixth test case is a positive overflow condition since the two most positive numbers are added.

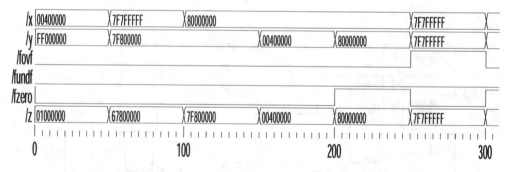

Figure 15-5 FADD simulation waveform — 1.

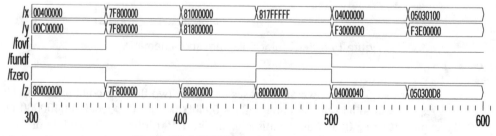

Figure 15-6 FADD simulation waveform — 2.

Figure 15-6 shows more test cases. The first test case adds 1.5 and –1.5 with the result as 0. The second test case adds the two most negative numbers. The overflow flag is set and the result is set to the most negative number. The fourth test case adds two small floating-point numbers (one is positive and the other is negative). Underflow occurs and the result is set to 0. Further verification is left as Exercise 15.2. The synthesis is left as Exercise 15.3.

15.3 FLOATING-POINT MULTIPLICATION

The value of a floating-point number x can be represented as $x(\text{man}) * 2^{x(\text{exp})}$. Multiplying floating-point numbers x and y can be represented as:

$$x(\text{man}) * y(\text{man}) * 2^{(x(\text{exp}) + y(\text{exp}))}$$

The multiplication can be described as the following algorithm:
1. Multiply mantissas. Set $z(\text{man})$ to be $x(\text{man}) * y(\text{man})$.
2. Add exponents. Set $z(\text{exp})$ to be $x(\text{exp}) + y(\text{exp})$.
3. Test for the special case of $z(\text{man})$. If $z(\text{man})$ is zero, set $z(\text{exp})$ to be –128. If $z(\text{man})$ requires to shift to normalize the result, $z(\text{man})$ is shifted right 1 or 2 bits, $z(\text{exp})$ is increased by 1 or 2 accordingly.
4. Check for $z(\text{exp})$ overflow or underflow. If $z(\text{exp})$ overflows, set $z(\text{exp})$ to be the most positive value when $z(\text{man})$ is positive, and set $z(\text{exp})$ to be the most negative value when $z(\text{man})$ is negative. If $z(\text{exp})$ underflows, z is set to 0 ($z(\text{exp})$ is set to –128, and $z(\text{man})$ is set to 0). If $z(\text{exp})$ does not overflow or overflow, z retains its value. Note that the number to be multiplied should be already normalized. This ensures that the product of the mantissa can be adjusted by shifting right 0, one, or two bit positions in normal situations. When underflow or overflow occurs, the mantissa and exponent will be set to appropriate values.

Figure 15-7 shows a simple example of multiplying –2.0 by –2.0. Both exponents are 0. The product of both mantissa has the value "0100" before the binary point. The product is shifted right two bit positions and the exponent is increased by 2. The result is 02000000 in hexadecimal after the implied bit is removed.

```
      10.000_0000_0000_0000_0000_0000          -2.0 exponent = 0
  ×   10.000_0000_0000_0000_0000_0000          -2.0 exponent = 0
  ─────────────────────────────────────────────────────────────────
    0100.000_0000_0000_0000_0000_0000_0000_0000_0000_0000_0000_0000

    0001.000_0000_0000_0000_0000_0000_0000_0000_0000_0000_0000_0000
```

The product is shifted right two bit positions. The exponent is increased by 2.

Figure 15-7 Floating-point multiplication example — 1.

Figure 15-8 illustrates the multiplication of 1.5 by 1.5. The binary value before the binary point is "0010". The product is shifted right one bit position so that the mantissa of the product is 1.0125 in decimal. The exponent is increased by 1. The product is 1.0125 × 2 which is equal to 2.25 in decimal. Multiplying 1.5 by 1.5 is 2.25 in decimal.

```
        01.100_0000_0000_0000_0000_0000              1.5  exponent = 0
  ×     01.100_0000_0000_0000_0000_0000              1.5  exponent = 0
```

```
    0010.010_0000_0000_0000_0000_0000_0000_0000_0000_0000_0000_0000

    0001.001_0000_0000_0000_0000_0000_0000_0000_0000_0000_0000_0000
```
The product is shifted right one bit positions. The exponent is increased by 1.

Figure 15-8 Floating-point multiplication example — 2.

Figure 15-9 shows multiplication of 1.0 by 1.0. The product of the mantissa before the binary point is "0001". There is no need to shift the product.

```
        01.000_0000_0000_0000_0000_0000              1.0  exponent = 0
  ×     01.000_0000_0000_0000_0000_0000              1.0  exponent = 0
```

```
    0001.000_0000_0000_0000_0000_0000_0000_0000_0000_0000_0000_0000
```

The product does not need to be shifted.

Figure 15-9 Floating-point multiplication example — 3.

The above floating-point multiplication algorithm can be described in the following VHDL code. Lines 1 to 3 reference IEEE library and packages. Line 6 declares normalized floating-point number inputs X and Y to be multiplied. In lines 7 to 9, three output signals FOVF, FUNDF, and FZERO are used to indicate the overflow, underflow, and zero conditions, respectively.

```
1    library IEEE;
2    use IEEE.std_logic_1164.all;
3    use IEEE.std_logic_arith.all;
4    entity FMULT is
5      port (
6        X, Y     : in  std_logic_vector(31 downto 0);
7        FOVF     : out std_logic; -- result overflow
8        FUNDF    : out std_logic; -- result underflow
```

```
9          FZERO    : out std_logic; -- result zero
10         Z        : out std_logic_vector(31 downto 0));
11    end FMULT;
```

Lines 13 to 18 declare signals. Line 19 to 20 declare a constant F0 as the value of the floating-point number 0.0. Lines 22 and 23 check whether the input floating-point number is zero. Lines 24 to 27 determine the mantissa with the implied bit added to form a 25-bit mantissa. Lines 28 to 29 rename signals. Line 30 generates the product. Lines 31 to 33 determine the number of bit positions to be shifted right. Lines 34 to 37 select the correct mantissa value for the product by shifting 0, one, or two bit positions based on the SHIFT value.

```
12    architecture RTL of FMULT is
13       signal Xexp, Yexp : std_logic_vector( 7 downto 0);
14       signal Xman, Yman : std_logic_vector(24 downto 0);
15       signal Zman       : std_logic_vector(23 downto 0);
16       signal PROD       : std_logic_vector(49 downto 0);
17       signal SHIFT      : std_logic_vector( 1 downto 0);
18       signal X0, Y0, Xsign, Ysign, Xzero, Yzero : std_logic;
19       constant F0 : std_logic_vector(31 downto 0) :=
20          "10000000" & "00000000" & "00000000" & "00000000";
21    begin
22      Xzero <= '1' when X = F0 else '0';
23      Yzero <= '1' when Y = F0 else '0';
24      X0 <= '0' when Xzero = '1' else not X(23);
25      Y0 <= '0' when Yzero = '1' else not Y(23);
26      Xman <= X(23) & X0 & X(22 downto 0);
27      Yman <= Y(23) & Y0 & Y(22 downto 0);
28      Xexp <= X(31 downto 24); Yexp <= Y(31 downto 24);
29      Xsign <= X(23); Ysign <= Y(23);
30      PROD <= signed(Xman) * signed(Yman);
31      SHIFT  <= "10" when (PROD(49) xor PROD(48)) = '1' else
32                "01" when (PROD(48) xor PROD(47)) = '1' else
33                "00";
34    with SHIFT select
35       Zman <= PROD(49) & PROD(47 downto 25) when "10",
36               PROD(48) & PROD(46 downto 24) when "01",
37               PROD(47) & PROD(45 downto 23) when others;
```

Lines 38 to 77 check the overflow, underflow, and zero conditions. Line 43 sets the default value for FOVF, FUNDF, and FZERO. Line 44 adds the exponents. Line 45 sets the exponent overflow and underflow flags to '0'. Line 46 checks for exponent overflow or underflow as the result of adding two exponents. Line 50 adjusts the exponent for the situations when the mantissa product needs to be shifted right. Lines 51 to 53 check the exponent overflow again. It is possible that the addition of exponents in line 44 does not cause an overflow. After adding 1 or 2, overflow occurs. Lines 54 to 57 adjust the underflow flag, if necessary. It is possible that the addition of exponents in line 44 results an underflow. After adding 1 or 2 in line 50, the underflow condition does not exist any more. The flag should be adjusted for this. Lines 59 to 73 set the corresponding values for zero, underflow, and overflow (positive or negative) conditions. If the result is not zero, underflow, or overflow, line 75 sets the Z output as the concatenation of Zexp and Zman.

```
38      expadd : process (Xexp, Yexp, SHIFT, Xzero, Yzero,
39                        Xsign, Ysign, Zman)
40        variable Zexp1, Zexp2 : std_logic_vector(7 downto 0);
41        variable EXPovf, EXPundf : std_logic;
42      begin
43        FOVF <= '0'; FUNDF <= '0'; FZERO <= '0';
44        Zexp1 := unsigned(Xexp) + unsigned(Yexp);
45        EXPovf := '0'; EXPundf := '0';
46        if ((Xexp(7) = Yexp(7)) and (Zexp1(7) /= Xexp(7))) or
47           (Zexp1 = "10000000") then
48          EXPovf := not Xexp(7); EXPundf := Xexp(7);
49        end if;
50        Zexp2 := unsigned(Zexp1) + unsigned("000000" & SHIFT);
51        if (EXPundf = '0') and ((EXPovf = '1') or
52           ((Zexp2(7) /= Zexp1(7)) and (Zexp1(7) = '0'))) then
53          EXPovf := '1';
54        elsif (((Zexp2(7) /= Zexp1(7)) and (Zexp1(7) = '0')) and
55              (EXPundf = '1')) or ((Zexp1 = "10000000") and
56              (SHIFT /= "00")) then -- reverse exp underflow
57          EXPundf := '0';
58        end if;
59        if (Xzero = '1') or (Yzero = '1') then -- result 0
60          FZERO <= '1';
61          Z      <= "10000000" & "00000000" & "0000000000000000";
62        elsif (EXPovf = '1') then -- overflow
63          FOVF   <= '1';
64          if (Xsign = Ysign) then -- set to most positive
65            Z    <= "01111111" & "01111111" & "1111111111111111";
66          else                      -- set to most negative
67            Z    <= "01111111" & "10000000" & "0000000000000000";
68          end if;
69        elsif (EXPundf = '1') or -- underflow
70              ((Zexp2 = "10000000") and
71              (Zman /= "0000000000000000000000000")) then
72          FUNDF <= '1'; FZERO <= '1';
73          Z      <= "10000000" & "00000000" & "0000000000000000";
74        else --------- normal situation
75          Z <= Zexp2 & Zman;
76        end if;
77      end process;
78    end RTL;
```

Figure 15-10 shows several test cases. The first three test case values are the same as in Figure 15-7, Figure 15-8, and Figure 15-9, respectively. Test cases 5, 6, and

7 have one value as 0.0 (represented in 8000000 floating-point format). The result is 0.0. Test case 8 multiplies the largest positive number by 2.0. The result is set to the most positive number since the overflow occurs.

Figure 15-11 shows more test cases. Test case 3 multiplies the most negative number by –2.0. The result is overflow to the most positive number. Test case 4 multiplies the most negative number of 1.5. The result is overflow to the most negative number. Test case 5 has the underflow result. Further verification is left as Exercise 15.4.

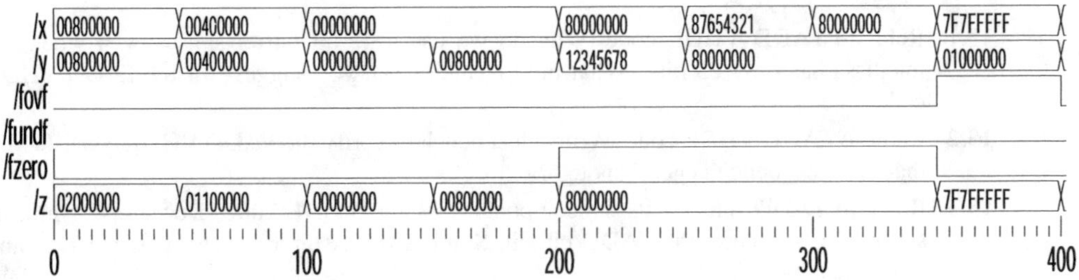

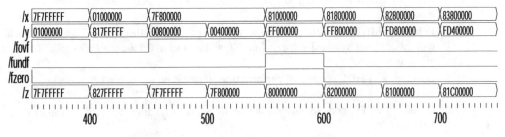

Figure 15-10 FMULT simulation waveform — 1.

Figure 15-11 FMULT simulation waveform — 2.

Other typical floating-point operations are normalization, rounding, integer to floating-point conversion, and floating-point to integer conversion. The normalization process is part of the floating-point addition. The rounding can be achieved by adding half of the least significant mantissa bit value. In this format, $1*2**(-24)$ is used. This requires the extended precision by having more bits for the mantissa; for example, 32 bits. The normal floating-point addition can then be used. The floating-point to integer conversion requires checking whether the exponent is within the range. If the exponent is greater than 30, overflow occurs since the integer can only have 32 bits. Converting an integer to a floating-point is similar to normalization that shifting is required. These are left as exercises. A microprocessor with a floating-point unit usually has various floating-point instructions. As a minimum, it will have the ADD,

SUB, NORM, F2I, ROUND instructions. They can be implemented with share hardware such as the barrel shifter. This is left as Exercise 15.12.

Note that the floating-point format used in this chapter is usually called the simple *2's complement floating-point format*. IEEE has published a floating-point format Standard 754. The format is slightly different. However, the concept is the same. These are left as Exercises 15.13 and 15.14.

15.4 EXERCISES

15.1 Refer to FADD VHDL code. Why do we use a 32-bit barrel shifter even when the mantissa has only 25 bits? What improvements can you suggest for the FADD VHDL code?

15.2 Refer to FADD VHDL code. Write a test bench to verify the FADD VHDL code. Justify that your test bench is good enough.

15.3 Refer to FADD and all its subcomponents as the VHDL code. Develop a synthesis strategy to synthesis the FADD circuit. Synthesize the circuit and report the area and timing results.

15.4 Refer to FMULT VHDL code. Write a test bench to verify the FMULT VHDL code. Justify that your test bench is good enough.

15.5 Refer to FMULT VHDL code. What improvements do you suggest for the FMULT VHDL code?

15.6 Refer to FMULT VHDL code. Synthesize the VHDL code and report the area and timing data.

15.7 Refer to FMULT VHDL code. How would you change the VHDL code so that the multiplier is not synthesized with line 30, and a custom 26×26 bit multiplier block is given as a macro?

15.8 Develop a VHDL code to perform the floating-point normalization. Verify and synthesize the VHDL code.

15.9 Develop a VHDL code to perform the floating-point rounding. Verify and synthesize the VHDL code.

15.10 Develop a VHDL code to convert a floating-point number to an integer. Verify and synthesize the VHDL code.

15.11 Develop a VHDL code to convert an integer to a floating-point number. Verify and synthesize the VHDL code.

15.12 Develop a VHDL code to do floating-point addition, subtraction, normalization, rounding, and integer to floating point conversion. Verify and synthesize the VHDL code.

15.13 Refer to 1985 IEEE Standard 754 floating-point format. Develop a floating-point addition circuit with VHDL code. Verify and synthesize the VHDL code.

15.14 Refer to 1985 IEEE Standard 754 floating-point format. Develop a floating-point multiplication circuit with VHDL code. Verify and synthesize the VHDL code.

Appendix A

Package PACK

```
1    library IEEE;
2    use IEEE.std_logic_1164.all;
3    package PACK is
4        function REDUCE_AND(DIN:in std_logic_vector) return
std_logic;
5        function REDUCE_OR(DIN:in std_logic_vector)  return
std_logic;
6        function REDUCE_XOR(DIN:in std_logic_vector) return
std_logic;
7        function COMP2 (DIN : in std_logic_vector)
8          return std_logic_vector ;
9        procedure COMP2 (
10         signal DIN   : in  std_logic_vector;
11         signal DOUT  : out std_logic_vector) ;
12       procedure DECODER24 (
13         signal DIN  : in  std_logic_vector(1 downto 0);
14         signal DOUT : out std_logic_vector(3 downto 0)) ;
15       procedure DECODER38 (
16         signal DIN  : in  std_logic_vector(2 downto 0);
17         signal DOUT : out std_logic_vector(7 downto 0)) ;
18       procedure INVTRIBUF(
19         signal DIN, EN : in  std_logic;
20         signal DOUT     : out std_logic) ;
21       procedure INVTRIBUF(
22         signal DIN  : in  std_logic_vector;
23         signal EN   : in  std_logic;
24         signal DOUT : out std_logic_vector) ;
25       procedure INVTRIBUFn(
26         signal DIN, ENn : in  std_logic;
27         signal DOUT : out std_logic) ;
28       procedure INVTRIBUFn(
29         signal DIN  : in  std_logic_vector;
30         signal ENn  : in  std_logic;
31         signal DOUT : out std_logic_vector) ;
32       procedure LATCH_C(
33         signal CLR, EN, D : in  std_logic;
```

```
34        signal Q          : out std_logic) ;
35     procedure LATCH_C(
36        signal CLR, EN : in  std_logic;
37        signal D        : in  std_logic_vector;
38        signal Q        : out std_logic_vector) ;
39     procedure LATCH(
40        signal DIN, EN : in  std_logic;
41        signal DOUT     : out std_logic) ;
42     procedure LATCH(
43        signal DIN : in  std_logic_vector;
44        signal EN   : in  std_logic;
45        signal DOUT : out std_logic_vector) ;
46     procedure LATCHn(
47        signal DIN, ENn : in  std_logic;
48        signal DOUT     : out std_logic) ;
49     procedure LATCHn(
50        signal DIN : in  std_logic_vector;
51        signal ENn  : in  std_logic;
52        signal DOUT : out std_logic_vector) ;
53     procedure LATCHR(
54        signal RSTn, EN, DIN : in  std_logic;
55        signal DOUT            : out std_logic) ;
56     procedure LATCHR(
57        signal RSTn, EN : in  std_logic;
58        signal DIN : in  std_logic_vector;
59        signal DOUT : out std_logic_vector) ;
60     procedure LATCHRn(
61        signal RSTn, ENn, DIN : in  std_logic;
62        signal DOUT : out std_logic) ;
63     procedure LATCHRn(
64        signal RSTn, ENn : in  std_logic;
65        signal DIN      : in  std_logic_vector;
66        signal DOUT     : out std_logic_vector) ;
67     procedure KMUX21(
68        signal SEL, DIN0, DIN1  : in  std_logic;
69        signal DOUT : out std_logic) ;
70     procedure KMUX21(
71        signal SEL  : in  std_logic;
72        signal DIN0, DIN1 : in  std_logic_vector;
73        signal DOUT : out std_logic_vector) ;
74     procedure KMUX31(
75        signal SEL                : in  std_logic_vector(1 downto 0);
76        signal DIN0, DIN1, DIN2 : in  std_logic;
77        signal DOUT              : out std_logic) ;
78     procedure KMUX31(
79        signal SEL  : in  std_logic_vector(1 downto 0);
80        signal DIN0, DIN1, DIN2 : in  std_logic_vector;
81        signal DOUT : out std_logic_vector) ;
82     procedure KMUX41(
83        signal DIN0, DIN1, DIN2, DIN3 : in  std_logic;
```

```vhdl
84        signal S0, S1                    : in  std_logic;
85        signal Y                         : out std_logic) ;
86     procedure KMUX41(
87        signal DIN0, DIN1, DIN2, DIN3 : in  std_logic_vector;
88        signal S0, S1 : in  std_logic;
89        signal Y   : out std_logic_vector) ;
90     procedure ODDONE(
91        signal DIN  : in  std_logic_vector;
92        signal DOUT : out std_logic) ;
93     procedure TRIBUF(
94        signal DIN, EN : in  std_logic;
95        signal DOUT    : out std_logic) ;
96     procedure TRIBUF(
97        signal DIN  : in  std_logic_vector;
98        signal EN   : in  std_logic;
99        signal DOUT : out std_logic_vector) ;
100    procedure TRIBUFn(
101       signal DIN, ENn  : in  std_logic;
102       signal DOUT : out std_logic) ;
103    procedure TRIBUFn(
104       signal DIN  : in  std_logic_vector;
105       signal ENn  : in  std_logic;
106       signal DOUT : out std_logic_vector) ;
107  end PACK;
108
109  package body PACK is
110     --------------------------------------------------------------------
111     function REDUCE_AND(DIN:in std_logic_vector) return std_logic is
112        variable result: std_logic;
113     begin
114       result := '1';
115       for i in DIN'range loop
116         result := result and DIN(i);
117       end loop;
118       return result;
119     end REDUCE_AND;
120     --------------------------------------------------------------------
121     function REDUCE_OR(DIN:in std_logic_vector) return std_logic is
122        variable result: std_logic;
123     begin
124       result := '0';
125       for i in DIN'range loop
126         result := result or DIN(i);
127       end loop;
128       return result;
129     end REDUCE_OR;
130     --------------------------------------------------------------------
```

```
----
131    function REDUCE_XOR(DIN:in std_logic_vector) return
std_logic is
132      variable result: std_logic;
133    begin
134      result := '0';
135      for i in DIN'range loop
136        result := result xor DIN(i);
137      end loop;
138      return result;
139    end REDUCE_XOR;
140    ----------------------------------------------------------------
----
141    function COMP2 (DIN : in std_logic_vector)
142      return std_logic_vector is
143      variable temp  : std_logic_vector(DIN'length-1 downto 0);
144      variable SEEN1 : std_logic;
145    begin
146      SEEN1 := '0';
147      for i in DIN'reverse_range loop
148        if (SEEN1 = '1') then
149          temp(i) := not DIN(i);
150        elsif (DIN(i) = '1') then
151          SEEN1 := '1';
152          temp(i) := '1';
153        else
154          temp(i) := '0';
155        end if;
156      end loop;
157      return temp;
158    end COMP2;
159    ----------------------------------------------------------------
----
160    procedure COMP2 (
161      signal DIN   : in  std_logic_vector;
162      signal DOUT  : out std_logic_vector) is
163      variable SEEN1 : std_logic;
164    begin
165      SEEN1 := '0';
166      for j in DIN'reverse_range loop
167        if (SEEN1 = '1') then
168          DOUT(j) <= not DIN(j);
169        else
170          if (DIN(j) = '1') then
171            SEEN1 := '1';
172          end if;
173          DOUT(j) <= DIN(j);
174        end if;
175      end loop;
176    end COMP2;
177    ----------------------------------------------------------------
```

```
----
178     procedure DECODER24 (
179        signal DIN  : in  std_logic_vector(1 downto 0);
180        signal DOUT : out std_logic_vector(3 downto 0)) is
181     begin
182        DOUT <= "0000";
183        case DIN is
184          when "00"   => DOUT(0) <= '1';
185          when "01"   => DOUT(1) <= '1';
186          when "10"   => DOUT(2) <= '1';
187          when "11"   => DOUT(3) <= '1';
188          when others => null;
189        end case;
190     end DECODER24;
191     ----------------------------------------------------------------
----
192     procedure DECODER38 (
193        signal DIN  : in  std_logic_vector(2 downto 0);
194        signal DOUT : out std_logic_vector(7 downto 0)) is
195     begin
196        DOUT <= "00000000";
197        case DIN is
198          when "000"  => DOUT(0) <= '1';
199          when "001"  => DOUT(1) <= '1';
200          when "010"  => DOUT(2) <= '1';
201          when "011"  => DOUT(3) <= '1';
202          when "100"  => DOUT(4) <= '1';
203          when "101"  => DOUT(5) <= '1';
204          when "110"  => DOUT(6) <= '1';
205          when "111"  => DOUT(7) <= '1';
206          when others => null;
207        end case;
208     end DECODER38;
209     ----------------------------------------------------------------
----
210     procedure INVTRIBUF(
211        signal DIN, EN  : in  std_logic;
212        signal DOUT     : out std_logic) is
213     begin
214        if (EN = '1') then
215          DOUT <= not DIN;
216        else
217          DOUT <= 'Z';
218        end if;
219     end INVTRIBUF;
220     ----------------------------------------------------------------
----
221     procedure INVTRIBUF(
222        signal DIN  : in  std_logic_vector;
223        signal EN   : in  std_logic;
224        signal DOUT : out std_logic_vector) is
```

```
225     begin
226       if (EN = '1') then
227         DOUT <= not DIN;
228       else
229         for i in DIN'low to DIN'high loop
230           DOUT(i) <= 'Z';
231         end loop;
232       end if;
233     end INVTRIBUF;
234     ----------------------------------------------------------------
235     procedure INVTRIBUFn(
236       signal DIN, ENn  : in  std_logic;
237       signal DOUT : out std_logic) is
238     begin
239       if (ENn = '0') then
240         DOUT <= not DIN;
241       else
242         DOUT <= 'Z';
243       end if;
244     end INVTRIBUFn;
245     ----------------------------------------------------------------
246     procedure INVTRIBUFn(
247       signal DIN  : in  std_logic_vector;
248       signal ENn  : in  std_logic;
249       signal DOUT : out std_logic_vector) is
250     begin
251       if (ENn = '0') then
252         DOUT <= not DIN;
253       else
254         for i in DIN'low to DIN'high loop
255           DOUT(i) <= 'Z';
256         end loop;
257       end if;
258     end INVTRIBUFn;
259     ----------------------------------------------------------------
260     procedure LATCH_C(
261       signal CLR, EN, D : in  std_logic;
262       signal Q          : out std_logic) is
263     begin
264       if (CLR = '0') then
265         Q     <= '0';
266       elsif (EN = '1') then
267         Q     <= D;
268       end if;
269     end LATCH_C;
270     ----------------------------------------------------------------
271     procedure LATCH_C(
```

```
272      signal CLR, EN : in  std_logic;
273      signal D        : in  std_logic_vector;
274      signal Q        : out std_logic_vector) is
275    begin
276      if (CLR = '0') then
277        Q    <= (Q'range => '0');
278      elsif (EN = '1') then
279        Q    <= D;
280      end if;
281    end LATCH_C;
282    ----------------------------------------------------------------
       ----
283    procedure LATCH(
284      signal DIN, EN : in  std_logic;
285      signal DOUT    : out std_logic) is
286    begin
287      if (EN = '1') then
288        DOUT <= DIN;
289      end if;
290    end LATCH;
291    ----------------------------------------------------------------
       ----
292    procedure LATCH(
293      signal DIN  : in  std_logic_vector;
294      signal EN   : in  std_logic;
295      signal DOUT : out std_logic_vector) is
296    begin
297      if (EN = '1') then
298        DOUT <= DIN;
299      end if;
300    end LATCH;
301    ----------------------------------------------------------------
       ----
302    procedure LATCHn(
303      signal DIN, ENn : in  std_logic;
304      signal DOUT     : out std_logic) is
305    begin
306      if (ENn = '0') then
307        DOUT <= DIN;
308      end if;
309    end LATCHn;
310    ----------------------------------------------------------------
       ----
311    procedure LATCHn(
312      signal DIN  : in  std_logic_vector;
313      signal ENn  : in  std_logic;
314      signal DOUT : out std_logic_vector) is
315    begin
316      if (ENn = '0') then
317        DOUT <= DIN;
318      end if;
```

```
319    end LATCHn;
320    ----------------------------------------------------------------
----
321    procedure LATCHR(
322      signal RSTn, EN, DIN : in  std_logic;
323      signal DOUT         : out std_logic) is
324    begin
325      if (RSTn = '0') then
326        DOUT <= '0';
327      elsif (EN = '1') then
328        DOUT <= DIN;
329      end if;
330    end LATCHR;
331    ----------------------------------------------------------------
----
332    procedure LATCHR(
333      signal RSTn, EN : in  std_logic;
334      signal DIN  : in  std_logic_vector;
335      signal DOUT : out std_logic_vector) is
336    begin
337      if (RSTn = '0') then
338        DOUT <= (DOUT'range => '0');
339      elsif (EN = '1') then
340        DOUT <= DIN;
341      end if;
342    end LATCHR;
343    ----------------------------------------------------------------
----
344    procedure LATCHRn(
345      signal RSTn, ENn, DIN : in  std_logic;
346      signal DOUT : out std_logic) is
347    begin
348      if (RSTn = '0') then
349        DOUT <= '0';
350      elsif (ENn = '0') then
351        DOUT <= DIN;
352      end if;
353    end LATCHRn;
354    ----------------------------------------------------------------
----
355    procedure LATCHRn(
356      signal RSTn, ENn : in  std_logic;
357      signal DIN       : in  std_logic_vector;
358      signal DOUT      : out std_logic_vector) is
359    begin
360      if (RSTn = '0') then
361        DOUT <= (DOUT'range => '0');
362      elsif (ENn = '0') then
363        DOUT <= DIN;
364      end if;
365    end LATCHRn;
```

```
366      -----------------------------------------------------------------
----
367      procedure KMUX21(
368        signal SEL, DIN0, DIN1  : in  std_logic;
369        signal DOUT : out std_logic) is
370      begin
371        case SEL is
372          when '0' => DOUT <= DIN0;
373          when others => DOUT <= DIN1;
374        end case;
375      end KMUX21;
376      -----------------------------------------------------------------
----
377      procedure KMUX21(
378        signal SEL  : in  std_logic;
379        signal DIN0, DIN1 : in  std_logic_vector;
380        signal DOUT : out std_logic_vector) is
381      begin
382        case SEL is
383          when '0' => DOUT <= DIN0;
384          when others => DOUT <= DIN1;
385        end case;
386      end KMUX21;
387      -----------------------------------------------------------------
----
388      procedure KMUX31(
389        signal SEL                : in  std_logic_vector(1 downto 0);
390        signal DIN0, DIN1, DIN2 : in  std_logic;
391        signal DOUT               : out std_logic) is
392      begin
393        case SEL is
394          when "00"   => DOUT <= DIN0;
395          when "01"   => DOUT <= DIN1;
396          when others => DOUT <= DIN2;
397        end case;
398      end KMUX31;
399      -----------------------------------------------------------------
----
400      procedure KMUX31(
401        signal SEL  : in  std_logic_vector(1 downto 0);
402        signal DIN0, DIN1, DIN2 : in  std_logic_vector;
403        signal DOUT : out std_logic_vector) is
404      begin
405        case SEL is
406          when "00"   => DOUT <= DIN0;
407          when "01"   => DOUT <= DIN1;
408          when others => DOUT <= DIN2;
409        end case;
410      end KMUX31;
411      -----------------------------------------------------------------
----
```

```
412    procedure KMUX41(
413      signal DIN0, DIN1, DIN2, DIN3 : in  std_logic;
414      signal S0, S1                 : in  std_logic;
415      signal Y                      : out std_logic) is
416        variable SEL : std_logic_vector(1 downto 0);
417    begin
418      SEL := S1 & S0;
419      case SEL is
420        when "00"    => Y <= DIN0;
421        when "01"    => Y <= DIN1;
422        when "10"    => Y <= DIN2;
423        when others => Y <= DIN3;
424      end case;
425    end KMUX41;
426    ----------------------------------------------------------------
427    procedure KMUX41(
428      signal DIN0, DIN1, DIN2, DIN3 : in  std_logic_vector;
429      signal S0, S1 : in  std_logic;
430      signal Y  : out std_logic_vector) is
431        variable SEL : std_logic_vector(1 downto 0);
432    begin
433      SEL := S1 & S0;
434      case SEL is
435        when "00"    => Y <= DIN0;
436        when "01"    => Y <= DIN1;
437        when "10"    => Y <= DIN2;
438        when others => Y <= DIN3;
439      end case;
440    end KMUX41;
441    ----------------------------------------------------------------
442    procedure ODDONE(
443      signal DIN  : in  std_logic_vector;
444      signal DOUT : out std_logic) is
445        variable temp : std_logic;
446    begin
447      temp := DIN(DIN'high);
448      for i in DIN'high-1 downto DIN'low loop
449        temp := temp xor DIN(i);
450      end loop;
451      DOUT <= temp;
452    end ODDONE;
453    ----------------------------------------------------------------
454    procedure TRIBUF(
455      signal DIN, EN : in  std_logic;
456      signal DOUT    : out std_logic) is
457    begin
458      if (EN = '1') then
459        DOUT <= DIN;
```

```
460        else
461          DOUT <= 'Z';
462        end if;
463      end TRIBUF;
464      -----------------------------------------------------------------
----
465      procedure TRIBUF(
466        signal DIN  : in  std_logic_vector;
467        signal EN   : in  std_logic;
468        signal DOUT : out std_logic_vector) is
469      begin
470        if (EN = '1') then
471          DOUT <= DIN;
472        else
473          for i in DIN'low to DIN'high loop
474            DOUT(i) <= 'Z';
475          end loop;
476        end if;
477      end TRIBUF;
478      -----------------------------------------------------------------
----
479      procedure TRIBUFn(
480        signal DIN, ENn  : in  std_logic;
481        signal DOUT : out std_logic) is
482      begin
483        if (ENn = '0') then
484          DOUT <= DIN;
485        else
486          DOUT <= 'Z';
487        end if;
488      end TRIBUFn;
489      -----------------------------------------------------------------
----
490      procedure TRIBUFn(
491        signal DIN  : in  std_logic_vector;
492        signal ENn  : in  std_logic;
493        signal DOUT : out std_logic_vector) is
494      begin
495        if (ENn = '0') then
496          DOUT <= DIN;
497        else
498          for i in DIN'low to DIN'high loop
499            DOUT(i) <= 'Z';
500          end loop;
501        end if;
502      end TRIBUFn;
503    end PACK;
```

Index

Symbols
(DP 251
.synopsys_dc.setup 40, 41
.synopsys_vss.setup 34

Numerics
2's Complement
    Multiplication 419
2's complementer 141

A
after 159
analyze 61
assert 58, 109
asynchronous counter 150
asynchronous FIFO 266
asynchronously reset 230

B
barrel shifter 132
Booth recoding table 420

C
Carry Look Ahead Adder 101
carry look ahead adder 101
carry look ahead unit 103
carry ripple adder 97
carry save adder 409
case 55
check number 391
check_design 43
CLA16 105
CLA16 block diagram 105
CLA64 113
clock buffer 222
clock buffer tree 222
clock delay 222

clock divider 242
clock skew 222, 230
Code Converter 71
combinational gates 22
commercial environments
    343
Comparator with Multiple
    Outputs 82
compile.common 41
configuration 98, 107, 414,
    461
configuration registers 196
conv_std_logic_vector 57
counter 150
countone 119
countone circuit 119
create_clock 61, 102
CY7C910 instruction
    summary table 350

D
define_design_lib 41
Design Analyzer 39
Design Compiler 39
design environment 32
Design Time 39
directory structure 33
dividend 445
Divider 451
divisor 445
DMA 193
DRAM 274
Dual Port RAM 251
dynamic RAM 274

E
elaborate 61, 76, 81, 95, 99,

379
encoder 68
equality checker 73
even parity 390, 391
exponen 467

F
FIF 266
first in first out (FIFO) 260
fix_hold 231
flip flop 4
Floating Point Addition 468
Floating Point Multiplication
    479
floating point number 467

G
gated clock 246
generate 80, 97, 98
generate carry 101
generic 97, 226
generic map 109
global set and reset (GSR)
    246
group_path 377

H
half adder 91
Hamming 390
hold timing error 231

I
Identifiers 2, 4, 32, 166,
    192, 196, 223, 251, 288,
    342, 391, 392, 393, 402,
    403, 404, 405, 406, 408,
    445

include 61
incrementer 139
instruction decoder 349
interrupt mask register 190
interrupt pending register 190
Interrupt registers 189
interrupt request register 191

J
JK flip flop 12

L
latch 15
leading zero 472
leading zero circuit 123
LEADZERO 130
Least negative 468
Least positive 468
LFSR 176
linear feedback shift register 176
link_library 41
literals 18
loop 108, 119

M
mantissa 467
microcontroller 341
Microprocessor registers 219
microprogram 363
military environment 343
Most negative 468
Most positive 468
multiplexer 23
Multiplication with Shift and Add 437
multiplier 408

N
normalized number 468

O
odd parity 391

P
parallel to serial converter 177
parity check 391
partial products 409
pipeline register 363
port map 129
prioritize_min_paths 233
propagate carry 101

Q
QuickVHDL 36
quotient 445
qvcom 36
qvlib 36
qvmap 36
qvsim 37

R
random access memory (RAM) 274
read 99, 109
redundancy ratio 390
Register Block Partitioning and Synthesis 202
remainder 445
remove_attribute 437
remove_design 43, 378
remove_license 43, 378
report_area 43
report_constraint 43
report_net 209
report_timing 43, 208, 378
Reset Circuitry 228
resource library 35
run_quick_vss 385

S
SDF 381
select 55
selector 23, 53
sensitivity list 8
serial to parallel converter 180
set_clock_skew 61, 233, 377, 379
set_dont_touch 95
set_dont_touch_network 196, 377
set_fix_hold 233, 379
set_input_delay 61, 67, 102, 377
set_operating_conditions 379
set_output_delay 61, 102, 377
set_scan_style 377
set_wire_load 379
setup timing error 230
sh date 130
signal chopper 143
signal extender 148
signal manipulator 143

signed 429
Simulation Environment 36
simulation environment 33
simulation tools 32
single error detecting code 390
slack 63
software reset 193
SRAM 274
stack pointer 347
std_logic_arith 56
std_logic_unsigned 57
symbol_library 41
Synchronization between Clock Domains 238
synchronous counter 154
Synchronous FIFO 260
synchronous reset 230
Synthesis Environment 39
synthesis technology library 45
synthesis tools 32
synthetic_library 41
systematic 390

T
target_library 41
Test Compiler 39
testvhd 385
three-state buffer 18
type declaration 22

U
ungroup 111, 379, 437
uniquify 95
unit adder 423
use configuration 109
use entity 109

V
VHDL analyzer 32
VHDL Compiler 39
VHDL Design Process for A Block 47
VHDL simulation tools 32
VHDL simulator 32
VHDL source code debugger 32
VHDL Synthesis Rules 26
VHDL synthesis tools 32
VHDL waveform viewer 32
vhdlan 35
vhdldbx 35

W
wait for 159
wait until 146
Wallace Tree 423
Wallace Tree Adders 423
wif2tab 376
working library 35
write 43, 129, 378
write_timing 380
writeline 129

Y
ytv2ctv 383

About the Author

K. C. CHANG, Ph.D.

K. C. Chang received his B.S. degree in Electrical Engineering from the National Taiwan University in 1979. His M.S. and Ph.D. degrees in computer science were earned at the University of Minnesota in 1984 and 1986, respectively. Later in 1986 he joined The Boeing Company where he teaches very high speed integrated circuit hardware description language (VHDL) and synthesis courses in addition to his regular responsibilities of computer-aided design algorithm development, VHDL, synthesis, and application specific integrated circuit (ASIC) design. He has designed several ASICs with VHDL and synthesis, including 32-bit digital signal processing (DSP) and reduced instruction set computer (RISC) microprocessors. He gives conferences, tutorials, and industry VHDL short courses on these same subjects. Dr. Chang has served as a reviewer of technical papers for the Association on Computer Machinery/Institute of Electrical and Electronics Engineers (ACM/IEEE) Design Automation Conference since 1987. He holds three US patents and has published 16 technical papers, including 3 papers in *IEEE Transactions on Computers* and *Transactions on Computer-Aided Design* journals. His first VHDL book entitled Digital Design and Modeling with VHDL and Synthesis (IEEE Computer Society Press, 1997) has been widely used both in academia and industry. He also earned his M.B.A. degree from the City University, Bellevue, Washington, in 1994. Since 1997, he has been an Associate Technical Fellow at Boeing and served as an Affiliate Associate Professor in the University of Washington Electrical Engineering Department.